AF531640

Pvt. Ltd.

Horticultural Crops — Varietal Wealth

H.P. Singh

Deputy Director General (Horticulture),
Indian Council of Agriculture Research,
New Delhi

V.A. Parthasarathy

Indian Institute of Spices Research
Calicut, Kerala

D. Prasath

Indian Institute of Spices Research,
Calicut, Kerala

2009

Studium Press (India) Pvt. Ltd.

Horticultural Crops — Varietal Wealth

ISBN: 978-93-80012-09-4

Citation:

Singh, H.P., Parthasarathy, V.A. and Prasad, D., (Eds.). 2009. ***Horticultural Crops — Varietal Wealth.*** *Studium Press (India) Pvt. Ltd. pp. 563*

Published by:

Studium Press (India) Pvt. Ltd.
4735/22, 2nd Floor, Prakash Deep Building
(Near Delhi Medical Association),
Ansari Road, Darya Ganj, New Delhi-110 002
Tel.: 23240257, 65150447; Fax: 91-11-23240273;
jngovil@gmail.com; jngovil@hotmail.com

Printed at:

Salasar Imaging Systems
C-7/5, Lawrence Road Indl. Area, Delhi - 110 035

Foreword

Indian horticulture has made rapid strides during the recent years and, there has been an appreciable growth in production and productivity leading to year round availability of horticultural produce. This development has been possible due to generation of appropriate technologies and focused attention of Government. The Indian Council of Agricultural Research (ICAR) has spearheaded innovations in horticulture through its institutional network comprising of Institutes, Directorates, National Research Centres and over 200 centres of All India Coordinated Research Projects across the country.

The improved technologies in form of varieties/hybrids have exhibited their impact on yield and yield-contributing characters resulting in enhanced yield, quality and profitability. However the information on cultivars has remained scattered and need for information on varieties has been strongly felt by the stakeholders. I am happy to note that information on varieties released for cultivation in horticultural crops has been compiled and is being brought out in form of a book entitled **"Horticultural Crops — Varietal Wealth".**

I compliment the authors for compiling valuable information on horticultural crops in the country, which could prove to be of use to farmers, scientists, researchers, policy makers and other stakeholders involved in improvement of horticulture sector.

Dated : 20th April, 2009
New Delhi

(MANGALA RAI)
Secretary,
Department of Agricultural Research & Education
and
Director General,
Indian Council of Agricultural Research,
Ministry of Agriculture, Krishi Bhavan,
New Delhi–110 001

Acknowledgement

We acknowledge the contribution of breeders from across the country working at 71 research centres in National Agricultural Research System, who made their sincere efforts in development of the cultivars. The book is first of its kind to bring all the information on the varieties developed in horticultural crops. The efforts have been made by many experts, Directors and coordinators in surfacing the information to put in proper perspective. Therefore, we sincerely thanks to Directors: BMC Reddy, (CISH), Dr T.A. More (CIAH), Dr Nazeer Ahmed (CITH), Dr G.V. Thomas (CPCRI), Dr S.K. Pandey (CPRI), Dr S. Edison (CTCRI), Dr (Mrs) Meenakshi Srinivas (IIHR), Dr V.A. Parthasarathy (IISR), Dr Mathura Rai (IIVR), Dr M.M. Mustaffa (NRCB), Dr M.G. Bhatt (NRC Cashew), Dr V.J. Shivankar (NRC Citrus),Dr P.G. Adsule (NRC Grapes), Dr K.K. Kumar (NRC Litchi), Dr S. Maiti (NRCM&AP), Dr M. Kochu Babu (NRC Oilpalm), Dr K.E. Lawande (NRC O & G), Dr R.P. Medhi (NRC Orchids), Dr V.T. Jadav (NRC Pomegranate), Dr B.B. Vasishta (NRC Seed Spices). Project CoordinatorsDr S.S. Sidhu (Floriculture), Dr P.S. Naik (Potato), Dr M.S. Palaniswamy (Tuber Crops), Dr S. Arul Raj (Plantatin Crops), Dr M. Anandraj (Spices) and Dr A.K. Misra (Subtropical Fruits) for their support in compilation of the information. We are thankful to all those who contributed directly or indirectly in bringing out the valuable publication. Financial support provided by various organizations in developing these cultivars is thankfully acknowledged.

We dedicate this piece of work to the Horticultural breeders who has done the commendable job to breed these varieties.

Contributors

Dr. B.M.C. Reddy: *Sub Tropical Fruits,* Director, Central Institute for Subtropical Horticulture Rehmankkhera, P. O. Kakori Lucknow – 227 107, Uttar Pradesh; *e-mail:* director@cish.ernet.in, *Tel:* 0522 - 2841022.

Dr. T.A. More: *Arid Fruits,* Director, Central Institute of Arid Horticulture, Sri Ganganagar Road NH - 15, Beechwal, Bikaner – 334 006, Rajasthan; *e-mail:* ciah@nic.in, *Tel:* 0151 - 2250960.

Prof. Nazeer Ahmed: *Temperate Fruits,* Director, Central Institute of Temperate Horticulture, Old Air Field, Rangreth – 190 005, J & K; *e-mail:* cith@nic.in, dnak59@rediffmail.com, *Tel:* 0194 - 2305044.

Dr. G.V. Thomas: *Coconut, Arecanut and Cocoa,* Director, Central Plantation Crops Research Institute, Kudlu P.O. Kasaragod – 671 124, Kerala; *e-mail:* georgevthomas@yahoo.com, *Tel:* 04994 - 232333.

Dr. S.K. Pandey: *Potato,* Director, Central Potato Research Institute, Shimla – 171 001, Himachal Pradesh; *e-mail:* dircpri@ernet.in, *Tel:* 0177 - 2625073.

Dr. S. Edison: *Tropical Tuber Crops,* Formerly Director, Central Tuber Crops Research Institute, Sreekariyam, Thiruvananthapuram – 695 017, Kerala; *e-mail:* ctcritvm@yahoo.com, *Tel:* 0471 - 2598431.

Dr. Meenakshi Srinivas: *Tropical Fruits, Vegetables and Flowers,* Formerly Director (Acting), Indian Institute of Horticultural Research, Hessaraghatta lake post, Bangalore – 560 089; *e-mail:* director@iihr.ernet.in, *Tel:* 080 - 2846633.

Dr. V.A. Parthasarathy: *Spices,* Director, Indian Institute of Spices Research, Marikunnu P.O. Kozhikode – 673 012, Kerala; *e-mail:* parthasarathy@spices.res.in, *Tel:* 0495 - 2730294.

Dr. Mathura Rai: *Vegetables,* Director, Indian Institute of Vegetable Research, Post Bag No. 01, P.O. Jakhin, Varanasi – 221 005, Uttar Pradesh; *e-mail:* directo@iivr.org.in, mathura.rai@gmail.com, *Tel:* 0542 - 2635236.

Dr. M.M. Mustafa: *Banana,* National Research Centre for Banana, Thogamalai Main Road, Thayanur P.O. Thiruchirapalli – 620 102, Tamil Nadu; *e-mail:* directornrcb@gmail.com, *Tel:* 0431 - 2618104.

Dr. M.G. Bhat: *Cashew,* Director, Directorate of Cashew Research, Puttur P.O. Dakshina Kannada – 574 202, Karnataka; *e-mail:* nrccaju86@yahoo.com, *Tel:* 08251 - 231530.

Dr. V.J. Shivankar: *Citrus,* Director, National Research Centre for Citrus, Post Box No. 464, Shankar Nagar P.O. Amravati Road, Nagpur – 440 010, Maharashtra; *e-mail:* Citrus9_ngp@sancharnet.in, *Tel:* 0712 - 2500813.

Dr. P.G. Adsule: *Grapes,* Director, National Research Centre for Grapes, P.B. No. 3, Manjri Farm Post, Solapur Road, Pune – 412 307, Maharashtra; *e-mail:* dirnrcg@gmail.com, *Tel:* 020 - 26915573.

Dr. K.K. Kumar: *Litchi,* Director, National Research Centre for Litchi, P.O. MIC Bela, Muzaffarpur – 842 005, Bihar; *e-mail:* nrclitchi@yahoo.co.in, *Tel:* 0621 - 2272933.

Dr. Satyabrata Maiti: *Medicinal and Aromatic Plants,* Director, Directorate for Medicinal & Aromatic Plants, Boriavi, Anand – 387 310, Gujarat; *e-mail:* satyabratamaiti@hotmail.com, *Tel:* 02692 - 271602.

Dr. M. Kochu Babu: *Oil Palm,* Director, Project Directorate of Oilpalm Research, Near Jawahar Navodaya Vidyalaya, Pedavegi, West Godavari – 534 450, Andhra Pradesh; *e-mail:* kochubabum@gmail.com, *Tel:* 08812 - 259409.

Dr. K.E. Lawande: *Onion and Garlic,* Director, Project Directorate of Onion and Garlic, Pune-Nasik Highway, Rajgurunagar, Pune – 410 505, Maharashtra; *e-mail:* kelawande@nrcog.res.in, *Tel:* 02135 - 222026.

Dr. R.P. Medhi: *Orchids,* Director, National Research Centre for Orchids, Pakyong, Gangtok – 737 106, Sikkim; *e-mail:* nrcorchids@rediffmail.com, *Tel:* 03592 - 257954.

Dr. V.T. Jadhav: *Pomegranate,* Director, National Research Centre on Pomegranate, NH-9, Bypass Road, Shelgi, Solapur – 413 006, Maharashtra; *e-mail:* nrcpomegranate@indiatimes.com, *Tel:* 0217 - 2374262.

Dr. B.B. Vasisht: *Seed Spices,* Formerly Director, National Research Centre on Seed Spices, Tabiji Farm, Ajmer – 305 206, Rajasthan; *e-mail:* director@nrcss.ernet.in, *Tel:* 0145 - 2443238.

Project Coordinators

Dr. A.S. Sidhu: *Ornamentals,* Project Coordinator Acting, AICRP on Floriculture, Division of Floriculture and Landscaping, Indian Agricultural Research Institute, Pusa, New Delhi – 110 012; *e-mail:* pcflori@gmail.com, *Tel:* 011 - 25843768.

Dr. P.S. Naik: *Potato,* Project Coordinator, AICRP on Potato, Shimla – 171 001, Himachal Pradesh; *e-mail:* Naikps1952@gmail.com, *Tel:* 0177 - 2624398.

Dr. M.S. Palaniswamy: *Tropical Tuber Crops,* Project Coordinator, AICRP on Tuber Crops, Central Tuber Crops Research Institute, Sreekariyam, Thiruvananthapuram – 695 017, Kerala; *e-mail:* aicrptc@yahoo.co.in, *Tel:* 0471 - 2590071.

Dr. S. Arul Raj: *Palms,* Project Coordinator, AICRP on Palms, Central Plantation Crops Research Institute, Kudlu P.O. Kasaragod – 671 124, Kerala; *e-mail:* aicrppalms@yahoo.com, *Tel:* 04994 - 232733.

Dr. M. Anandaraj: *Spices,* Project Coordinator, AICRP on Spices, Indian Institute of Spices Research, Marikunnu P.O. Kozhikode – 673 012, Kerala; *e-mail:* anandaraj@spices.res.in, *Tel:* 0495 - 2731794.

Dr. A.K. Misra: *Sub Tropical Fruits,* Project Coordinator, Central Institute for Subtropical Horticulture, Rahmankhera, Lucknow – 227 107, Uttar Pradesh; *e-mail:* pcstf@cishlko.org, *Tel:* 0522 - 2841115.

Acronyms

Name of Institute/ university	*Expansion*
AAU	Anand Agricultural University
AAU	Assam Agricultural University
AC&RI	Agricultural Colege & Research Institute
AES	Agricultural Experimental Station
AICRP	All India Coordinated Research Project
AINRP on M&AP	All India Network Research Project Medicinal & Aromatic Plants
ANGRAU	Acharya NG Ranga Agricultural University
ARS	Agricultural Research Station
ARU	Agricultural Research Unit
BAC	Bihar Agriculture College
BAU	Birsa Agriculture University
BCKV	Bidhan Chandra Krishi Viswa Vidyalaya
CARI	Central Agricultural Research Institute
CCSHAU	Chaudhary Charan Singh Haryana Agricultural University
CHES	Central Horticultural Experiment Station
CIAH	Central Institute for Arid Horticulture
CISH	Central Institute for Subtropical Horticulture
CPCRI	Central Plantation Crops Research Institute
CPRI	Central Potato Research Institute
CRC	Cardamom Research Centre
CRS	Coconut Research Station
CSAUAT	Chandra Sekhar Azad University of Agriculture & Technology
CTCRI	Central Tuber Crops Research Institute
DARL	Defence Agricultural Research Laboratory
DBSKKV	Dr. Balasaheb Sawant Konkan Krishi Vidyapeeth
FRS	Fruit Research Station

GAU	Gujarat Agricultural University
GBPAUT	Govind Ballabh Pant University of Agriculture and Technology
HARP	Horticulture and Agroforestry Research Programme
HAU	Haryana Agricultural University
HC&RI	Horticultural College & Research Institute
HPAU	Himachal Pradesh Agricultural University
HPKV	Himachal Pradesh Krishi Vishva Vidyalaya
HRS	Horticultural Research Station
IARI	Indian Agricultural Research Institute
ICAR R.C. NEH Region	ICAR Research Complex for NEH Region
ICRI	Indian Cardamom Research Institute
IGAU	Indira Gandhi Agricultural University
IGKV	Indira Gandhi Krishi Vishwa Vidyalaya
IIHR	Indian Institute of Horticultural Research
IISR	Indian Institute of Spices Research
IIVR	Indian Institute of Vegetable Research
JNKVV	Jawaharlal Nehru Krishi Vishwa Vidyalaya
KAU	Kerala Agricultural University
KKV	Konkan Krishi Vidyapeeth
MPKV	Mahatma Phule Krishi Vidyapeeth
NAU	Navsari Agricultural University
NBPGR	National Bureau of Plant Genetic Resources
NDUAT	Narendra Deva University of Agriculture and Technology
NHRDF	National Horticultural Research and Development Foundation
NRC	National Research Centre
NRCOG	National Research Centre for Onion and Garlic
OUAT	Orissa University of Agriculture & Technology
PAU	Punjab Agricultural University
PDKV	Dr. Panjabrao Deshmukh Krishi Vidyapeeth
RARS	Regional Agricultural Research Station
RAU	Rajendra Agricultural University
RS	Regional Station
SKN	Shri Karan Narendra College of Agriculture
SKUA & T	Sher-e-Kashmir University of Agriculture & Technology

SKUAS & T	Sher-e-Kashmir University of Agricultural Sciences & Technology
SKUAST(K)	Sher-e-Kashmir University of Agricultural Sciences & Technology-Kashmir
SVBPUA & T	Sardar Vallabha Bhai Patel University of Agriculture & Technology
TNAU	Tamil Nadu Agricultural University
TNAU RS	Tamil Nadu Agricultural University Research Station
UAS	University of Agricultural Sciences
UBKVV	Uttar Banga Krishi Viswa Vidyalaya
VPKAS	Vivekananda Parvatiya Krishi Anusandhan Sansthan
VRS	Vegetable Research Station
YSPUHF	Dr. Y. S. Parmar University of Horticulture and Forestry

Preface

Indian agriculture has made rapid strides, converting the country from food scarcity to sufficiency. But the challenges ahead in 21st century are much greater than before. The growing population has to be fed and surplus has to be produced in the scenario of declining land, water and threat of climate change. The horticulture, which includes fruits, vegetables, spices, flowers, and medicinal and aromatic plants, has proved beyond doubt its potentiality for gainful diversification. Initiatives taken by the Government and other stakeholders have impacted the development in terms of increased production, productivity and availability of horticultural produce. The emerging trend worldwide and also in the country is indicative of a paradigm shift in dietary needs of the people with rise in the income, which demand for more horticultural produce. Since the growing of horticultural crops is rewarding to the farmers in terms of returns per unit area, the sector is expected to contribute significantly for food and nutritional security, employment opportunity and poverty alleviation.

Past trend in development of horticulture has been satisfying in terms of technological adoptions leading to unprecedented growth in horticulture. Production and export have increased manifolds and high-tech production system utilizing new cultivars and technology is now the reality. The Indian Council of Agricultural Research, an apex body for National Agricultural Research System, has played a pivotal role in technology-led growth. Currently, horticultural research in the country is done through 10 Institutes, 6 Directorates, 7 National Research Centres and 13 All India Coordinated Research Projects having 251 centres located across the country. State Agricultural Universities, a few of them exclusively dedicated to horticulture have significantly contributed for the development of location – and situation-specific cultivars and technologies.

During the last six decades, since 1950, over 1,500 varieties for about 129 horticultural crops have been developed. However, information on the varieties have been scattered which has been an impediment in effective its utilization and many of the varieties remained unnoticed. Therefore, compilation of cultivars developed and recommended for the growing in different parts of the country needed an immediate attention.

When we look at the compilation, we are fascinated to see the efforts made by the breeders to develop the cultivars having traits needed to improve the quality and productivity, besides resistance to biotic and abiotic stresses. This Information is vital not only for breeders but also for students as well as farmers. This is the first collected information on varietal wealth of horticultural crops, methology and value of the cultivars. This book shall be of great values to all those who are concerned with horticulture.

We are highly grateful to Dr. Mangala Rai, Secretary (DARE)&DG, ICAR, for his inspiration and guidance. We also express our thanks to Dr. P. Rethinam, formerly Executive Director, APCC, Jakarta, for peer review of the manuscript. Last but not the least, we are thankful to all those who have contributed in bringing out this book and also to all those who developed the cultivars.

Editors

Contents

Introduction

Horticulture has emerged as the best option to ensure livelihood and food security as well as to ameliorate the "Green Revolution Fatigue" that is gradually pervading Indian agriculture. It brings in sustainability, profitability, nutritional/job security, women empowerment and food safety to the conventional, input-intensive cop husbandry (Table 1). National goal of achieving 4.0% growth in agriculture, can be achieved through the major contribution in growth from horticulture, *i.e.* 6.0 %. Horticulture, including fruits, vegetables, tuber crops, medicinal and aromatic plants, spices, plantation crops and floriculture, has proved beyond doubt its potential for its fruitful diversification in agriculture in an effective land-use planning.

The horticulture sector has been contributing significantly to the GDP in agriculture (28.5% form 13.08% area). Among horticultural crops, contribution of banana to agricultural GDP is highest (1.99%). Horticultural crops form an important part of wholesome food containing carbohydrates, proteins, minerals, dietary fibres and vitamins in adequate amount. The economic importance of horticultural crops such as fruits, vegetables, flowers, plantation crops and spices has been increasing over the years due to the increasing domestic and international demand. Horticultural crop produces more edible energy and protein per unit area and time than many other food crops. For the small and marginal farmers, horticultural crops fits well into multiple cropping systems and has contributed to increasing cropping intensity in subtropical areas. The high profitability of horticultural crops as cash crops has made

Table 1. Change in cropping pattern in India during 1990–2004
(Unit: '000 ha)

Commodity	1990–95	1995–2000	2000–04	1990–04
Rice	150.1	1873.3	-2590.0	-566.6
Wheat	843.9	719.0	760.0	2322.9
Coarse cereals	-5365.4	11.1	-300.0	-5654.3
Cereals	-4500.0	2000.0	-3000.0	-5500.0
Pulses	-2380.0	-1930.0	2120.0	-2190.0
Foodgrains	-6830.0	40.0	-890.0	-7680.0
Oilseeds	1810.0	-3190.0	4230.0	2850.0
Cotton	1595.3	-505.3	390.0	1480.0
Jute and mesta	-90.0	90.0	-120.0	-120.0
Sugarcane	461.4	172.6	-680.0	-46.0
Fruits	483.0	512.0	1095.0	2090.0
Vegetables	-258.0	915.0	506.0	1163.0
Fruits and vegetables	225.0	1427.0	1601.0	3253.0
Horticulture	900.0	1856.0	4514.0	7270.0

Source: Mittal, S. 2007. ICRIER Working Paper 197

it a viable commercial enterprise also. The wide range of horticultural crops suitable to diversified agroclimatic conditions provide ample opportunities to farmers to adopt multiple cropping systems for minimizing risk of crop failure and maximizing their farm income.

The rapid technological advances in varietal improvement, biotechnology, agro-techniques, plant protection, storage and processing, etc. have been made by 10 central institutes, 27 regional centres, 6 directorates, 8 national research centres and 14 AICRPs with 300 centres and three full-fledged horticultural universities existing across the country. The research efforts have recorded impressive achievements in terms of developing new cultivars and production technologies, which have demonstrated 200–300% increase in yield, improving the quality of produce, besides reducing post-harvest losses. Many value-added products have been developed, which have walked towards commercial adoption. Thus, horticulture will play a key role in poverty alleviation and comprehensive rural development. Recognizing the importance of Horticulture sector in the growth of Indian agriculture, the National Horticulture Mission has been launched in 2004 with a total budget of Rs 6,500 crore with the objective of doubling horticultural production from 149 to 300 million tonnes

Table 2. Projected demand of horticultural crops during 2020-21

Commodity	Production			Growth rate (%)
	1998-99	**2011-12**	**2020-2021**	
Fruits	44.04	81.00	98.00	7.8
Vegetables	87.53	185.00	220.00	9.2
Spices	2.91	5.50	650.00	8.0
Coconut	10.27	20.00	24.00	8.4
Cashew nut	0.46	1.70	2.00	25.1
Cocoa, other	3.00	6.80	9.50	11.1
Total	**146.82**	**300.00**	**360.00**	**9.0**

Source: Singh, H.P. 2007. *Indian Horticulture* 52(4)

by 2011–12 and 360 million tonnes by 2020 (Table 2). It has been recognized that horticulture will bring prosperity to rural India like Information Technology and Biotechnology brought in the urban and semi-urban areas. India is poised to be world leader in horticultural research and development.

Area and Production

The area and production of horticultural crops were 13.43 million ha (mha) and 97.83 million t (mt) respectively during 1991–92 which increased to 20.09 mha and 20.07 mmt during 2007–08. The expansion of horticulture in non-traditional areas resulted in increase in area under fruits from 3.8 mha in 1999 to 5.34 mha in 2006 with an increased production of about 10 mmt. The area under vegetable cultivation also showed a similar increase of 1.54 mha from 1999 to 2006 with an increased production of 19.3 mmt. Even though the area under flowers has not shown any tremendous increase over a decade the production has gone up to 27.7 mt (both loose and cut flowers) during 2005–06 as compared to 0.52 mt in 1999 (Table 3). This is mainly because of protected and precision farming techniques. Among the horticultural crops fruits recorded a three-fold increase in area and four-fold increase in production over the last five decades. India is the largest producer of mango, banana, coconut, cashew nut, ginger, turmeric and black pepper, and the second largest producer of fruits and vegetables. India accounts for 10% of the total world fruit production, ranking first in production of banana, sapota and acid lime. In vegetables, India is the second largest producer with 113.0 mt, contributing to 14.4% of the total world production. India is first in cauliflower and pea production, second in onion and third

Table 3. Change in area under horticultural crops in India during 1990–2004

(Unit: '000 ha)

Commodity	**1990–95**	**1995–2000**	**2000–04**	**1990–04**	**Change in area (%)**
Fruits	483	512	1095	2090	28.75
Vegetables	-258	915	506	1163	16.00
Plantation Crops	435	129	240	804	11.06
Spices	211	284	2655	3150	43.33
Flowers	29	16	18	63	0.87
Total	900	1856	4514	7270	100.00

Source: Mittal, S. 2007. ICRIER Working Paper 197

Table 4. Productivity of major horticultural crops in India *vis-s-vis* other countries

Crop	**National (tonnes/ha)**	**Worlds highest (tonnes/ha)**	**Country**	**Increase over India (%)**
Potato	17.05	43.87	Netherlands	157
Banana	34.00	52.54	Costa Rica	54.5
Mango	6.75	12.50	Brazil	85
Citrus	8.40	25.00	Brazil	297.6
Coconut: copra	0.51	0.61	Indonesia	20.0
Cassava	31.44	31.44	India	–
Sweet potato	8.94	24.24	Japan	171
Grape	25.69	25.69	India	–
Onion	13.41	51.19	USA	73.80
Garlic	4.93	17.71	USA	72.20
Apple	7.68	84.01	Austria	993.88
Pear	5.59	283.19	Austria	4966.01
Tomato	17.08	26.69	USA	70.45
Eggplant	16.08	17.48	Japan	34.7
Chilli	9.18	14.40	Spain	44.5
Okra	9.59	0.47	Jordan	17.78
Peas	9.14	8.35	Lithuania	20
Black pepper	0.315	2.95	Malaysia	836
Cardamom	0.175	0.250	Guatemala	42.8
Ginger	3.47	–	India	–
Turmeric	3.92	–	India	–

in cabbage production in the world. Even though India is leading in productivity of some horticultural crops like grapes, cassava, ginger and turmeric, still there is a scope to increase the productivity in other horticultural crops compared to other countries (Table 4). The area and production of flowers, spices and condiments, mushrooms, coconut and cashew also indicated three-fold increases in last few decades.

Diversification in Horticulture

There has been a visible shift towards horticulture as evidenced from Table 3. As a vibrant sector in agriculture, horticulture provides avenues for diversification, enhanced returns per unit area, better land- and water-use and opportunities for employment generation. The major fruits grown in India are mango, banana, citrus, guava, pineapple, grape and papaya in the tropical and subtropical regions, and apple, pear, plum, peach and walnut in temperate region. Apart from these, sapota, annona, ber, pomegranate, litchi, peach, pear, plum, apricot and walnut are grown on a sizeable area. A number of other fruits, such as jackfruit, phalsa, mulberry, aonla, bael, fig, date palm, etc. are also grown in different regions. A number of tropical, subtropical fruits like avocado, macadamia nut, mangosteen and rambutan, though introduced in the country, have yet to be commercially exploited. There is also a need to give priority to nuts such as pistachio nut which is suitable for temperate regions of the country. More than 50 vegetable crops are grown in the country under diversified agroclimatic conditions. Cut and loose flowers, medicinal and aromatic plants, spices and plantation crops amenable to various agroclimatic conditions and cropping systems as components may give rise to diversified value-added products. Diversification and value-addition will be the key words in the Indian horticulture in the 21st century. Thus, India is at the brink of a "*Golden Revolution*" in horticulture.

Genetic Enhancement

In order to ensure food security to the growing population, food production has to be increased by improving the production potential of crops through genetic enhancement. The natural resources in the country are diminishing at a rapid pace causing great concern. Concerted efforts are therefore, made for documentation, molecular characterization, conservation and utilization of important plant genetic resources. India is endowed with a large germplasm pool scientifically conserved in various horticultural institutes besides

NBPGR, New Delhi. About 36, 108 germplasm accessions of cultivated, wild and related genera (fruits – 5,980, vegetables – 18,480, ornamentals – 2,743, spices and plantation crops – 8,196 and medicinal aromatic crops – 709) are being maintained at various research institutes. Till date, 1,530 improved high-yielding, high quality coupled with disease and peat resistance varieties and hybrids have been released by various institutes/universities for cultivation in diverse agroclimatic conditions of the country (Tables 5, 6, 7, 8 and Figs 1, 2 and 3). As a result, productivity of banana and potato has gone up three times each and cassava two times. In the last 4 years itself, the country has recorded 8% increase in potato production. Regular-

Table 5. High-yielding hybrids of fruit crops

Fruit crop	Varieties/hybrids
Mango	Mallika, Amrapali, CISH M2, Ratna, Arka Puneet, Arka Aruna, Arka Anmol, Arka Neelkiran, Sindhu (pulp making), Dashehari 51 (regular-bearing, high yielding), Pusa Surya (red peel, good shelf-life)
Banana	FHIA-01, FHIA-03, CO-2, H-1 and H-2
Grape	Arkavati, Arka Kanchan, Arka Hans, Arka Shyam, Arka Neelmani, Tas-e-Ganesh, Sharad Seedless, Sonaka and Dilkush
Acid lime	Pramalini, Vikdram, PKM -1, Sai Sarbati and Jai Devi
Guava	CISH-G-1, CISH-G-2, ISH-G-3, Arka Mridula, Arka Amulya, Lalit (Pink peel), Safed Jam, Kohir Safeda
Papaya	H-39, Cp-81, CO-1-7, Surya, Pusa Delicious, Pusa Majesty, Pusa giant, Pusa, Dwarf, Pusa Nanha
Litchi	Swarna Rupa, Sabour madhu, Sabour Priya
Pomergranate	Hybrid Ruby, Ganesh, Hyti (red arils), Arakta and Bhagwa
Ber	Gomah Kirti, Umran, Gola and Kaithalli
Aonla	Chakaiya, Krishna, Kanchn, NA-6, NA-7 and NA-10
Beal	NB-5, NB-9, CISH B-1, CISH B-2, Pant Aparna, Pant Shivani
Apple	Sunehari, Lal Ambri, Chaubatia Princes, Amrich, Amred and Royal Ambri
Custard apple	Arka Sahan
Sapota	Kalipatti, Gupthi, Cricket Ball, PKM-2 and DHS
Kiwifruit	Abbott, Allison, Bruno, Hayward and Monty
Peach	Shan-e-Punjab and Flordasun (low-chilling)

Table 6. Vegetables and spices varieties resistance to pests and diseases

Crop	Varieties	Resistant to
Tomato	Arka Vardan	Root-knot nematode
	Arka Alok, Arka Abha, Arka shreshta, Arka Abhijit	Bacterial wilt
	Arka Ashish	Powdery mildew
Brinjal	Pusa Purple Long	Shoot borer and little leaf
	Pusa Purple Cluster	Little leaf
	Arka Nidhi, Arka Keshav, Arka Neelakanth	Bacterial wilt
Chilli	NP46A	Tolerant to thrips
	PMR 57/88K	Powdery mildew
	Pant C1	Tolerant to mosaic and leaf curl
	Musalwadi	Tolerant to die back and powdery mildew
	Jawahar 218	Tolerant to leaf curl and fruit rot
	Punjab Lal, Pusa Sadabahar	TMV, CMV and leaf curl virus
	Ujjwala	Bacterial wilt
Cabbage	Pusa Mukta, K1	Black rot
	Pusa Drumhead	Black leg
Cucumber	Poinsett	Downy mildew, powdery mildew, anthracnose, angular leaf spot
Bottle gourd	Punjab Komal	CMV
Watermelon	Arka Manik	Powdery mildew, downy mildew and anthracnose
Muskmelon	Arka Rajhans	Powdery mildew
Okra	Punjab Padmini, Parbhani Kranti, Arka Abhay, Arka Anamika	Yellow-vein mosaic
Pumpkin	Arka Suryamukh	Fruit fly
Frenchbean	Arka Bold	Rust
Pea	FC1, Jawahar Matar 5, Jawahar Peas 83, JP4 (JM6), Arka Ajit	Powdery mildew
Onion	Arka Kalyan, Arka Kirthiman	Purple blotch, basla rot and thrips
Amaranthus	Arka Suguna	White rust

Table 6. *Contd.*

Crop	Varieties	Resistant to
Black pepper	IISR Shakthi Pournami	*Phytophthroa capcisi* *Melodogyne* spp.
Cardamom	IISR Vijetha IISR Avinash	Mosaic Rhizome rot
Ginger	Mahima Maran	*Meloidogyne* spp. Soft rot
Cumin	Gujarat Cumin 4	Wilt

Table 7. Varieties of horticultural crops (based on breeding methods)

	Introd-uction	Selection	Hybridi-zation	Muta-tion	NA	Total
			Fruits			
Mango		5	23			28
Banana		1	3			4
Papaya		12	4		3	19
Grape	19	3	13			35
Pomegranate		6	4			10
Guava		4	7			11
Citrus	1	4	1			6
Sapota		4	4			8
Avacado		1				1
Passionfruit			1			1
Litchi		2	3			5
Anona		1	1			2
Jack fruit		3				3
Ber		2	1			3
Aonla		6				6
Fig	4					4
Karonda		3				3
Tamarind		3				3
Apple		1	14	1		16
Pear		3			3	6
Peach		1				1
Plum		2				2
Almond		4				4
Apricot				3		3
Walnut		2				2

Table 7. ***Contd.***

	Introd-uction	Selection	Hybridi-zation	Muta-tion	NA	Total
			Vegetable Crops			
Tomato	5	30	43	2	6	86
Brinjal	1	41	39	1	2	84
Chilli		38	15	1		54
Sweet pepper	3	5	2			10
Okra	1	6	20	2	1	30
Bitter gourd		13	5	1		19
Pumpkin		14	2			16
Squash	2	1	1			4
Bottle gourd		14	9		6	29
Muskmelon		13	8		1	22
Watermelon	5	6	5			16
Ash gourd		7				7
Cucumber	2	10	6		1	19
Long melon		2				2
Pointed gourd		4				4
Ivy gourd		3				3
Ridge gourd		6	2	1	1	10
Sponge gourd		8				8
Round gourd		2	1			3
Sanke gourd		6	1	1		8
Sanpmelon		3				3
Kachri		2				2
Kakdi		2				2
Khejri		1				1
Cabbage	3	5	1			9
Cauliflower		14	3		2	19
Broccoli		5			1	6
Knol khol	3	1				4
Carrot	1	5	3		1	10
Beet root		1				1
Radish	1	4	1		5	11
Frenchbean	3	8	8	1	1	21
Vegetable cowpea		10	9			19
Vegetable Dolichos bean		2	3		3	8

Table 7. ***Contd.***

	Introd-uction	Selection	Hybridi-zation	Muta-tion	NA	Total
Garden pea	6	1	16		1	24
Clusterbean		1	1		2	4
Amaranthus		10			5	15
Spinach		1	1			2
Moringa		1	1			2
Potato	1	2	44			47
Cassava	1	8	8			17
Sweet potato	2	25	7		3	37
Greater yam		7	1			8
White yam		3				3
Lesser yam		3				3
Elephant-foot yam		4	1			5
Colocasia (taro, tania, bunda)		18	1			19
Chinese potato/coleus			4			4
Yam bean		1				1
Onion	2	41	6			49
Garlic		21				21
		Flower Crops				
Tube rose			4			4
Chrysanthemum		4	11	3		18
Marigold		1			2	3
Gladiolus		9	25	1	12	47
Rose		73	62	4	45	184
Carnation				1		1
Jasminum		7		1		8
China aster		2	2			4
Orchids			5			5
Bougainvillea		3	3			6
Gerbera		2				2
		Plantation Crops				
Coconut		19	11			30
Arecanut	7		1			8
Cocoa		8	4			12
Cashewnut		18	8			26

Table 7. *Contd.*

	Introd-uction	Selection	Hybridi-zation	Muta-tion	NA	Total
		Spices				
Black pepper		12	4			16
Cardamom		12	1			13
Ginger		7				7
Turmeric		20		3		23
Cinnamon		6				6
Nutmeg		3				3
Currey leaf						0
Coriander		22		1		23
Cumin		8		1		9
Fennel		10				10
Fenugreek		13		2	2	17
Ajowain		2				2
Dill		2				2
Nigella		1				1
Anise		1				1
Celery		1				1
		Medicinal Crops				
Opium poppy		6				6
Asalio		1				1
Safed musli		1				1
Periwinkle		1				1
Senna		2				2
Yam		2				2
Foxglove		1				1
Yellow harned poppy		1				1
Liquorice		1				1
Sarpaganda		1				1
Long pepper		1				1
Isabgol		3		1		4
Solanum lacinatum		1				1
Khasi kateri		1		1		2
Aswagandha		1	1			2
Betlevine		5				5

Table 7. ***Contd.***

	Introd-uction	Selection	Hybridi-zation	Muta-tion	NA	Total
Egyptian henbane		1				1
Thyme		1				1
		Aromatic Crops				
Lemon grass		1				1
Palmarosa		2				2
Spearmint		1				1
Mushakbala		1				1
Geranium		1				1
Total	**73**	**820**	**495**	**33**	**109**	**1,530**

Table 8. Varieties of horticultural crops (based on year of release)

Crop	1951–70	1971–80	1981–90	1991–2000	2001–08	NA	Total
			Fruits				
Mango		3	11	11	2	1	28
Banana			1	2	1		4
Papaya		2	9	3		5	19
Grape	8	4	4	16	3		35
Pomegranate	1		4	3	2		10
Guava				3	2	6	11
Citrus	1		1	3		1	6
Sapota		2	1	3	2		8
Avacado				1			1
Passionfruit		1					1
Litchi		1		2		2	5
Anona				1	1		2
Jack fruit				2	1		3
Ber				1	2		3
Aonla				1	1	4	6
Fig			3	1			4
Karonda						3	3
Tamarind					3		3
Apple		5	8	2	1		16
Pear				2	1	3	6

Table 8. ***Contd.***

Crop	1951–70	1971–80	1981–90	1991–2000	2001–08	NA	Total
Peach				1			1
Plum				2			2
Almond				4			4
Apricot			3				3
Walnut					2		2
			Vegetable Crops				
Tomato	5	11	16	21	20	13	86
Brinjal	3	8	25	24	13	11	84
Chilli		3	15	13	15	8	54
Sweet pepper		3	3	2	2		10
Okra	2	2	7	5	10	4	30
Bitter gourd		2	4	5	5	3	19
Pumpkin		2	4	1	6	3	16
Squash		1	2			1	4
Bottle gourd		1	5	4	10	9	29
Muskmelon	1	4	5	5	5	2	22
Watermelon		4	2	3	3	4	16
Ash gourd		1	1		5		7
Cucumber		2	3	1	6	7	19
Long melon			1	1			2
Pointed gourd				2	2		4
Ivy gourd					3		3
Ridge gourd		2	1	1	5	1	10
Sponge gourd		1	1	1	5		8
Round gourd		1	1		1		3
Sanke gourd		2	2	2	2		8
Sanpmelon				2	1		3
Kachri				2			2
Kakdi				2			2
Khejri				1			1
Cabbage	2	3	1	3			9
Cauliflower	2	6	2	4	4	1	19
Broccoli				2	3	1	6
Knol khol		3			1		4

Table 8. ***Contd.***

Crop	1951–70	1971–80	1981–90	1991–2000	2001–08	NA	Total
Carrot	1	1	2	2	1	3	10
Beetroot				1			1
Radish	1	1	2	3	1	3	11
Frenchbean		2	3	6	6	4	21
Vegetable cowpea			2	6	8	3	19
Vegetable dolichos bean			2	1	2	3	8
Garden pea		3	4	4	9	4	24
Clusterbean				1		3	4
Amaranthus	1	1	2	4	2	5	15
Spinach				1	1		2
Moringa			1	1			2
Potato	17	6	4	9	11		47
Cassava		6	3	6	2		17
Sweet potato		6	4	8	16	3	37
Greater yam				2	3	3	8
White yam				2	1		3
Lesser yam				1	1	1	3
Elephant-foot yam					3	2	5
Colocasia (taro)				3	6	10	19
Chinese potato				2	2		4
Yam bean				1			1
Onion	2	13	5	10	8	11	49
Garlic			5	4	2	10	21
			Flower Crops				
Tuberose				1	3		4
Chrysanthe-mum		3	3	9	3		18
Marigold			1			2	3
Gladiolus		7	7	29	3	1	47
Rose	51	67	39	17	10		184
Carnation					1		1
Jasminum		3	1	4			8
China aster				4			4

Table 8. ***Contd.***

Crop	1951–70	1971–80	1981–90	1991–2000	2001–08	NA	Total
Orchids					5		5
Bougainvillea		6					6
Gerbera				2			2
			Plantation Crops				
Coconut			9	9	12		30
Arecanut		1	2	3	2		8
Cocoa					2	10	12
Cashew nut		3	9	13	1		26
			Spices				
Black pepper	1		2	9	4		16
Cardamom			1	8	4		13
Ginger			2	3	2		7
Turmeric			7	12	4		23
Cinnamon				5	1		6
Nutmeg				1	2		3
Currey leaf							0
Coriander		2	6	8	7		23
Cumin	2		2	4	1		9
Fennel	1	1	2	1	5		10
Fenugreek			3	6	6	2	17
Ajowain					2		2
Dill					2		2
Nigella					1		1
Anise					1		1
Celery					1		1
			Medicinal Plants				
Opium poppy			3	3			6
Asalio				1			1
Safed musli					1		1
Periwinkle					1		1
Senna				1	1		2
Yam		2					2
Foxglove				1			1
Yellow harned poppy				1			1

Table 8. *Contd.*

Crop	1951–70	1971–80	1981–90	1991–2000	2001–08	NA	Total
Liquorice			1				1
Sarpaganda						1	1
Long pepper				1			1
Isabgol		1	2	1			4
Solanum lacinatum				1			1
Khasi kateri			1	1			2
Aswagandha			1	1			2
Betlevine				2	3		5
Egyptian henbane						1	1
Thyme				1	1		
Aromatic Crops							
Lemon grass				1			1
Palmarosa			1			1	2
Spearmint				1			1
Mushakbala				1			1
Geranium			1				1
	102	**215**	**291**	**421**	**329**	**172**	**1,530**

NA — Details not available

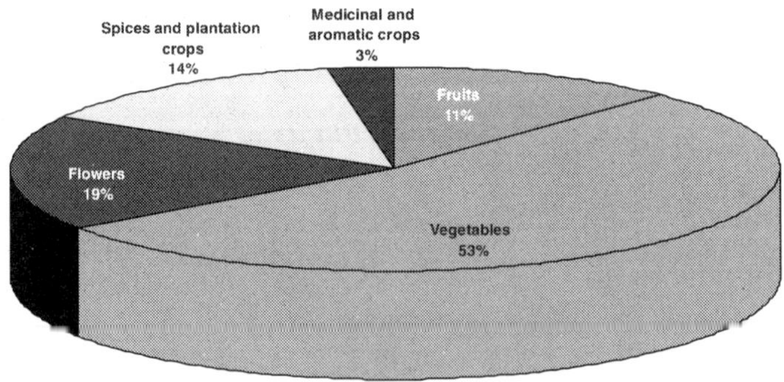

Fig 1. Varieties of horticultural crops

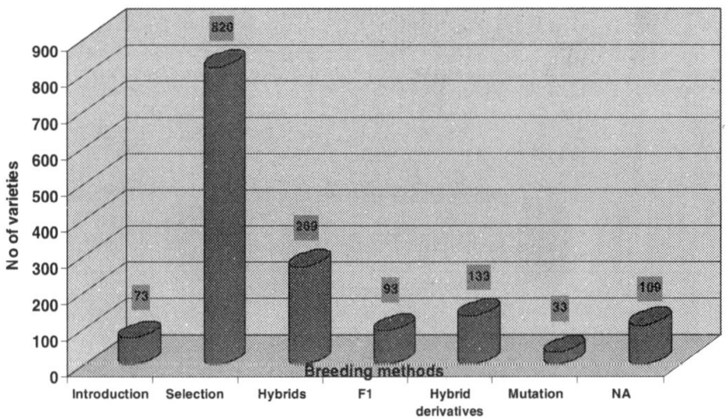

Fig 2. Varieties and breeding methods

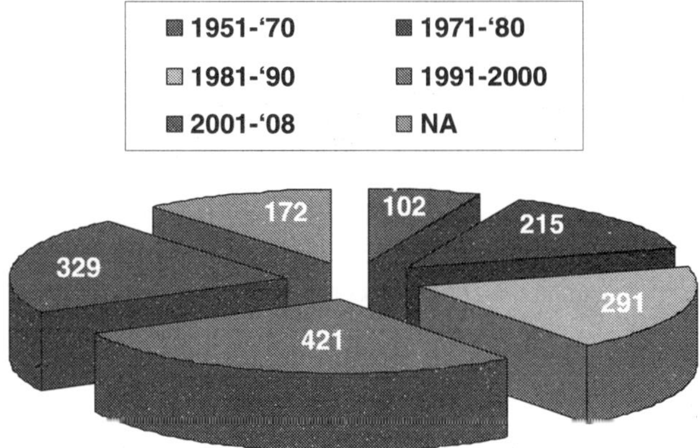

Fig 3. Varieties and year of release

bearing mango hybrid, export quality grapes, multi-disease resistant vegetable hybrids, high-value spices and tuber crops of industrial uses have been developed. Tomato varieties resistant to tomato leaf curl virus, bacterial wilt and *Fusarium* wilt have been developed. Improved varieties have revolutionized the Horticulture sector. For instance, high-yielding Gauri, Sankar and Sree Bhadra sweet potatoes have brought a revolution by minimizing malnutrition, improving nutritional security in the Kandhamal district in Orissa.

Genetic Resources of Horticulture Crops

India is endowed with varied agroclimatic conditions, which are congenial for growing a wide range of horticultural crops. Almost all types of horticulture crops can be grown in one or the other region of the country. One hundred and ninety species of economically important crops have been listed out comprising 109 fruits, 54 vegetables and 27 spices and condiments, among others. India is the largest producer of fruits and vegetables in the world. Horticultural crops have attained 50% growth in production during the past 10–15 years. These crops form a subset of agro-biodiversity, which include horticultural plant species or their wild gene pool, having genetic material of actual or potential value. The Horticultural Genetic Resources also include improved and obsolete varieties, populations, landraces, genetic stocks and breeding materials of crop plants and their wild relatives. Among horticultural genetic resources, landraces and populations are heterogenous group of plants, which are able to survive and sustain under varying agroclimatic conditions. However, introduction of improved varieties over a period of time has replaced innumerable local varieties and landraces, which result in genetic erosion and genetic vulnerability to biotic and abiotic stresses, leading to crop losses. For example, a high-yielding and yellow-vein mosaic resistant okra 'Pusa Sawani' has replaced most of the local landraces in India. Hence, cultivar replacement is considered to be one of the main causes of genetic erosion.

India is the centre of diversity for many horticultural crops, which were shown to have all kinds of endemic varieties, alleles and even, Linnean species. Moreover, these centres are characterized by repository of dominant genes and recessive genes in the periphery. Among horticultural crops, most of the main fruits and vegetables originated from the Near-Eastern Asia, Central Asia and Mediterranean centres, while root and tuber crops and tropical fruit trees are from the Central American and Andean centres. India is known as the land of spices, being the origin of black pepper and cardamom and also for ginger, turmeric, *Garcinia* and *Myristica* with maximum diversity. Rich diversity also occurs in India for medicinal and aromatic plants and traditional

Table 1: India — Centre of Diversity of Horticultural Crops

Fruits	*Primary centre:* mango, citrus, jack fruit, bael, aonla, ber, khejri, jamun, tamarind, phalsa, lasora, karonda, woodapple, pilu, bilimbi, garcinia
	Secondary centre: banana, pomegranate, mulberry, malus, pyrus, prunus, rubus
Vegetables	*Primary centre:* brinjal, smooth guard, ridge guard, cucumber, parwal, amaranthus, basella, sword bean, winged bean, kundru, dolichos bean, indian lettuce, drumstick
	Secondary centre: cowpea, okra, chilli, pumpkin and brassicas
Ornamentals	*Flowers:* orchids, rhododendrons, mask rose, begonia, balsam, globe amaranth, glory lily, foxtail lily, primula, blue poppy, lotus, water lily, clematis, tulip
	Trees: kachnar, amaltas, pink cassia, butea, Indian coral tree, pride of India, scarlet cordia, yellow silk cotton tree, karanj, tecomella, tulip tree, chalta, sita ashok, arjun, michelia, kadamba, maulsari
	Shrubs and climbers: jasmine, ixora, hamiltonia, clerodendron, crossandra, plumbago, tabernamontana, trachlospermum, passiflora, clitoria, porana, gloriosa, clematis
Plantation crops, spices and condiments	Pepper, greater galanga, bengal cardamom, anethum sowa, ajowain, cinnamon, cumin, *curcuma* spp., curry leaf, long pepper, betelvine, long pepper, ginger, Indian cinnamon, Indian tamarind, kokum and tamarind
Tuber crops	Greater yam, lesser yam, potato yam, elephant-foot yam, yam bean, winged bean, alocasia, giant taro, colocasia
Medicinal and aromatic plants	Muskdana, belladonna, jamalgota, Malabar grass, rosha grass, citronella grass, lemon grass, datura, puskarmul, jasmine, saya, isabgol, patchouli, sarpagandha, sandal wood, costus, nuxvomica, Indian almond, vetiver, kutaki, ban-kakri, asparagus, atees, vatsnabh, Indian ginseng, ashoka, arjuna, bijayasal, kurchi, neem, guruchi, lodhara

knowledge associated with their uses, particularly in the Western Ghats and North-Eastern region. Besides, Indian subcontinent is rich repository for ornamental trees, shrubs, climbers, herbs and succulents. Wild relatives are also available in plenty in the rain forest habitats (Table 1).

The germplasm of horticultural crops is being conserved both under *in situ* and *ex-situ* gene banks. The germplasm that are being conserved under different ICAR institutes are given in Tables 2–7.

Table 2: Genetic Resources of Horticultural Crops

Crop	No. of Accessions
Fruits	5,980
Vegetables	18,480
Flowers	2,743
Spices and plantation crops	8,196
Medicinal and aromatic plants	709
Total	**36,108**

Table 3: Flower Crops

Crop	IIHR	IARI	NRCO	NBPGR	Total
Tuberose	11				11
Chrysanthemum	54				54
Marigold					0
Gladiolus	47				47
Rose	274				274
Carnation	45				45
Jasminum	13				13
China aster	4				4
Orchids			2205		2205
Bougainvillea					0
Anthurium	60				60
Gerbera	18				18
Crossandra	12				12
Total					**2743**

Table 4: Fruit Crops

Crop	IIHR	CIAH	CAZRI	CISH	CITH	NRCB	NRCC	NBPGR	NRCG	NRCP	Total
Mango	453	52		718							1,223
Banana	75					341					416
Papaya	16			26							42
Grape	106								444		550
Pomegranate	20	190			12			170		187	579
Guava	70			105							175
Citrus	114						602				716
Sapota	33	7									40
Avacado	5										5
Passionfruit	11										11
Litchi											0
Anona	16	9									25
Jack fruit	34										34
Ber		395									395
Aonla		64									64
Fig	9	8						2			19
Karonda		48									48
Tamarind		25									25
Date palm		55									55
Bael		36									36
Phalsa		8									8

Table 4: ***Contd.***

Crop	IIHR	CIAH	CAZRI	CISH	CITH	NRCB	NRCC	NBPGR	NRCG	NRCP	Total
Jamun	21	50									71
Apple					236			165			401
Pear					40			70			110
Peach					22			47			69
Plum					34			37			71
Almond					26			60			86
Apricot					56			47			103
Walnut					150			110			260
Cherry					26			2			28
Chesnuts					0			33			33
Kiwi fruit					6			15			21
Olive					21						21
Pecan nut					6			11			17
Persimon					2			11			13
Pistachi nut					1	1		2			3
Ribes								26			26
Rubes								45			45
Strawberry					56	40		96			152
Total											**5980**

Table 5: Vegetable Crops

Crop	IIHR	IARI	IIVR	CIAH	CTCRI	CPRI	NRCOG	NBPGR	ICAR	RC Barabani
Tomato	500		1200							1,700
Brinjal	280		600	30						910
Chillies	230		307	132						669
Sweet Pepper	40	150	80							270
Bhendi			470	2						472
Paprika	50	100	-							150
Bitter gourd			267	41						308
Pumpkin			124	21						145
Squash			-							0
Bottle gourd			190							190
Musk melon			84	60						144
Water melon			-							0
Ash gourd			105							105
Cucumber	50		400							450
Long melon			20							20
Pointed gourd			215							215
Ivy gourd			42	1						43
Ridge gourd			-	57						57
Sponge gourd			-	15						15
Round gourd			-	10						10

Table 5: ***Contd.***

Crop	IIHR	IARI	IIVR	CIAH	CTCRI	CPRI	NRCOG	NBPGR	ICAR	RC Barabani
Sanke gourd			-							0
Sanp melon			185	82						267
Kachri			-	558						558
Kakdi			-	18						18
Khejri			-							0
Cho-cho			-						75	75
Cabbage		195	-							195
Cauliflower		200	134							334
Broccoli			-							0
Knol khol			-							0
Carrot			75							75
Beetroot			-							0
Radish			53							53
Frenchbean	129		65							194
Vegetable cowpea	235		266							501
Vegetable Dolichos bean	176		400							576
Garden pea	60		425							485
Cluster bean				7						7
Amaranthus										0
Spinach										0

Table 5: ***Contd.***

Crop	IIHR	IARI	IIVR	CIAH	CTCRI	CPRI	NRCOG	NBPGR	ICAR	RC Barabani
Moringa				41						41
Potato						3020				3,020
Cassava					1669					1,669
Sweet potato					1032					992
Yams					679					679
Greater yam										0
White yam										0
Lesser yam										0
Elephant-foot yam										0
Ariods					1171					1,171
Colocasia (taro)										0
Colacasia (bunda)										0
Chinese potato					90					90
Xanthosoma										0
Coleous					2					2
Yam bean					63					0
Minor tubers					197					185
All tuber crops					0					707
Onion							480			480
Garlic							825			825
Total										**18,480**

Table 6: Spices and Plantation Crops

Crop	CPCRI	IISR	NRCSS	NRCC	NRCOP	NBPGR	Total
Coconut	365						365
Arecanut	153						153
Cocoa	230						230
Cashew				509			509
Oilpalm					110		110
Black pepper		2327					2327
Cardamom		436					436
Ginger		684					684
Turmeric		1040					1040
Cinnamon		408					408
Nutmeg		484					484
Clove		233					233
Cassia							0
Garcinia		86					86
Allspices		2					2
Vanilla		82					82
Coriander			275				275
Cumin			135				135
Fennel			259				259
Fenugreek			204				204
Ajowain			79				79
Dill			61				61
Nigella			12				12
Anise			7				7
Celery			11				11
Caraway			4				4
							8196

Table 7: Medicinal and Aromatic Crops

	NRCMAP	IIHR	Total
Senna	5		5
Safed musli	54		54
Isabgol	9		9
Aswagandha	11	151	162
Betlevine		104	104
Lemon grass	13		13
Aloe vera	55	40	95
Coleus		25	25
Andrographis paniculata	59		59
Asparagus sp.	47		47
Commiphora	67		67
Datura sp.	2		2
Evolvulus alsinoides	3		3
Phyllanthus sp.	13		13
Tinospora cordifolia	38		38
Tribulus terrestris	7		7
Urginea indica	6		6
	389	**320**	**709**

Varieties of Horticultural Crops — Status of Last Five Decades

FRUIT CROPS

Tropical and Subtropical Fruits

MANGO

Botanical name: *Mangifera indica*
Family: Anacardiaceae

Mango, most important tropical fruit of the world, is called as the 'King of fruits'. Indo-Burma-Siam region and the Philippines are considered to be the probable places of origin of mango. The crop has been under cultivation for more than 4,000 years in India. The fully ripe fruit is not only considered as a delicious table fruit, but also used in preparation of jam, jelly, squash, syrups and nectarines. The raw fruits, before ripening, can be used in preparation of pickles, chutneys etc. India continues to be the largest mango-producing country in the world, accounting for more than 50% of the total world production. It is grown in about 1.60 million ha in India with an annual production of 10.78 million tonnes. Although India is the largest producer, its productivity is only 9 tonnes/ha as against 30 tonnes/ha in Israel.

The mango fruit is a drupe; when mature. The fruits vary in their size: 10–25 cm long, 7–12 cm in diameter, and weigh up to 2.5 kg. The ripe fruit is variably coloured — yellow, orange and red, red on

MANGO

Variety	Year of release	Breeding method	Pedigree/ parentage	Important traits	Institutes/ universities
Dashehari 51	1998	Selection	Clonal selection from Dashehari	A clone of Dashehari with regular bearing habit	CISH, Lucknow
Ambika	2000	Hybridization	Amrapali × Janardhan Pasand	Colour is bright yellow with dark red blush, flesh dark yellow, firm with scanty fibre. TSS is 21° Brix. It is late in ripening. The hybrid has potential of export and internal market because of its attractive fruit colour.	CISH, Lucknow
Al Fazali	1981	Hybridization	Alphonso × Fazali	Good quality and late maturing, free from malformation and fruit fly	BAC, RAU, Sabour
Sundar Langra	1981	Hybridization	Langra × Sunder Pasand	Moderate and regular bearing	BAC, RAU, Sabour
Safari	1989	Hybridization	Gulabkhas × Bombai	Medium fruit size with red blush on base	BAC, RAU, Sabour
Jawahar	1989	Hybridization	Gulabkhas × Mahmud Bahar	Semidwarf, precocious and regular bearer	BAC, RAU, Sabour
Menaka	1991	Selection	Seedling selection of Gulabkhas	Late-maturing, attractive fruit colour (red blush on basal side of fruits) good yield with moderate storage life	BAC, RAU, Sabour

MANGO : *Contd.*

Variety	**Year of release**	**Breeding method**	**Pedigree/ parentage**	**Important traits**	**Institutes/ universities**
Subhash	1991	Selection	Chance seedling selection from Zardalu	Similar to langra, medium size, attractive yellow colour of zardalu, TSS 24%, acidity 0.294%	BAC, RAU, Sabour
Neelphonso	1986	Hybridization	Neelum × Alphonso	Regular bearer, dwarfer, late maturing, suitable for table and juice	AES, NAU, Paria
Neeleshan	1986	Hybridization	Neelum × Baneshan	Dwarf, late-maturing with good keeping quality	AES, NAU, Paria
Neeleshwari	1986	Hybridization	Neelum × Dashehari	Dwarf, late maturing with non fibrous, firm texture and very good fruit quality	AES, NAU, Paria
Manjeera	1985	Hybridization	Rumani × Neelum	Regular, heavy bearing in bunches and suitable for high density planting	FRS, Sangareddy
AU Rumani	1985	Hybridization	Rumani × Neelum	Fruit large, heavy yielder and fibreless	FRS, Sangareddy
Mallika	1971	Hybridization	Neelum × Dashehari	Heavy bearer, big fruit size and good quality fruits	IARI, New Delhi
Amrapali	1979	Hybridization	Dashehari × Neelum	Regular bearer, dwarf suitable for high density planting	IARI, New Delhi

MANGO : *Contd.*

Variety	Year of release	Breeding method	Pedigree/ parentage	Important traits	Institutes/ universities
Pusa Arunima	2002	Hybridization	Amrapali × Sensation	Regular bearer, high quality, good yield, attractive colour	IARI, New Delhi
Pusa Surya	2002	Selection	Selection from Eldon	Regular bear, high yield with good quality fruits, attractive peel colour	IARI, New Delhi
Arka Aruna	1992	Hybridization	Banganapalli × Alphonso	Dwarf, regular, suitable for high- density planting and free from spongy tissue	IIHR, Bangalore
Arka Anmol	1996	Hybridization	Alphonos × Janardhan Pasand	High quality and free from spongy tissue	IIHR, Bangalore
Arka Puneet	1992	Hybridization	Alphonso × Banganapalli	Red blush, high-keeping quality and free from spongy tissue	IIHR, Bangalore
Arka Neelkiran	1992	Hybridization	Alphonso × Neelum	Regular bearer, late harvesting, free from spongy tissue	IIHR, Bangalore
Ratna	1981	Hybridization	Neelum × Alphonso	Regular bearer and free from spongy tissue	RFRS, Vengurla
Sindhu	1992	Hybridization	Ratna × Alphonso	Regular bearer and free from spongy tissue, early maturity, seedless	RFRS, Vengurla

MANGO : ***Contd.***

Variety	Year of release	Breeding method	Pedigree/ parentage	Important traits	Institutes/ universities
Konkan Ruchi	1999	Hybridization	Neelum × Alphonso	Regular bearer, high-yielding and suitable for pickle making	RFRS, Vengurla
PKM-1	1980	Hybridization	Chinnasuvarnarekha × Neelum	Regular bearer, high yielding and bearing in clusters	HC&RI, TNAU, Periyakulam
PKM-2	1982	Hybridization	Neelum × Mulgoa	Regular bearer with excellent fruit quality	HC&RI, TNAU, Periyakulam
Paiyur 1	1992	Selection	Clonal selection from Neelum from Karukkanchavadi, Dharmapuri district	Trees are dwarf, low spreading and suitable for high density planting. This is a regular-bearer with a mean yield of 22.3 kg/tree (8929 kg/ha) nine years after planting. Fruits are medium long, ovate with yellow skin. Fruits have excellent taste with good keeping quality	HRS, TNAU, Paiyur
Sai-Sugandh	—	Hybridization	—	Fruit colour golden yellow on ripening, average weight 380 g, TSS 22%, aroma pleasant, sweet taste, free from spongy tissue, stone weevil, mono-embryonic, suitable for export and table purpose	MPKV, Rahuri

the side facing the sun and yellow where shaded; green usually indicates that fruit is not yet ripe. The chromosome number is n = 20 and 2n = 40.

Botany

Tree, leaves alternate, petiolate, entire, coracious or membranous. Flowers are borne on panicle, small, penta or tetramerous, polygamous in terminal panicles, the bracts are deciduous, petals 4–5, rarely 6, free/adnate to the disc. Stamens 5, rarely 10–12 inserted within the disc or on it, free or connate at base; disc is swollen, 4–5 lobed or pedicelate or absent. Ovary sessile, always one celled, oblique or globose, style lateral. Fruits is fleshy drupe; stone compressed and fibrous.

Pollination mechanism/floral biology

- Primarily terminal but axillary and multiple panicle may also arise from axillary buds frequently.
- Flowers are male and hermaphrodite.
- Total number of flowers vary from 1,000 to 6,000.
- Sex ratio is a varietal feature and is also greatly influenced by environment.
- Flowers start opening in the morning and the anthesis is completed generally by the noon.
- Maximum flowers open between 9 and 10 AM. Dehiscence takes place after the flowers have opened.
- Although receptivity of stigma continues up to 72 h after anthesis, it is most receptive for about 6 h after anthesis.
- Highly cross-pollinated crop.
- Pollination takes place by housefly and honeybees (entomophilous).

BANANA

Botanical name: *Musa* spp.
Family: Musaceae

Banana is most important fruit crop in India, next to mango, having about 30% of the total cultivated area under fruit crops. Being relatively a cheap commodity, banana plays a major role in satisfying the nutritional requirement of the poor and thus is rightly called as “poor man’s apple”. It is a fast-growing, herbaceous perennial plant,

BANANA

Variety	Year of release	Breeding method	Pedigree/ parentage	Important traits	Institutes/ universities
Udhayam	2005-06	Selection from Kenthali	Single plant	The average bunch weight is 37 kg having a potential up to 45–50 kg. Crop duration is 13 months and produces high-yield in ratoons also. Exhibits field tolerance to Sigatoka leaf spot diseases and nematodes infestation. It has cylindrical bunch with well-spaced hands, hence suitable for long distance *transportation* and less fruit damage during transit. The fruits are high in sugar with 31° Brix with harmonious blend of acidity and sweetness, hence suitable for processing into value-added products like figs, banana juice, wine, etc. Suitable for Tamil Nadu, Andhra Pradesh, Kerala, Bihar, West Bengal and Tripura	National Research Centre for Banana, Trichy, Tamil Nadu
BRS-1	1998-99	Hybridization	Agniswar × Pisang Lilin	It is a triploid (AAB) hybrid of pome type. It has a short cropping cycle, resistance to leaf spot, *Fusarium* wilt and burrowing nematode (*Radopholus similis*). Medium statured with 14–16 kg bunch. Elongated fruits turn	Banana Research Station, Kannara, Kerala

BANANA : *Contd.*

Variety	Year of release	Breeding method	Pedigree/ parentage	Important traits	Institutes/ universities
				attractive golden yellow on ripening. Fruits are slightly acidic which vanishes upon full ripening with high sugar content. It has faster ratoonability, completing 4 crop cycles in 3 years. Suitable for Kerala	
BRS-2	1998-99	Hybridization	Vannan × Pisang Lilin	It belongs to Mysore subgroup, Poovan type. It is a medium-statured plant growing up to 7–8 feet. Crop-cycle is short with bunch coming to harvest in 11–12 months. Average weight of bunch ranges from 15–20 kg with short, stout, dark green Poovan like fruits, which are arranged very compactly. Fruits are slightly acidic with pleasant sweet-sour aroma. Tolerance to leaf spot diseases and nematodes. Suitable for Kerala.	Banana Research Station, Kannara, Kerala
CO-1 (AAB)	1985-86	Hybridization	Ladan (AAB) × Clone Sawar (BB) ↓ F_1 × Kadali	It is a Pome hybrid with AAB genome developed by 3-way sequential crosses. It retains the typical apple flavor of Virupakshi even when grown in plains contrary to Virupakshi, which develops aroma only when grown at higher altitudes.	TNAU, Coimbatore

arising from underground rhizomes. Leaves are, oblong or elliptic, becoming as much as 9 feet long and wide. Approximately 44 leaves appear before the inflorescence. The inflorescence is a spike originating from the tip of the corm. Female flowers, with inferior ovaries, occupy the lower 5–15 rows on stalk, with neuter or hermaphrodite flowers in the centre, and males at the top. Male flowers and bracts are shed one day after opening. The flowers consist of three each sepals and petals, two each outer whorl stamen and inner stammen, and a tri-locular, inferior ovary. The adaxial petal is not opposed by an inner whorl stamen, resulting in only two stamen in this whorl. Fruits epigynous berry, and are borne in hands of up to 20 fruits with 5–13 hands per spike. Fruits appear as angled, slender, green fingers during growth, reaching harvest maturity 90–120 days after flower opening.

Floral Biology

- Banana blooms year-round, but most flowering is concentrated from February to August.
- The longevity of female flowers in more than male flowers and opening of both types of flowers occurs at any time during the day, but mostly in the morning.
- The natural volume and rate of female flowers are significantly higher than male flowers.
- Cavendish group of bananas are triploids and therefore completely sterile, fruit is set parthenocarpically.
- Floral morphology suggests that wild bananas are bat pollinated in their native range.
- Insects such as bumble bees (*Bombus eximius* and *B. montivolans*), honey bees (*Apis cerana* and *A. florea*), and wasps (*Vespa mandarinia*) are primary floral visitors, showing a preference for female flowers.

PAPAYA

Botanical name: *Carica papaya* L.
Family: Caricaceae

Papaya is an important fruit growing in tropical and subtropical regions. Its fruits are valued for high nutritive value with higher production potential. Papaya has a wide range of adaptability with high returns per unit area. Its demand is increasing not only for table purposes but also for its nutraceutical value coupled with low calorific value. Papaya cultivation in the country has largely been remained in peri-urban areas, catering to the need of urban people. The production is intensive characterized by chemical based inputs and is highly remunerative. Papaya has emerged as an industrial fruit crop in the recent times with perceptible increase in area and production. Apart from its delicious fruit, the demand for proteolytic enzyme papain obtained from the milky latex of partially mature fruit acted as a major factor for significant increase in papaya cultivation in many parts of the world.

The somatic chromosome number in dicotyledonous genus *Carica*, is 2n = 18. It is normally a small-unbranched, soft-wooded tree but sometimes branches arise when apical bud is damaged. The stem is hollow with soft fleshy tissue, covered with smooth grey bark marked externally by numerous large orbicular leaf scars. Leaves are very large up to 75 cm across and deeply divided into about 7 palmate lobes again being pinnately lobed. They usually form a crown over the central trunk. Most *Carica* spp. are dioecious, except for *C. papaya,* which is characterized by various flower types and three primary, polygamous sexual types, *viz.* pistillate (female; mm), staminate (male; M1m) and hermaphrodite (M2m). The 5-petalled flowers of papaya are fleshy, waxy, cream to yellow in colour, and slightly fragrant. Flowers are borne singly or on cymose inflorescences in leaf axils. Staminate trees produce long pendulous male inflorescences bearing 10 stamens in each flower, while pistillate trees bear one or two flowers at each leaf axil, with the absence of stamens and a large ovary with numerous ovules. Hermaphrodite trees normally bear one to several bisexual flowers characterized by an elongated, slender ovary and usually 10 stamens. Papaya can be grouped into either dioecious or gynodioecious. The dioecious consists of female and male trees, while gynodioecious female and hermaphroditic trees. In gynodioecious group, pollen for fertilization of the female flowers is derived from the bisexual flowers of hermaphrodite trees.

Floral Biology

- The *C. papaya* flowers throughout the year in India, inflorescence axillary.
- Takes 21 days from bud emergence to anthesis, peak anthesis being between 5 and 6 AM.
- Stigma receptivity is maximum on the day of anthesis.
- Anther dehiscence is completed 18–36 h before flowers are opened and stigma became receptive a day before the flowers opened, remaining receptive for 6 days.
- Pollens from hermaphrodite flowers are inferior to staminate flowers

PAPAYA

Variety	Year of release	Breeding method	Pedigree/ parentage	Important traits	Institutes/ universities
CO 1	1972	Selection	Selection from Ranchi type	Dioecious type, flesh is orange-yellow	TNAU, Coimbatore
CO 2	1979	Selection	Pure line selection from local type	Dioecious type, fruits are oblate and large in size, flesh orange-yellow, dual pupose variety (for fruits and papain extraction)	TNAU, Coimbatore
CO 3	1983	Hybridization	CO 2 × Sunrise Solo	Gynodioecious type, highly suitable for table purpose, fruits are pyriform in shape, flesh is red	TNAU, Coimbatore
CO 4	1983	Hybridization	CO 1 × Washington	Dioecious type, purple petiole and stem	TNAU, Coimbatore
CO 5	1985	Selection	Washington	Dioecious type suitable exclusively for papain production, pink petiole	TNAU, Coimbatore
CO 6	1986	Selection	Giant	Dioecious type, fruits large, dual purpose variety	TNAU, Coimbatore
CO 7	1997	Hybridization	Pusa Delicious × CO 3	It is superior to CO 3 in terms of fruit size and quality and gynodioecious type	TNAU, Coimbatore
Coorg Honey Dew	—	Selection	Honey Dew	Gynodioecious type, fruits oblong, plants dwarf and high-yielding	IIHR, Bangalore

PAPAYA : *Contd.*

Variety	Year of release	Breeding method	Pedigree/ parentage	Important traits	Institutes/ universities
Pink Flesh Sweet	1992	Selection	—	Gynodioecious type	IIHR, Bangalore
Surya	1994	Hybridization	Sunrise Solo × Pink Flesh Sweet	Gynodioecious type	IIHR, Bangalore
Pusa Delicious	1982	Selection	Line: 1-15	Gynodioecious type, fruits round, medium in shape, flesh deep orange	Pusa, Samastipur, Bihar
Pusa Majesty	1982	Selection	Line: 22-3	Gynodioecious line, fruits round, good keeping quality	Pusa, Samastipur, Bihar
Pusa Dwarf	1982	Selection	1-45 (D)	Dioecious type, more precocious in bearing, bears at 30 cm height	Pusa, Samastipur, Bihar
Pusa Giant	1982	Selection	1-45 (V)	Dioecious type, fruits are attractively big sized	Pusa, Samastipur, Bihar
Pusa Nanha	1982	Selection	Mutant Dwarf	Dioecious type ultra dwarf variety	Pusa, Samastipur, Bihar

PAPAYA : *Contd.*

Variety	Year of release	Breeding method	Pedigree/ parentage	Important traits	Institutes/ universities
Pant 1	1984	Selection	Seedling selection	Dwarf plant, heavy-yielder, plant start bearing from 40–45 cm above soil, fruit weight 1–1.5 kg, resistant to anthracnose	GBPAUT, Pant Nagar
Pant 2	1984	Selection	Seedling selection	Medium sized plants, tolerant to frost and wet feet condition, vigorous, bear at 60-90 cm height, fruit are medium to large	GBPAUT, Pant Nagar
Pant 3	1984	Selection	Seedling selection	Medium height, bearing first flower at a height of 115–130 cm from the ground on strong system, fruit small to medium (0.5–0.9 kg), exellent quality, tolerant to frost and water logging	GBPAUT, Pant Nagar
Punjab Sweet	—	—	—	Dioecious type, trees start bearing at 75 cm height, fruits yellow fleshed, round to oval	PAU, Ludhiana

- Papaya is pollinated primarily by wind and small insects
- Honey bees and butterflies are more effective pollinating agents.

GRAPE

Botanical name: *Vitis vinifera*
Family: Vitaceae

Grape is a deciduous crop. Its natural habitat is temperate climate. Presently, its cultivation is concentrated in peninsular India, accounting for 90% of the total area. Major grape-growing states are Maharashtra, Karnataka, Andhra Pradesh, Tamil Nadu, Punjab, Haryana, Delhi, western Uttar Pradesh, Rajasthan and Madhya Pradesh.The grape is non-climacteric fruit that grows on perennial and deciduous woody vines. Grapes can be eaten raw or used for making jam, juice, jelly, vinegar, wine, grape seed extracts and grape seed oil. Grapes are also used in some kinds of candy. Seedlessness is a highly desirable subjective quality in table grape selection, and seedless cultivars now make up overwhelming majority of table grape plantings. Because grapevines are vegetatively propagated by cuttings, the lack of seeds does not present a problem for reproduction. It is desirable to use a seeded-variety as female parent or rescue embryos early in development using tissue culture techniques. The source for seedlessness trait are Thompson Seedless, Russian Seedless, and Black Monukka, all being cultivars of *Vitis vinifera*.

Vigorous, deciduous vines growing 60–100 feet in the wild. In contrast to most other grapes, have a tight, non-shedding bark, warty shoots and unbranched tendrils. The slightly lobed, two and-a-half to 5 inch leaves are rounded to broadly ovate with coarsely serrate edges and an acuminate point. Dark green above and green tinged yellow beneath, the leaves are glossy on both sides, becoming firm and subglabrous at maturity. Flowers are dioecious, with male and female flowers on different plants. The small, greenish flowers are borne in short, dense panicles. It appears that both wind and insects play a role in the pollination of female flowers. The flowers have five sepals (calyx) forming the outer part. The corolla also called the cap or calyptra, is made up of five petals joined at the top, as the catyptra opens it falls off. There are five stamens opposite the petals. The ovary consist of carpels having two locules. In each locule there are two ovules. The fruit is borne in small, loose clusters of 3–40 grapes, quite unlike the large, tight bunches characteristic of European and American grapes. The round, fruits have a thick and tough skin, containing up to 5 hard, oblong seeds.

GRAPE

Variety	Year of release	Breeding method	Pedigree/ parentage	Important traits	Institutes/ universities
Perlette	1969	Introduction	Regina de Vignette × Sultana Marble from UC, Davis, California, USA	Early ripening, seedless berries	IARI, New Delhi
Pusa Seedless	1970	Clonal selection	Thompson Seedless	Good yield, bold attractive bunches with uniform ripening	IARI, New Delhi
Pusa Navrang	1997	Hybridization	Madelien Angevine × Rubi Red	Vines moderately vigour, anthracnose resistant, a prolific-bearer, early ripener. Berries teinturier dark red, suited for juice and port wine	IARI, New Delhi
Pusa Urvashi	1997	Hybridization	Banqui Abhyad × Perlette	Vines moderately vigored, prolific-bearer, bunches medium well-filled, berries spherical transparent greenish yellow. Seedless suitable for table and raisin making	IARI, New Delhi
Arka Kanchan	1980	Hybridization	Anab-E-Shahi × Queen of Vineyards	Good yield, muscat flavour, suited for port wine making	IIHR, Bangalore
Arka Hans	1980	Hybridization	Banglore Blue × Anab-e-Shahi	Pleasant foxy flavour, suited to table wine	IIHR, Bangalore

GRAPE : *Contd.*

Variety	Year of release	Breeding method	Pedigree/ parentage	Important traits	Institutes/ universities
Arka Shyam	1980	Hybridization	Banglore Blue × Black Champa	Bluish black berries with mild foxy flavour, good for juice and desert wine	IIHR, Bangalore
Arka Vathi	1980	Hybridization	Black Champa × Thompson Seedless	Greenish seedless berries with thin skin suited for raisin making and also for table wines	IIHR, Bangalore
Arka Neelmani	1991	Hybridization	Black Champa × Thompson Seedless	Moderate vigorous and good yielder, black seedless berries good for juice and table wine	IIHR, Bangalore
Arka Chitra	1994	Hybridization	Angur Kalan × Anab-e-Shahi	Prolific-yielder on moderate vigorous vines, berries bold, seeded, golden yellow with pink blush	IIHR, Bangalore
Arka Trishna	1994	Hybridization	Banglore Blue × Convent Large Black	Small bunches with deep tan coloured berries, suitable for juice and red dessert wine	IIHR, Bangalore
Arka Krishna	1994	Hybridization	Black Champa × Thompson Seedless	Berries dark red, seedless clusters medium size, good for juice and table purposes	IIHR, Bangalore

GRAPE : *Contd*

Variety	Year of release	Breeding method	Pedigree/ parentage	Important traits	Institutes/ universities
Arka Soma	1994	Hybridization	Anab-e-Shahi × Queen of Vineyards	Bunches medium-sized cylindrical with bold white seeded berries with Muscat flavour, good for dessert wines	IIHR, Bangalore
Arka Majestic	1994	Hybridization	Angur Kalan × Black Champa	Vines moderate vigorous, but prolific-bearers. Berries deep red, bold, seeded with crisp pulp, suited for table purposes	IIHR, Bangalore
Arka Shweta	1994	Hybridization	Anab-e-Shahi × Thompson Seedless	Vines moderately vigorous, bunches medium cylindrical, berries greenish yellow and seedless, suited for table purposes	IIHR, Bangalore
Cheema Sahebi	1964	Selection	Selection from OP seedlings of Pandri Sahebi (Sel-7)	High-yielding, uniform late ripening	College of Agriculture, Pune
Madhu Angoor	1998	Clonal selection	Probably, from Carolina Black Rose	Vines tolerant to downy mildew, bunches medium large, bold dark red seeded berries, mild Muscat flavoured, suitable for table purpose	ANGRAU, Hyderabad

GRAPE : *Contd.*

Variety	Year of release	Breeding method	Pedigree/ parentage	Important traits	Institutes/ universities
Thompson Seedless	1936	Introduction	Origin from Asia Minor, named in USA and introduced from Australia	Seedless grapes has multiple use fresh table, raisin and in wine-making	—
Muscat of Hamburg (Gulabi)	1936	Introduction	Origin from England but introduced from Australia	Seeded coloured table grape with muscat flavour	—
Kishmish Chernyi	1964	Introduction	Introduced from Erstwhile USSR (Uzbekistan)	Black seedless grape, consistent yield and good table grape quality	—
Beauty Seedless	1964	Introduction	USA	Black seedless and early ripening	—
Flame seedless	1993	Introduction	Origin from USA, introduced from Australia	Early ripening, muscat flavoured and red coloured, crisp seedless berries	—
Red Globe	1985	Introduction	USA	Bold, red coloured seeded grape with firm pulp and thick skin and has excellent keeping quality	—
Crimson Seedless	2002	Introduction	USA	Naturally bold seedless coloured berries with excellent shelf-life	—

GRAPE : *Contd*

Variety	Year of release	Breeding method	Pedigree/ parentage	Important traits	Institutes/ universities
Italia	2002	Introduction	Originally from Italy, but introduced from USA	Greenish white seeded bold berries with pleasant muscat flavour	—
Cabernet Sauvignon	1982	Introduction	Introduced from France	Small cylindrical bunches with small round black purple berries suitable for table wine making	—
Chenin Blanc	1987	Introduction	Introduced from France	White seeded medium-sized, round berries. Bunches winged and moderate size	—
Sauvignon Blanc	1987	Introduction	Introduced from Blanc	White seeded small berry, bunches medium compact and often winged. Suitable for table wine making	—
Shiraz	1993	Introduction	Introduced from both France and Australia	Bunches cylindrical, medium-sized, berries medium size, round and medium-late-maturing, suitable for rose wine or blending to provide colour	—
Zinfandel	1995	Introduction	Introduced from France	Bunches medium size, moderately compact, dark purple small berries, suited for rose wines	—

GRAPE : *Contd.*

Variety	Year of release	Breeding method	Pedigree/ parentage	Important traits	Institutes/ universities
Malbec	1995	Introduction	Introduced from France	Red wine variety, bunch size medium, berries small to medium spherical	—
Ugni Blanc	1995	Introduction	Introduced from France	White wine variety used for blending with other varietals to improve aroma. Clusters compact and medium and winged	—
Grenache	1995	Introduction	Introduction from France	Red wine variety mostly used for blending clusters variable in size and shape a late maturing variety	—
Centennial Seedless	1985	Introduction	Introduced from USA and also from Australia	White seedless, naturally bold and early ripening with attractive bunch and berry having mild muscat flavour	—
Fantasy seedless	2001	Introduction	Introduced from USA	Dark bluish black seedless crisp naturally bold berries, having neutral flavour	—

Floral Biology

Anthesis starts early in the morning and continues beyond 5 PM, the peak anthesis being between 8 and 10 AM under South Indian conditions. The peak anthesis is between 7 and 9 AM in north Indian conditions. Maximum number of flowers complete anthesis before noon and later it declines sharply. The stigma receptivity is characterized by the presence of sugary secretion on stigma giving it a bright appearance. Production of pollen-grains per anther is comparatively higher in Muscat, Pachadraksha and Kishmish than either in Anab-e-shahi or Bangalore Blue. The *V. vinifera* have perfect flowers, but some native North American species have imperfect flowers. Flowers are self-pollinated under South Indian condition.

POMEGRANATE

Botanical name: *Punica granatum*
Family: Lythraceae

Pomegranate is a bush or small tree. It has since earliest times occupied a position of importance alongside grape and fig. Romanites considered pomegranate as the apple in the garden of Eden. According to Bible, King Solomon possessed an orchard of pomegranates. When the children in Israel, wandering in the wilderness, sighed for abandoned comforts of Egypt, the cooling pomegranates were remembered longingly. Centuries later, prophet Muhammad remarked, "Eat the pomegranate, for it purges the system of envy and hatred".

Its seeds have an astringent smell and sweet-sour taste. A sweet and fresh syrup, known as Grenadine, is made from its juice. Pomegranate syrup, used in Middle Eastern cooking, has an intense concentrated flavour. The seeds are used in Indian cooking as a souring agent. Crushed seeds are sprinkled in some of the Middle Eastern cuisines. Its seeds are also used in gargles and in the Indian medicines. They are good to ease fevers and to counteract diarrhoea.

The plant growing 5 or 7 m (16 or 23 feet) in height, has elliptic to lance-shaped, bright-green leaves about 75 mm (3 inches) long and handsome axillary orange-red flowers borne towards the ends of branchlets. The calyx is tubular and persistent. It has 5–7 lobes; the petals are lance-shaped, inserted between calyx lobes. The ovary is embedded in calyx tube and contains several compartments in two series, one above the other. The fruit is of a large orange size, obscurely six-sided, with a smooth, leathery skin that ranges from brownish-yellow to red; within, it is divided into several chambers, containing many thin, transparent vesicles of reddish, juicy pulp,

POMEGRANATE

Variety	Year of release	Breeding method	Pedigree/ parentage	Important traits	Institutes/ universities
Ganesh	1954	Clonal selection	Open-pollinated seedlings of Alandi	Fruit weight – 234.36 g, fruit size medium, rind thickness – 0.33 cm, arils in 100 g – 371, aril colour – pink, juice – 78%, TSS – 15.5° Brix, acidity – 0.50%, fruit colour – greenish yellow with orange tinge, yield 12.21 kg/tree	MPKV, Rahuri
G-137	1984	Clonal selection	Ganesh	Fruit weight – 289.82 g, fruit size large, rind thickness – 0.45 cm, arils in 100 g – 238, aril colour – dark pink, juice – 88.7%, TSS – 16.0° Brix, acidity – 0.39%, fruit colour – yellowish pink, yield – 11.97 kg/tree	MPKV, Rahuri
Mridula	1994	Hybridization	Ganesh × Gul-e-Shah Red	Fruit weight (g) 178.21, fruit size medium, rind thickness – 0.29 cm, aril colour – blood red, TSS – 15.28° Brix, acidity – 0.46%, yield – 14.36 kg/tree	MPKV, Rahuri
Phule Arakta	2003	Hybridization	Ganesh × Gul-e-Shah Red	Av. Fruit weight 182.70 g, fruit size medium, rind thickness – 0.32 cm, aril colour dark red, TSS – 15.97° Brix, acidity – 0.45%, yield – 29.83 kg/tree	MPKV, Rahuri

POMEGRANATE : *Contd.*

Variety	Year of release	Breeding method	Pedigree/ parentage	Important traits	Institutes/ universities
Bhagwa	2003	Clonal selection	—	Av. Fruit weight – 405.97 g, fruit size large, rind thickness – 0.40 cm, aril colour dark red, TSS – 15.4° Brix, acidity – 0.37%, fruit colour – dark saffron, yield – 30.38 kg/tree	MPKV, Rahuri
CO 1	1983	Clonal selection	—	Fruit weight – 340 g, soft seeded, pink arils, juicy, TSS 16° Brix	TNAU, Coimbatore
Yercaud 1	1985	Clonal selection	Acc. No. 455	Fruit weight – 350 to 400 g, soft seeded, deep purple arils, juicy, TSS- 15.2° Brix	HRS, TNAU, Yercaud
Ruby	1997	Hybridization	3-way cross (Ganesh × Kabul) × Yercud × (Ganesh × Gulsha Rose Pink)	Fruit weight – 270 g, skin colour – reddish with green streaks, rind thickness–0.24 cm, weight of arils – 37.2 g/100 g, TSS – 16° Brix, soft seeded, high yielding–16-18 tonnes/ha	IIHR, Bangalore
Amlidana	1999	Hybridization	Ganesh × Nana	Fruit weight – 120 g, arils highly acidic – 4.8%, short tree statured, suitable for *Anardana*	IIHR, Bangalore
Jyothi	1985	Selection	Selection from open-pollinated seedlings of Bessein seedless and Dholka	Fruit medium to lager sized, with yellowish red, more fleshy and pink aril, sweet fruit, soft seeded, moderate yielder	UAS, Dharwad

each surrounding an angular, elongated seed. The fruit is eaten fresh, while juice is the source of grenadine syrup, used in flavourings and liqueurs. The seeds are extracted from the peel, pith and membranes, then dried. When dry, they are small and dark-red to black in colour and slightly sticky.

GUAVA

Botanical name: *Psidium guajava*
Family: Myrtaceae

Guava is highly productive. It is a nutritious fruit of tropical and subtropical regions. It ranks fifth important fruit crop in India. It is a rich source of vitamin C and pectin. It is available at a cheaper cost, both as fresh and processed forms. It gives fruits twice a year both in rainy and winter seasons. Most of the commercial varieties are diploids, the chromosome number being 2n = 22. The seedless varieties are triploids. The somatic chromosome number of triploid is 2n = 33.

Botany

A shrub or small tree, with smooth, light reddish-brown, with pubescent 4-angled young branches, its trunk normally attains a diameter of about 25 cm. Leaves are opposite, ovate — elliptic or oblong — elliptic, acute — acuminate, pubescent beneath, often rather brittle, prominently nerved, lateral nerves 10–20 pairs, blades mostly 7–15 cm long and 3–5 cm wide. Flowers are hermaphrodite, solitary or 2–4 together in leaf axils, peduncle about 1–2 cm long, pubescent, calyx 4–5 lobed, about 6–8 mm long, petals white 10 15 cm long, fugacious usually 4/5, obovate, slightly concave, stamens numerous, white, style 10–12 mm long, stigma peltate. Fruits globose, ovoid, or pyriform, many seeded, pulp granular, juicy seeds, yellowish, reniform.

Floral Biology

- The cracking of calyx occurs nearly 24 h before flower opening.
- The calyx splits anywhere between 13 and 26 h before flower opening.
- The peak anthesis is found to be between 5.00 and 6.30 AM in most of the varieties.
- In north India, anthesis takes place between 6.00 and 7.30 AM.

GUAVA

Variety	Year of release	Breeding method	Pedigree/ parentage	Important traits	Institutes/ universities
Lalit	1999	Selection	Selection from half-sib population of Apple colour	Saffron yellow coloured fruits with red blush, flesh firm and pink with good blend of sugar and acid. It is suitable for both table and processing purposes. The pink colour in the beverage remains stable for more than a year in storage	CISH, Lucknow
Shweta	2005	Selection	Selection from half-sib population of Apple colour	High-yielding, attractive guava variety with globose fruits, medium-sized, creamy white exocarp with red spots or blush, snow-white flesh, high TSS (12.5–13.2%) and vitamin C (300 mg/100 g pulp) with good keeping quality	CISH, Lucknow
Arka Amulya	1992	Hybridization	Allahabad Safeda × Seedless	Heavy yield, few and soft seeds, white pulp and high TSS	IIHR, Bangalore
Arka Mirdual	1992	Selection	Selection from open pollinated seedlings of Allahabad Safeda	Round fruits, skin yellow and smooth, white flesh, 12° Brix TSS, soft seeds, good keeping quality, suitable for jelly making	IIHR, Bangalore

GUAVA : *Contd.*

Variety	Year of release	Breeding method	Pedigree/ parentage	Important traits	Institutes/ universities
Hybrid-1	—	Hybridization	Sardar × Allahabad Safeda	High yield with soft seeds	BAC, Sabour
Hybrid-2	—	Hybridization	Chittidar × Allahabad	High yield with soft seeds	BAC, Sabour
TRY (G) 1	2005	Selection	Selection from elite mother plant, origin unknown	Off season bearing, shiny greenish yellow fruit with desirable aroma, high TSS (10° Brix), drought and sodicity tolerant, high yielder (40.52 kg/tree)	TNAU, Coimbatore
Safed Jam	—	Hybridization	Allahabad Safeda × Kohir	Fruits large, few and soft seeds, good keeping quality	FRS, Sangareddy
Kohir Safeda	—	Hybridization	Kohir × Allahabad Safeda	White fleshed, fruits large, few soft seeds	FRS, Sangareddy
Hisar Safeda	—	Hybridization	Allahabad Safeda × Seedless	Fruits round, 92 g, pulp creamy white with few soft seeds	CCSHAU, Hisar
Hisar Surkha	—	Hybridization	Apple colour × Banarasi Surkha	Fruits round, 86 g, pulp pink	CCSHAU, Hisar

- The dehiscence of anthesis starts 15–30 min after anthesis in all the varieties and continues up to 2 h.
- The stigma receptivity is maximum on the day of anthesis
- The stigma is receptive 2 days before dehiscence and up to 4 days afterwards
- Receptivity of stigma continued up to 48 h, it remained maximum for 2–3 h after dehiscence.
- Guava is primarily a safe fruitful, although some strains seem to produce more fruits when cross-pollinated with another variety
- Bees and other insects visit its flowers.

CITRUS

Botanical name: *Citrus* spp.
Family: Rutaceae

Commonly grown citrus fruits belong to three genera: *Citrus, Fortunella* and *Poncirus*. All these genera are closely related and have intergeneric fertility. Citrus production in India is ranked third after banana and mango. There are several species and varieties of Citrus, mandarin finds its premier position in Punjab, Maharashtra, Karnataka, North-Eastern states, Tamil Nadu, Kerala, Uttar Pradesh and West Bengal, In India, total area under citrus cultivation is 0.56 mha with annual production of 4.58 million tonnes. In India, Punjab ranks first in area and production. Fruits are rich in nutritive and medicinal value.

Limes and lemons are commercially-grown in tropical and subtropical regions of India. Of them, acidlime (*Citrus aurantifolia*) is the third important fruit after mandarins and sweet oranges. On the other hand, lemons (*C. limon*) are cultivated to a limited extent. India ranks fifth among major lime and lemon-producing countries in the world. India is perhaps the largest producer of acid lime in the world. It is cultivated in almost all the states, Andhra Pradesh, Maharashtra, Tamil Nadu, Karnataka, Gujarat, Bihar and Himachal Pradesh being major producing states. Lemons are lesser popular than limes in India. They are cultivated to a considerable extent commercially in Punjab, Rajasthan, and *tarai* region of Uttar Pradesh. Besides acid lime, sweet lime (*C. limettioides*), Tahiti lime (*C. latifolia*) and Rangpur lime (*C. limonia*) are also cultivated on a limited scale in India. Sweet lime, indigenous to India, is an important citrus fruit in north India. Tahiti lime grows well in Karnataka and Tamil Nadu. However, Sweet and Tahiti limes could not replace acidlime under commercial cultivation. Rangpur lime is mostly cultivated for rootstock purpose.

CITRUS

Variety	Year of release	Breeding method	Pedigree/ parentage	Important traits	Institutes/ universities
Jai Devi (PKM 1)	1990	Selection	Selection from Kadayam type of Thirunelveli district of Tamil Nadu	Big-sized fruits (52 g), high ascorbic acid (34.29 mg/100 g of fruit juice) and TSS (8.0° Brix). High yielder, less susceptible to leaf minor and citrus butterfly	HC&RI, TNAU, Periyakulam
Rasraj	1998	Hybridization (*C. aurantifolia* × *C. limon*)	Inter-specific polyembryonic hybrid	Resistant to bacterial canker, fruits yellow, average weight 55 g, rind is thicker than lime, 6% acidity, 8° Brix TSS	IIHR, Bangalore
Sai-Sharbati	—	Selection	Selection from local germplasm, Western Maharashtra	Fruits smooth, uniform, skin thin, TSS, juice, acidity high, high yielder, tolerant to canker and tristeza	MPKV, Rahuri
Vikram	1996	Selection	Selection from local Kagzi Lime	Bears fruits in clusters of 5–10 during June–July and November–December and off season fruits during September and May–June and high-yielder. Fruit weight 45.94 g	MAU, Parbhani, Maharashtra
Pramalini	1998	Selection	Selection from local Kagzi Lime	Cluster bearing cultivar (3–7 fruits per cluster, fruit weight 46.96 g, thin peel thickness, high juice content (57.72%) and high yielder in December and off season fruits during September and May–June	MAU, Parbhani, Maharashtra

CITRUS : *Contd.*

Variety	Year of release	Breeding method	Pedigree/ parentage	Important traits	Institutes/ universities
Kinnow	1956	Introduction	Riverside Citrus Experiment station, California, USA	Tree vigorous, large and symmetrical. Fruit medium, deep yellowish orange colour, smooth surface, rind thick, peels tough and leathery, pulp deep yellowish orange, very juicy, high sugar, acid moderate, very rich flavour, 25–28 seeds	Abohar

Floral Biology

- In subtropical zones with cool winter temperatures, most Citrus species flower once a year during early spring, while flower induction is initiated in early January.
- In tropics and coastal areas, flowering takes place several times a year.
- Flowers are borne singly in axils of leaves/in short axillary, corymbose racemes.
- Inflorescence develops from dormant axillary vegetative buds
- Flowers have 4–8 thick linear petals and a 4–5 lobed calyx.
- Anthers surround the pistil or at near the level of stigma.
- Consist of two locules that detusce longitudinally.
- About 6–14 carpels join to each other and to a central axis
- Anthers dehisce prior to opening of flower.
- Stigma is receptive for one to a few days before anthesis and for several days afterward.
- Normal anthers are bright yellow at maturity. Pollen is of sticky, adherent type — entomophilous (honeybees, mites and thrips).
- Self-pollination can easily occur because of nearness of anthers to stigma.

SAPOTA

Botanical name: *Manilkara zapota* syn. *Achras sapota*
Family: Sapotaceae

India is considered to be the largest producer of sapota in the world. It is mainly cultivated in coastal areas of Maharashtra, Karnataka, Tamil Nadu, Kerala, Uttar Pradesh, West Bengal, Punjab and Haryana. It is a small to medium-sized, evergreen tree with a spreading crown and straight stem. The leaves are elliptic obovate and bears flowers solitary or in fascicles, white or pale yellow in colour. The origin of sapota is considered as Tropical America. The chromosome number is n = 13.

Botany

- Tall, medium or small tree
- Leaves — elliptic lanceolate, ovate lanceolate
- Flowers — yellow or white in colour

- Stamens and pistil — six, perfect, filaments short, epipetalous, parallel or obliquely placed
- Roundish, ovate, oval or elliptic fruits — berry

Floral Biology and Pollination Mechanism

- Flowers appear more or less throughout the year.
- Time taken for emergence of a flower bud to its opening — 36 days.
- Flower opening under Coimbatore conditions — 4.00–4.30 AM.
- Sapota is protogynous — Style and stigma grow out of the bud about 2 days before flower opening.
- Stigma is receptive 12 h after opening, peak being between 8.00 and 10.00 AM.
- Wind pollinated.
- Thrips (*Thrips hawaiiensis* Morgan and *Haplothrips tenuipennis* Bagnall) are the principle pollinators of sapodilla in India.
- Three flowering seasons, *viz.* June–August; October–December and March.
- Anthesis in long oval is by 7.00 PM and by midnight, the flowers are fully open.
- Fruit setting is highest under natural pollination, followed by cross-pollination.

Emasculation, Pollination and Fruit Setting

- Flower buds are emasculated and bagged between 4 and 5 PM one day before stigma protrudes out of the bud.
- Stamens from flower buds which are due to shed their pollen in early hours of the next day are collected the previous evening.
- Sapota trees produce many numbers of flowers, but the number of fruits reaching maturity is very small.
- The major period of fruit drop occurs during first five weeks following fruit setting and as little as 1.6% of flowers produced by a tree may develop into fruit.
- Shape of fruit is markedly affected by cross-pollination.
- The self-pollinated fruits are oblong to oval, whereas the cross-pollinated ones are round.
- The fruit setting is highest under natural pollination followed by cross-pollination.

SAPOTA

Variety	Year of release	Breeding method	Pedigree/ parentage	Important traits	Institutes/ universities
CO 1	1972	Hybridization	Cricket ball × Oval	The fruit shape is long oval, medium-sized with total soluble solids of 18° Brix. Each fruit weighs about 125 g. A yield of 175–200 kg/tree can be obtained in a year. The fruit is bigger than Oval and sweeter than Cricket Ball. The flesh is granular in texture and reddish brown in colour	TNAU, Coimbatore
CO 2	1974	Selection	Clonal selection from Baramasi	Fruits are round, medium sized, skin outer surface cinnamon brown in colour, inner surface yellowish green, flesh soft, juicy, slightly gritty, light brown in colour, aroma slight, taste sweet, seed medium sized, black, obovate in shape, beaked, suture distinct, slightly adhering to flesh and placed centrally in the fruit. Yield 175 kg/tree or 11.8 tonnes/ha/year	TNAU, Coimbatore
CO 3	2000	Hybridization	Cricket ball × Vavilavalasa	Highly suitable for high density planting. It bears fruits all through the year with a peak during February–June and September–October. Under high density planting system, it yields 40–50 tonnes/ha. It is suitable for growing in plains up to an elevation of 100 m MSL under wide range of soil condition	TNAU, Coimbatore

SAPOTA : *Contd.*

Variety	Year of release	Breeding method	Pedigree/ parentage	Important traits	Institutes/ universities
PKM 1	1981	Selection	Clonal selection from Guthi	A mean yield of 3,547 fruits (236 kg) can be obtained per tree per year. The tree is dwarf and adaptable to Southern and Central districts. Fruits are of two shapes, *viz.* round (rarely) and oval (maximum). High average yielder with medium size fruits. Average fruit weight is 100 g.	HC&RI, TNAU, Periyakulam
PKM 2	1992	Hybridization	Guthi × Kirtibarthi	This hybrid is a high yielder recording a mean yield of 80 kg of fruits/tree in during fifth year after planting. The yield increase is 42.76% over PKM.1 variety. The fruits are bigger in size (95 g)	HC&RI, TNAU, Periyakulam
PKM 3	1994	Hybridization	Guthi × Cricket ball	The variety is adaptable to tropical plains of Tami Nadu and yields 14 tonnes/ha. The fruits bear in clusters with oval shaped large fruits. The vertical growth habit of tree allows high density planting. The fruits mature earlier than other varieties in this season thus fetching higher return. The variety is tolerant to leaf spot and leaf webber.	HC&RI, TNAU, Periyakulam

SAPOTA : *Contd.*

Variety	Year of release	Breeding method	Pedigree/ parentage	Important traits	Institutes/ universities
PKM (Sa) 4	2003	Selection	Open pollinated clone of PKM 1	High yielder (20.8 tonnes/ha), spindle shaped fruits suitable for dry flakes production, attractive pulp with light pinkish honey brown colour, 24–25° Brix TSS.	HC&RI, TNAU, Periyakulam
PKM (Sa) 5	2007	Selection	Selection from open-pollinated seedlings maintained in a private orchard at Virudunagar of Tamil Nadu	High yielder (18.70 tonnes/ha), high TSS (25.5° Brix) and suitable for preparation of dry flakes, milk shake powder and fruit jam, oval shaped attractive fruits with smooth, light brown skin, flesh is crisp and retains coppery brown colour.	HC&RI, TNAU, Periyakulam

AVOCADO

Botanical name: *Persea americana*
Family: Lauraceae

Avocado, also known as butter pear or alligator pear, is a tree native to Mexico, South America and Central America. The name "avocado" also refers to the fruit (technically a large berry) of tree that contains an egg-shaped pit (hard seed casing). Avocados are a commercially valuable crop whose trees and fruits are cultivated in tropical climates throughout the world, producing a green-skinned, pear-shaped fruits that ripen after harvesting. Trees are partially self-pollinating and often are propagated through grafting to maintain a predictable quality and quantity of fruits.

The fruits are more or less round to egg- or pear-shaped, typically the size of a temperate-zone pear is larger, on the outside bright green to green-brown (or almost black) in colour. The fruit has a markedly higher fat content than most other fruits, mostly monounsaturated fat. A ripe avocado yields to a gentle pressure when held in palm of hand and squeezed. The flesh is typically greenish-yellow to golden-yellow when ripe. The flesh oxidizes and turns brown quickly after exposure to air. To prevent this, lime or lemon juice can be added to avocados after they are peeled. The avocado is very popular in vegetarian cuisine, making an excellent substitute for meats in sandwiches and salads because of its high fat content. The fruit is not sweet, but fatty, distinctly yet subtly flavored, and of smooth, almost creamy texture.

Ecological Races

Mexican race: Small fruits weighing less than 250 g and ripening 6-8 months after flowering oil content is up to 30%, resistant to cold temperature.

Guatemalan race: Fruits are fairly large, weighing up to 600 g and borne on stalks. Fruits ripen 9–12 months after flowering. Oil content ranges between 8 and 15%.

West Indian race: Fruits are medium-sized; require 6–9 months for ripening from the date of flowering. The oil content ranging between 3 and 10%, least resistant to cold temperature.

Botany

Avocado is a dense, evergreen tree, shedding many leaves in early spring, reaches 80 feet in height. Its leaves are alternate, glossy, elliptic and dark green with paler veins. They normally remain on tree for 2–3 years. The leaves of West Indian varieties are scentless, while Guatemalan types are rarely anise-scented. The leaves are high in oils and slow to compost and may collect in mounds beneath trees. Avocado flowers terminal panicles of 200–300 small yellow-green blooms. Each panicle produces only 1–3 fruits. Perfect, sepals about 5 mm long, with petals usually a bit longer. Three sepals opposite to inner stamen. Petals are opposite to staminode. Pistil has an ovary with ovule, stigma in highly papillate superior ovary, containing one ovule with placenta arranged basally. Stamens are two types. Two outer whorls appear identical and their pollen sacs open inward, pollen sacs of the inner whorl open outward. The stigma and transmitting tissue secrete a substance containing lipids and carbohydrates that is assumed to promote pollen germination and tube growth. Abnormalities in avocado floral structures are common. It is reported that most common abnormality is staminoidy, the conversion of other flower parts to stamens. pistilloidy, *i.e.* the conversion of other flower parts to pistils. The flesh of avocado fruit is deep green near the skin, becoming yellowish nearer the single large, inedible ovoid seed. The flesh is hard when harvested but softens to a buttery texture.

Floral Biology

- Flowers are protogynous and exhibit a unique mechanism of synchronous dichogamy for alteration of sexes.
- Each avocado flower opens twice, the first time as a functional female (stage I) and the second time as functional male (stage II).
- Pistil ripens earlier than the stamen — protogynous.
- Based on pattern of opening of flower the avocado cultivars are classified into two types A and B.
- In type A, flower opens first as functional female during the morning hours, close near midday and reopens as functional male in the afternoon of the following day.
- The type B cultivars exhibit the first opening of female flowers in the afternoon and the second opening (functional male) occurs in morning of the following day.
- As a result of synchronous dichogamy, cross-pollination has the best opportunity for fertilization.
- Thus, every morning type A pistils can be fertilized by type B pollen and in the afternoon the type B pistils can be crossed by the A pollen.

AVOCADO

Variety	Year of release	Breeding method	Pedigree/ parentage	Important traits	Institutes/ universities
TKD 1	1997	Selection	This variety (PA 2) is a selection from germplasm pool	The fruits are dark green, round-shaped and medium sized. Trees are semi-spreading, suitable for high density planting. Fruits mature earlier in the harvesting season. The tree yields 264 kg of fruits/tree/year (26.4 tonnes/ha). This is recommended for lower Pulneys, Shevroys and other hills of Tamil Nadu under irrigated and rainfed conditions.	TNAU, Coimbatore

PASSION FRUIT

Variety	Year of release	Breeding method	Pedigree/ parentage	Important traits	Institutes/ universities
Kaveri	1978	Hybridization	Hybrid of Purple × Yellow	High-yielding, ovoid to round, purple dotted large-size fruits (80–90 g), 30–35% juice, 3% acidity, tolerant to collar rot, wilt, brown leaf spot and nematodes.	IIHR, Bangalore

- The flowers are entomophilies and dependent on honey bees for pollination.
- The higher the fertility rate and the larger the pollen-grains.
- The RH may play some role in prolonging the receptivity of pistil.
- Various pollinating agents visit the its flowers for nectar and pollen. These include honey bee, various species of wild bees, wasps, flies, and hummingbirds.

PASSION FRUIT

Botanical name: *Passiflora edulis*
Family: Passifloraceae

Passionfruit is a plant cultivated commercially in frost-free areas. It is native to South America and widely grown in India, New Zealand, the Caribbean, Brazil, Ecuador, California, Southern Florida, Hawaii, Australia, East Africa, Israel and South Africa. The passion fruit is round to oval, yellow or dark purple at maturity, with a soft to firm, juicy interior filled with numerous seeds. The fruit can be used to eat or for its juice, which is often added to other fruit juices to enhance the aroma.

Two types of passionfruit have greatly different exterior appearances. The Bright yellow variety of passionfruit, which is also known as the Golden Passionfruit, can grow up to the size of a grapefruit, has a smooth, glossy, light and airy rind. It is used as a rootstock for purple passion fruit in Australia. The dark purple passionfruit (for example, in Kenya) is smaller than a lemon, with a dry, wrinkled rind at maturity. Its purple varieties have traces of cyanogenic glycosides in the skin, and hence are mildly poisonous. However, thick, hard skin is hardly edible, and if boiled (to make jam), cyanide molecules are destroyed at high temperatures. These forms of *Passiflora edulis* have been found to be different species. They occur in different climates in nature and bloom at different times of day. The purple fruited species is self-fertile and yellow fruited species, despite claims to the contrary, is self-sterile. It requires two clones for pollenization.

The passionfruit vine is a shallow-rooted, woody, perennial, climbing by means of tendrils. The alternate, evergreen leaves, deeply 3-lobed when mature, are finely toothed, 7.5–20 cm long, deep-green and glossy above, paler and dull beneath, and, like the young stems and tendrils, tinged with red or purple, especially in yellow form. A single, fragrant flower, 5–7.5 cm wide, is borne at each node on new growth. The bloom, clasped by 3 large, green, leaflike bracts, consists of 5 greenish-white sepals, 5 white petals, a fringelike corona of straight, white-tipped rays, rich purple at the base, also 5 stamens with large anthers, the ovary, and triple-branched style forming a prominent central structure. The flower of yellow is the more showy, with more intense colour. The nearly round or ovoid fruit, 4–7.5 cm wide, has a tough rind, smooth, waxy, ranging in hue from dark-purple with faint, fine white specks, to light-yellow or pumpkin-colour. It is 3 mm thick, adhering to a 6 mm layer of white pith. Within is a cavity more or less filled with an aromatic mass of double-walled, membranous sacs filled with orange-colored, pulpy juice and as many as 250 small, hard, dark-brown or black, pitted seeds. The flavor is appealing, musky, guava-like, subacid to acid.

LITCHI

Botanical name: *Litchi chinensis*
Family: Sapindaceae

A tropical fruit native to southern China, litchi is commonly found in Madagascar, India, Bangladesh, Pakistan, southern Taiwan, northern Vietnam, Indonesia, Thailand, the Philippines, and Southern Africa. It is a medium-sized, evergreen tree, reaching 15–20 m tall, with alternate pinnate leaves, each leaf 15–25 cm long, with 2–8 lateral leaflets 5–10 cm long; the terminal leaflet is absent. The newly-emerging young leaves are bright coppery-red at first, before turning green as they expand to full size. The flowers are small, greenish-white or yellowish-white, produced in panicles up to 30 cm long. The fruit is a drupe, 3–4 cm long and 3 cm in diameter. The outside is covered by a pink-red, roughly-textured rind that is inedible but easily removed. The inside consists of a layer of sweet, translucent white flesh, rich in vitamin C, with a texture somewhat similar to that of a grape. The edible flesh consists of a highly developed aril enveloping the seed. The center contains a single glossy brown nut-like seed, 2 cm long and 1–1.5 cm in diameter. The seed, similar to a buckeye seed, is not poisonous but should not be eaten. The fruit matures from July to October, about 100 days after flowering. There are two subspecies:

- *Litchi chinensis* subsp. *chinensis*. China, Indochina (Vietnam, Laos and Cambodia). Leaves with 4–8 (rarely 2) leaflets.
- *Litchi chinensis* subsp. *philippinensis* (Radlk.) Leenh. Philippines, Indonesia. Leaves with 2–4 (rarely 6) leaflets.

Breeding objectives are to develop varieties with dwarf stature and regular-bearer. Though an ideal litchi cultivar for modern requirements is lacking, some points are to be borne in mind while

LITCHI

Variety	Year of release	Breeding method	Pedigree/ parentage	Important traits	Institutes/ universities
Swarna Roopa	—	Clonal selection	From cultivars planted at research station (CHES, Ranchi)	Attractive deep-pink fruit colour, small seed and high TSS/acid ratio. Fruits are highly resistant to cracking. Fruits mature a week later than the late cultivar China	CHES, Ranchi
CHES-2	—	Selection	Bombaia	Late-maturing, inside canopy fruit bearing habit, fruits free from sunburn and cracking	CHES, Ranchi
Sabour Madhu (H-105)	1997	Hybridization	Purbi × Bedana	Higher number of fruits (24) per panicle and ripen 8 days later than another late maturing cultivar, Kasba, higher TSS and aril percentage than Purbi. Fruit shape resembles Purbi	Regional Fruit Research Station, Sabour, Bihar
Sabour Priya (H-73)	1997	Hybridization	Purbi × Bedana	Better fruit quality than Purbi, fruit shape has combination of both the parents. The fruit weight is higher than the better parent (Purbi)	Regional Fruit Research Station, Sabour, Bihar
Saharanpur Selection	1980	Seedling selection	Local cultivar	Late maturing, fruit TSS around 19.8%, very low fruit cracking (2% only)	GBPUAT, Pantnagar

developing a superior cultivar. They are large fruit size (>20 g) having small seed size or higher percentage of aborted seeds, good skin colour, fruit quality and high keeping quality.

ANNONA

Botanical name: *Annona reticulata*
Family: Annonacae

Edible fruits of genus *Annona* are collectively known as annonaceous fruits. Of the 40 genera of its family, genus *Annona* has 120 species, 6 of them having pomological significance. Annona fruits are syncarpia formed by fusion of pistil and receptacle into a large fleshy aggregate fruit. Annonaceous fruits have morphological affinity for each other but each type is unique in its taste, flavour, pulp colour and texture.

Among annonaceous fruits, custard-apple is most favourite in India. Its plants come up unattended in parts of Andhra Pradesh, Assam, Bihar, Karnataka, Maharashtra, Madhya Pradesh, Orissa, Rajasthan and Tamil Nadu as a scrub or hedge plant. Of late, custard-apple has gained commercial significance and exclusive orchards are emerging in Andhra Pradesh and Maharashtra.

Other annonas are cultivated on a limited scale. Bullock's heart is more commonly found in south India than in north India. It is usually associated with gardens and compounds and not commercial orchards. Cherimoya is mostly restricted to Assam and hills of south India. Atemoya and sour sop are cultivated in some gardens as miscellaneous fruits. Atemoya, cherimoya and ilama also provide excellent opportunities for a large-scale exploitation in India.

The custard-apple is believed to be a native of the West Indies but it was carried in early times through Central America to southern Mexico. Tree is erect, with a rounded or spreading crown and trunk 25–35 cm thick. Height ranges from 4.5 to 10 m. The leaves are deciduous, alternate, oblong or narrow-lanceolate, 10–20 cm long, 25 cm wide, with conspicuous veins. Flowers, in drooping clusters, are fragrant, slender, with 3 outer fleshy, narrow petals 23 cm long; light-green externally and pale-yellow with a dark-red or purple spot on the inside at the base. The flowers never fully open.

The compound fruit, 8–16 cm in diameter, may be symmetrically heart-shaped, lopsided, or irregular; or nearly round, or oblate, with a deep or shallow depression at the base. The skin, thin but tough, may be yellow or brownish when ripe, with a pink, reddish or brownish-red blush, and faintly, moderately, or distinctly reticulated. There is a thick, cream-white layer of custardlike, somewhat granular,

ANNONA

Variety	Year of release	Breeding method	Pedigree/ parentage	Important traits	Institutes/ universities
Arka Sahan	1995	Interspecific hybridization	Progeny of Island gem (*Annona atemoya*) × Mammoth (*A. squomosa*)	Waxy bloom skin, light green in colour, moderate thick with large flat eyes, creamy white colour flesh with mild pleasant aroma, 12 tonnes/ha yield.	IIHR, Bangalore
APK (Ca) 1 (*Annona reticulata)*	2003	Selection	Clonal selection from a high-yielding type in Courtallam, Tamil Nadu	High yield in black soils, drought tolerant, suitable for both rainfed and irrigated conditions, high-yielding (7,300 kg/ha)	TNAU, Coimbatore

flesh beneath the skin surrounding the concolorous moderately juicy segments, in many of which there is a single, hard, dark brown or black, glossy seed, oblong, smooth, less than 1.25 cm long. Actual seed counts have been 55, 60 and 76. A pointed, fibrous, central core, attached to the thick stem, extends more than halfway through the fruit. The flavour is sweet and agreeable though without the distinct character of the cherimoya, sugar apple, or atemoya.

JACK FRUIT

Botanical name: *Artocarpus heterophyllus* Lam
Family: M*oraceae*

Jackfruit is popularly known as poor man's food in the eastern and southern parts of India. A rich source of vitamin A, C, and minerals, it also supplies carbohydrates. Tender jackfruits are popularly used as vegetable. The skin of its fruits and leaves are excellent cattle feed. Its timber is valued for furniture making since it is rarely attacked by white ants. The latex from the bark contains resin. Pickles and dehydrated leather are its preserved delicacies. Canning of flakes can be done. They can be bottled and served after mixing with honey and sugar. Nectar is prepared from its pulp. The rind rich in pectin, can be used for making jelly. The flakes, seeds, sterile flowers, skin and core contain calcium pectate 4.6, 1.6, 3.7, 3.2 and 2.1% respectively. They are considered as a good sources of pectin.

In south India, jackfruit is a popular food ranking next to mango and banana in total annual production. It is believed to be indigenous to the rain forests of the Western Ghats of India. It spreads early on to other parts of India, Southeast Asia, the East Indies and ultimately the Philippines. It is often planted in central and eastern Africa and is fairly popular in Brazil and Surinam.

The tree is 9–21 m tall, with evergreen, alternate, glossy, somewhat leathery leaves to 22.5 cm long, oval on mature wood, sometimes oblong or deeply lobed on young shoots. All parts contain white latex. Short, stout flowering twigs emerge from the trunk and large branches, or even from the soil-covered base of very old trees. The tree is monoecious: tiny male flowers are borne in oblong clusters, 5–10 cm in length; female flower clusters are elliptic or rounded. Fruits 20–90 cm long and 15–50 cm wide. They weigh from 4.5–20 or 50 kg. The "rind' or exterior of compound or aggregate fruit is green or yellow when ripe and composed of numerous hard, cone-like points attached to a thick and rubbery, pale-yellow or whitish wall. The interior consists of large "bulbs" (fully-developed perianths) of yellow, banana-flavored flesh, massed among narrow ribbons of thin, tough

JACK FRUIT

Variety	Year of release	Breeding method	Pedigree/ parentage	Important traits	Institutes/ universities
PLR 1	1992	Selection	Clonal selection from Panikkankuppam local of Panruti taluk, South Arcot district	It is high yielding with medium height, less spreading and suitable for high density planting. In addition to regular bearing season (April–June), it gives yield during off-season (November–December) also. The annual yield per tree is about 80 fruits weighing around 900 kg. The average fruit weight is 12 kg containing 115–120 carpels	TNAU, Coimbatore
PLR (J) 2	2007	Selection	Clonal selection from Pathirakkotai local	High-yielding (95–110 fruits/tree/year), good quality bigger-sized fruits, highly palatable and edible flakes, less incidence to major pest and diseases	TNAU, Coimbatore
PPI 1	1996	Selection	Clonal selection from Mulagumoodu local near Pechiparai	It is a medium tall tree yields 107 fruits per tree per annum (weighing 1,818 kg) which accounts for 40.8% more yield than local. The tree bears twice annually (April–June and November-December) and produces high quality crisp carpels with more TSS and ascorbic acid content	TNAU, Coimbatore

undeveloped perianths (or perigones), and a central, pithy core. Each bulb encloses a smooth, oval, light-brown "seed" (endocarp) covered by a thin white membrane (exocarp). The seed is 2–4 cm long and 1.25–2 cm thick and is white and crisp within. There may be 100 or up to 500 seeds in a single fruit.

BER

Botanical name: *Ziziphus mauritiana*
Family: Rhamnaceae

Ber or Indian jujube is indigenous to India. Its fruits are rich in vitamin C, A and B complex. The composition varies in different varieties. Its leaves contain 5.6% digestible crude protein and 49.7% total digestible nutrients, making it a nutritive fodder for animals. Haryana, Punjab, Uttar Pradesh, Rajasthan, Gujarat, Madhya Pradesh, Bihar, Maharashtra, Andhra Pradesh and Tamil Nadu are major ber-growing states. Hisar, Rohtak, Jind, Panipat, Mohindergarh and Gurgaon (Haryana), Bharatpur, Jaipur and Jodhpur districts (Rajasthan), Sangrur and Patiala districts (Punjab), Banaskantha and Sabarmati (Gujarat), Bijapur and Bellary (Karnataka), and Tirunelveli, Ramanathapuram, Dharmapuri and Salem (Tamil Nadu) are ber-growing areas in India.

The jujube is a small, deciduous tree. Naturally, it is a drooping tree and often thorny with branches growing in a zig-zag pattern. The wood is very hard and strong. Jujube cultivars vary in size and conformation, with some being very narrow in habit and others being more widespread. After 30 years of growth, its trees can be 30 feet tall with a crown diameter of up to 15 feet. Plants send up suckers from their roots, and these suckers can appear many feet from the mother plant. Currently, these root suckers must be controlled by mowing or hoeing.

The small, ovate or oval leaves are 1–2 inches long and a shiny bright green. In autumn, leaves turn bright yellow before falling. There are usually two spines at the base of each leaf. Some spines may be hooked while others are long daggers. Virtually thornless cultivars are known. As the growing season commences, each node of a woody branch produces one to ten branchlets. Most of these are deciduous, falling from the plant in autumn. Flowers inconspicuous, 1/5 inch diameter, white to greenish-yellow flowers are somewhat fragrant and produced in large numbers in the leaf axils. The flowering period extends over several months from late spring into summer. However, individual flowers are receptive to pollen for only one day or less. Pollination needs of the jujube are not clearly defined, but

BER

Variety	Year of release	Breeding method	Pedigree/ parentage	Important traits	Institutes/ universities
Goma Kirti	1997	Selection	Clonal selection of Umran	Early maturity than Umran ber	CHES, Godhra
Thar Sevika	2007	Hybridization	Seb × Katha	Early-maturing and high-yielding	CIAH, Bikaner
Thar Bhubraj	2007	Selection	Selection from Bhusavar area of Bharatpur district of Rajasthan	Early and high-yielding	CIAH, Bikaner

appear to be done by ants or other insects and possibly by the wind. Most jujube cultivars produce fruits without cross-pollination. The jujube is well protected from late spring frosts by delayed budding until all chance of cold weather has passed.

The fruit is a drupe, varying from round to elongate and from cherry-size to plum-size depending on cultivar. It has a thin, edible skin surrounding whitish flesh of sweet, agreeable flavor. The single hard stone contains two seeds. The immature fruit is green in color, but as it ripens it goes through a yellow-green stage with mahogany-colored spots appearing on skin as fruit ripen further. The fully mature fruit is entirely red. Shortly after becoming fully red, the fruit begins to soften and wrinkle. The fruit can be eaten after it becomes wrinkled, but most people prefer them during the interval between the yellow-green stage and the full red stage. At this stage the flesh is crisp and sweet, reminiscent of an apple. Under dry conditions jujubes lose moisture, shrivel and become spongy inside.

AONLA

Botanical name: *Emblica officinalis*
Family: Euphorbiaceae

Aonla has originated from Asian continent particularly from south and central India, Sri Lanka and Malaysia. The principal economic value of amla is its vitamin C content, which ranges from 400 to 1,300 mg/100 g of fruit. Vitamin C is basically an antioxidant, which slows down the senescence of cells, enhancing the rejuvenation of cells in all living beings. So, aonla came to light very recently and which is one of the main reasons for its addition in many preparations for Ayurvedic and other systems of medicine. The vitamin C is not denatured or lost even after drying. This is very essential since it is mainly used as dried product. Apart from the vitamin C and its function as antioxidants, aonla improves hair growth, acts as diuretic, pungent, laxative etc. It is cultivated more common in India, particularly in Uttar Pardesh, semi-arid regions of Tamil Nadu, Maharashtra, Gujarat, Rajasthan, Andra Pradesh and Karnataka.

A deciduous tree, small to medium in size, its average height being 5.5 m. The bark is usually light brown to black, coming off in thin strips or flakes, exposing the fresh surface of a different colour underneath the older bark; the average girth of the main stem is 70 cm; in most cases, the main trunk is divided into 2–7 scaffolds very near the base. Leaves, 10–13 mm long, 3 mm wide, closely set in

AONLA

Variety	Year of release	Breeding method	Pedigree/ parentage	Important traits	Institutes/ universities
Goma Aishwarya	2007	Selection	A clonal selection of NA-7	High yielding than NA-7	CHES, Godhra
BSR 1	1995	Selection	It is a selection from Thimbam local	A high-yielder produces 155 kg of fruits per tree per year. The fruits are medium-sized (27 g) and born in clusters. The fruits contain more flesh, high ascorbic acid content (610 mg/ 100 g). Each fruit has six segments; the juice has a TSS of 18.1° Brix	TNAU, Coimbatore
Krishna (NA 4)	—	Selection	Probably chance seedling of Banarasi cultivar	Moderate bearer, fruits medium to large in size, flattened, conical, angular, basin papillate in shape, skin smooth, yellowish in colour with red blush on the exposed surface	NDUAT, Faizabad
Kanchan (NA 5)	—	Selection	Seedling selection from Cahkaiya	Heavy and regular bearer, medium sized fruits, higher fibre content	NDUAT, Faizabad
NA 6	—	Selection	Seedling selection from Cahkaiya	Prolific and heavy bearer, ideal for preserve and candy owing to low fibre content	NDUAT, Faizabad
NA 7	—	Selection	Seedling selection from Francis	Precocious, prolific and regular bearer, ideal for preparation of products	NDUAT, Faizabad

pinnate fashion, making its branches feathery in general appearance. The leaves develop after fruit setting. Flowers, unisexual, pale-green, 4–5 mm in length, borne in leaf-axils in clusters of 6–10; staminate flowers, tubular at the base, having a very small stalk, gamosepalous, having 6 lobes at the top; stamens 1–3, polyandrous, filaments 2 mm long; pistillate flowers, fewer, having a gamopetalous corolla arid a two-branched style; both staminate and pistillate flowers are borne on the same branch, but the staminate flowers occur towards the apices of small branches. Fruits, fleshy, almost depressed to globose. 2.14 cm in diameter, 5.68 g in weight, 4.92 ml in volume, primrose-yellow 601/2. The stone of fruit, six-ribbed, splitting into three segments, each containing usually two seeds; seeds 4–5 mm long, 2–3 mm wide, each weighing 572 mg, 590 microlitres in volume, citron green 793/3.

Aonla is a deciduous tree and the emergence of new shoots starts in the beginning of April. The shoots are light red and turn green after 2 or 3 days. After about 15 days, the small lateral twigs, which are 3–3.5 cm long, give out two rows of leaves on each side. They have the appearance of compound leaves, with about one hundred leaflets on each. Small circular green flowers also appear in the axils of these small leaves at the same time. The flowering season is from mid-April to first week of May under Sanwara (Himachal Pradesh) conditions. The flowering reaches its peak during April-end. The fruiting season is exceptionally long. The fruit in this area become fit for harvesting in December. They can be retained on the tree up to March without any significant loss in quality or yield. The picking of fruits is generally done by villagers in February and March.

FIG

Botanical name: *Ficus carica*
Family: Moraceae

Fig is believed to be indigenous to Western Asia and to have been distributed by man throughout the Mediterranean area. Remnants of figs have been found in excavations of sites traced to at least 5,000 B.C. It is a deciduous tree, 50 feet tall, but grows more typically to a height of 10–30 feet. The branches are muscular and twisting, spreading wider. Fig wood is weak and decays rapidly. The trunk often bears large nodal tumors, where branches are shed or removed. The twigs are terete and pithy rather than woody. The sap contains copious milky latex that is irritating to human skin. Fig trees often grow as a multiple-branched shrub, especially where subjected to frequent frost damage. They may be espaliered, but only where roots may be restricted, as in containers.

FIG

Variety	Year of release	Breeding method	Pedigree/ parentage	Important traits	Institutes/ universities
YCD 1 Timla Fig	1993	Introduction	—	Fruits attractive reddish-purple and large in size, 7.0 cm in diameter, each weighing 100–200 g. High-yielding, 4,000 fruits/tree. The bearing is throughout the year excepting winters	HRS, TNAU, Yercaud
Deanna	1986	Introduction	USA	Plants are vigorous, fruits produced parthenocarpically, fruits pyriform in shape, large (80 g), fruit skin colour golden-yellow, eye medium tight with straw yellow pulp which has a mean TSS of 18° Brix. Fruits are suitable for drying	—
Conadria	1986	Introduction	USA	Plants are moderate in vigour, fruits produced parthenocarpically, fruits pyriform in shape, medium-sized (45 g), fruit skin colour green, eye medium tight with a light pink pulp which has a mean TSS of 18° Brix	—
Excel	1986	Introduction	USA	Short stature plants with compact canopy, fruits produced partheno-carpically, they are ovoid in shpe, small (35 g), fruit skin colour lemon yellow, eye tight with pale yellow pulp which has a mean TSS of 20° Brix	—

Fig leaves are bright green, single, alternate and large (1 foot length). They are more or less deeply lobed with 1–5 sinuses, rough hairy on the upper surface and soft hairy on the underside. In summer, their foliage lends a beautiful tropical feeling. The tiny flowers of fig are not visible, clustered inside the green "fruits", technically a synconium. Pollinating insects gain access to flowers through an opening at the apex of synconium. In common, fig flowers are all female and need no pollination. There are 3 other types, caprifig which has male and female flowers requiring visits by a tiny wasp, *Blastophaga grossorum*; the Smyrna fig, needing cross-pollination by caprifigs in order to develop normally; and the San Pedro fig which is intermediate, its first crop independent like the common fig, its second crop dependent on pollination.

The common fig bears a first crop, called the breba crop, in the spring on last season's growth. The second crop is borne in fall on the new growth and is known as the main crop. In cold climates the breba crop is often destroyed by spring frosts. The matured "fruit" has a tough peel (pure green, green suffused with brown, brown or purple), often cracking upon ripeness, and exposing the pulp beneath. The interior is a white inner rind containing a seed mass bound with jelly-like flesh. The edible seeds are numerous and generally hollow, unless pollinated. Pollinated seeds provide the characteristic nutty taste of dried figs.

KARONDA

Botanical name: *Carissa carandas* Auct.
Family: Apocynaceae

Karonda is a hardy, evergreen, spiny and indigenous shrub. Widely grown in India, it is found wild in Bihar, West Bengal and south India. It is grown commonly as a hedge plant. Regular plantations of karonda are very common in Varanasi district of Uttar Pradesh. Fruits are sour and astringent in taste. They are richest source of iron containing good amount of vitamin C. Very useful to cure anaemia, its fruits have antiscorbutic properties also. Matured fruit contain high amount of pectin and, therefore, besides being used for making pickle, it can be exploited for making jelly, jam, squash, syrup and chutney, which are of great demand in international market.

It is native and common throughout India, Burma and Malacca and dry areas of Ceylon. It is commonly cultivated in these areas as a hedge and for its fruit. Its fruits are marketed in villages.

KARONDA

Variety	Year of release	Breeding method	Pedigree/ parentage	Important traits	Institutes/ universities
Pant Manohar	1998	Selection	Seedling selection	Medium sized, dense bushes, fruit size 2.13 cm × 1.69 cm, flesh 88.27%, TSS 3.92%, acidity 1.82%, dark brown colour	GBPUAT, Pantnagar
Pant Sudarshan	1998	Selection	Seedling selection	Fruit size 2.16 cm × 1.69 cm, colour dark brown, flesh 88.47%, TSS 3.45%, acidity 1.89%	GBPUAT, Pantnagar
Pant Suvarna	1998	Selection	Seedling selection	Plant upright growing and sparse, fruit size 2.16 × 1.66 cm, colour dark brown at ripening	GBPUAT, Pantnagar

This is a rank-growing, straggly, woody, climbing shrub, usually growing to 3–5 m high, sometimes ascending to tops of tall trees; and rich in white, gummy latex. The branches, numerous and spreading, forming dense masses, are set with sharp thorns, simple or forked, up to 5 cm long, in pairs in axils of leaves. The leaves are evergreen, opposite, oval or elliptic, 2.5–7.5 cm long; dark green, leathery, glossy on the upper surface, lighter green and dull on underside. The fragrant flowers are tubular with 5 hairy lobes which are twisted to the left in the bud instead of to right as in other species. They are white, often tinged with pink, and borne in terminal clusters of 2–12. The fruit, in clusters of 3–10, is oblong, broad-ovoid or round, 1.25–2.5 cm long; has fairly thin but tough, purplish-red skin turning dark-purple or nearly black when ripe; smooth, glossy; enclosing very acid to fairly sweet, often bitter, juicy, red or pink, juicy pulp, exuding flecks of latex. There may be 2–8 small, flat brown seeds.

TAMARIND

Botanical name: *Tamarindus indica*
Family: Fabaceae

Tamarind is an evergreen tree It belong to pea family native to tropical Africa. It is widely cultivated in other regions as an ornamental and for its edible fruits. The word tamarind literally means '*date of India*'. Tamarind, a native of East Africa, is now grown extensively in India, South East Asia and the West Indies. It is a semi-evergreen, tropical tree that grows to about 24 m (80 feet) tall and has long drooping branches with alternate, pinnately compound (feather-formed) leaves; the leaflets are about 2 cm (0.75 inch) long. The yellow flowers, about 2.5 cm across, with a red stripe are borne in small clusters. The dark brown fruit is a plump pod 7.5–24 cm long that does not split open. It contains 1–12 large, flat seeds embedded in a soft, brownish pulp. This pulp has a high tartaric acid content, that imparts for its sourness.

This portion of fruit is widely used in Orient in foods, beverages, and medicines. It is a standard ingredient in curries, chutneys, soups and several other dishes of India and South-East Asia. The juice is made into a refreshing drink in both the Middle East and the West Indies. Tamarind is a good laxative and an antiseptic. It is used for tummy upsets and for the treatment of ulcers. Over-ripe fruits can be used to clean copper and brass.

TAMARIND

Variety	Year of release	Breeding method	Pedigree/ parentage	Important traits	Institutes/ universities
PKM 1	1992	Clonal selection	Clonal selection from local type Endapuli	High-yielding, bears fruits in clusters (3–7), fruits are characteristically semi circle in shape, high pulp recovery (39%) with high tartaric acid and ascorbic acid content	HC&RI, TNAU, Periyakulam
DTS 1	1996	Clonal selection	—	High-yield potential and starts bearing within 5–6 years of planting	UAS, Dharwad

MANILLA TAMARIND (*Pithecellobium dulce*)

Variety	Year of release	Breeding method	Pedigree/ parentage	Important traits	Institutes/ universities
PKM (MT) 1	2008	Selection	Selection from open pollinated seedlings in Soolakkarai	Campact droping canopy amenable for high density planting, cluster-bearing, spirally twisted fruits with clear constrictions, pale yellow pods, white aril and black seeds, TSS 18° Brix, suitable for candy preparation, high yield (79 kg/tree)	HC&RI, TNAU, Periyakulam

TEMPERATE FRUITS

APPLE

Botanical name: *Malus domestica*
Family: Rosaceae (sub Family: Pomoideae)

Apple is most widely cultivated fruit. Its tree is small and deciduous, reaching 5–12 m tall, with a broad, often densely twiggy crown. The leaves are alternately arranged simple ovals 5–12 cm long and 3–6 cm broad on a 2–5 cm petiole with an acute tip, serrated margin and a slightly downy underside. Flowers are produced in spring simultaneously with budding of leaves. The flowers are white with a pink tinge that gradually fades, five-petaled, and 2.5–3.5 cm in diameter. The fruit matures in autumn, and is typically 5–9 cm in diameter. The centre of fruit contains five carpels arranged in a five-point star, each carpel containing 1–3 seeds. Apples are very nutritious.

The tree originated from Central Asia, where its wild ancestor is still found today. There are more than 7,500 known cultivars of apples, resulting in range of desired characteristics. Cultivars vary in their yield and the ultimate size of the tree, even when grown on the same rootstock.

Floral Biology

- Flowers of apple cultivars and seedlings vary considerably in their size and petal shape, and in colour from white to deep pink.
- Borne in cymose clusters on fairly short pedicles on spur type growth, but in some instances from terminal or lateral buds of previous season's growth.
- Five petals, a calyx of five sepals, 20 stamens and pistil which divides into five styles.
- Ovary has five carpels, each usually containing two ovules — in most cases seed content is 10 but some cultivars have more.

Pollination

- Mechanical transfer of pollen from anthers to stigmas and is a pre-requisite to fertilization of ovules and development of seeds and fruits.
- Cross-pollination by insects, particularly bees.

APPLE

Variety	Year of release	Breeding method	Pedigree/ parentage	Important traits	Institutes/ universities
Lal Ambri	1973	Hybridization	Red Delicious × Ambri	Upright tree with moderate vigour; free from incidence of scab	SKUAST(K), Srinagar
Sunheri	1973	Hybridization	Ambri × Golden Delicious	Upright tree with moderate vigour; free from incidence of scab	SKUAST(K), Srinagar
Firdous	1995	Hybridization	Golden Delicious × Rome Beauty × *Malus floribunda*	Resistant to scab powdery mildew, Alternaria, mid season crop, fruits of good taste large size, average yield – 7.20 tonnes/ha	SKUAST(K), Srinagar
Shireen	1995	Hybridization	Lord Lam bourne × Melba XR-1274078	Resistant to scab, Alternaria and powdery mildew, mid-season crop	SKUAST(K), Srinagar
Akber	2001	Hybridization	Ambri × Cox's orange Pippin	Upright, good yielding resistant to scab and powdery mildew	SKUAST(K), Srinagar
Amred	1978	Hybridization	Red Delicious × Ambri 157	Tree tall, spreading and open, maturity in second week of September, fruit medium in size, shape is conical, bright red strips over colour on barium yellow ground. Aromatic, crisp and juicy. Keeping quality is good. Low incidence of powdery mildew, sooty blotch and apple scab recorded	Dr YSPUHF, Solan, Himachal Pradesh

APPLE : *Contd.*

Variety	Year of release	Breeding method	Pedigree/ parentage	Important traits	Institutes/ universities
Ambstar-king	1978	Hybridization	Starking Delicious × Ambri 81	Vigorous with open canopy. Fruit matures in second week of September, medium in size. Shape round conical. Red streaks over chrome yellow ground colour. Crisp and juicy. Tolerant to scab	Dr YSPUHF, Solan, Himachal Pradesh
Ambrich	1978	Hybridization	Rich-a-Red × Ambri 15	Semi dwarf trees with spreading and drooping canopy. Semi spur type. Fruit maturity in second week of September. Fruit medium in size. Shape is round conical. Crisp, sub-acidic and juicy flesh. Moderately susceptible to powdery mildew, sooty blotch and fly speck. Tolerant to scab	Dr YSPUHF, Solan, Himachal Pradesh
Ambroyal	1986	Hybridization	Starking Delicious × Ambri 84	Semi dwarf trees with spreading canopy. Fruit matures in third week of September. Fruit size is medium, shape is conical. Red streaks on yellow ground colour. Soft and juicy flesh	Dr YSPUHF, Solan, Himachal Pradesh
Chaubattia Anupam	1984	Hybridization	Early Shanburry × Red Delicious	Vigorous trees with upright canopy. Somewhat early maturing. Fruits are medium to large. Somewhat	Directorate of Horticulture and Food

APPLE : *Contd.*

Variety	Year of release	Breeding method	Pedigree/ parentage	Important traits	Institutes/ universities
				conical in shape. Bright red streaks with red blush on pale ground colour. Crisp and aromatic flesh. Good keeping quality	Processing, Chaubattia, Uttarakhand
Chaubattia Princess	1984	Hybridization	Early Shanburry × Red Delicious	Trees are medium in vigour with upright canopy. Medium in size and conical in shape. Deep red streaks on pale back ground. Flesh crisp and juicy. Keeping quality good	Directorate of Horticulture and Food Processing, Chaubattia, Uttarakhand
Chaubattia Swarnima	1984	Hybridization	Benoni × Red Delicious	Fruit matures in first to second week of July. Red coloured with some strips over pale coloured skin. Juicy and slightly tart in taste. Good keeping quality	Directorate of Horticulture and Food Processing, Chaubattia, Uttarakhand
Chaubattia Alanker	1984	Hybridization	Fanny × Red Delicious	Second week of July is the ripening time. Round in shape. Red coloured strips over yellow ground colour skin	Directorate of Horticulture and Food Processing, Chaubattia, Uttarakhand

APPLE : *Contd.*

Variety	Year of release	Breeding method	Pedigree/ parentage	Important traits	Institutes/ universities
Chaubattia Anurag	1934	Hybridization	Esopus Spitzenberg × Red Delicious	Matures in third week of July to first week of September. Fruit medium in size, conical and slightly flattened at base. The skin is yellowish in colour with bright red colour at top. Flesh is juicy and sweet	Directorate of Horticulture and Food Processing, Chaubattia, Uttarakhand
Chaubattia Agrim	1984	Mutation	Induced mutation of (budwood irradiation with Gamma rays) Early Shanburry	Fruits are small, egg shaped slightly conical. Yellow coloured with slight red tinge. Flesh is dirty yellowish white coloured. Sour to sweet in taste. Matures in last week of June.	Directorate of Horticulture and Food Processing, Chaubattia, Uttarakhand
KKL 1	1987	Selection	Selection from Parlin's Beauty	The tree is medium size with a heavy yield of 22 tonnes/ha. It is found best adapted to warm winter conditions prevailing in Kodaikanal hills. It is a mid season variety. Fruits are available during July–August. The fruits are tasty with a TSS of 13° Brix. The tree yields 400 fruits in a year.	TNAU, Coimbatore

- Most flowers are admirably designed to facilitate pollination, but in a few genotypes, such as 'Delicious', the stamens are long and styles short, which allow the bees to visit flowers without making contact with stigmas leads to a considerable loss of crop.
- Before crossing, flowers of seed parent are usually emasculated at the balloon stage.
- Flicking off the sepals, petals and stamens with fingernails.
- Styles are left sticking up from the receptacle and ovary.
- Two flowers per cluster are emasculated and others removed.
- Un-emasculated flowers producing fruits can be identified, because fruits from emasculated flowers have no calyx.
- Pollination can be achieved by dipping a small, soft brush into a vial containing the pollen and lightly brushing the stigmas.

PEAR

Botanical name: *Pyrus communis*
Family: Rosaceae (sub Family: Pomoideae)

Pear is a pomaceous fruit produced by a tree of genus *Pyrus*. The term "pyriform" is sometimes used to describe something which is "pear-shaped". Fruit is composed of receptacle or upper end of flower-stalk (so-called calyx tube) greatly dilated, and enclosing within its cellular flesh the five cartilaginous carpels which constitute the "core" and are really the true fruit. From the upper rim of receptacle are given off the five sepals, the five petals, and the very numerous stamens. Another major relative of the pear (and thus the apple) is the quince.

Pears are consumed fresh, canned, as juice, and dried. The juice can also be used in jellies and jams, usually in combination with other fruits or berries. Fermented pear juice is called perry. Pears are rich in vitamin B2, vitamin C, E1, copper, potassium and are 100% free from bad cholesterol. Pear juice is also known to be the first juice introduced to infants because of its hypoallergenic properties.

Botany

- Deciduous tree up to 10 m in height.
- Grey erect trunk, with its bark breaking into stripes.
- Ovate leaves till 10 cm, bright green above and glabrous.

PEAR

Variety	Year of release	Breeding method	Pedigree/ parentage	Important traits	Institutes/ universities
Pant Pear 3	2000	Selection	Clonal selection	Tree medium-sized, mid maturing variety, high yield, medium-sized fruit, pyriform, flesh soft, TSS 15%, thin skin	GBPUAT, Pantnagar
Pant Pear 17	2000	Selection	Clonal selection	Tree medium sized, high yielding, late maturing, large fruit, round to pyriform, thin skin, soft flesh, TSS 14.5%	GBPUAT, Pantnagar
Pant Pear 18	2002	Selection	Clonal selection	Medium sized, early maturing, large round fruited, hard, juicy flesh, TSS 13%	GBPUAT, Pantnagar
Punjab Nector	—	—	—	Low chilling, semi-soft, average fruit weight 136 g, average yield 80 kg/tree	PAU, Ludhiana
Punjab Gold				Semi-soft, 14.5% TSS, average fruit weight 166 g, average yield 80 kg/tree	PAU, Ludhiana
Punjab Soft				Soft variety of Asian type pear, medium-sized fruits (128 g), average yield 86 kg/tree	PAU, Ludhiana

- White or rosy flowers, till 1.5 cm wide, gathered in corymbs from 3–7.
- The fruit is an edible pome.

Floral biology

- Blossoms — white, rarely pinkish and borne in umbel like racemes.
- Flower consists of five petals and sepals and 20–30 stamens with anthers that are usually red.
- Styles vary from two to five and are free but closely constricted at base.
- The ovary has five locules with two ovules per locule with a maximum seed set of ten.
- A greater percentage of the larger basal flowers set fruit compared with the smaller terminal ones and the king blossom.
- The king blossom is usually allowed to open and perhaps even the second blossom; then one or more of the remaining blossoms are emasculated when they are at the balloon stage of development.
- The remaining flowers removed at the time of emasculation
- The lowest mean daily temperature for start of flowering in pears is 9°C.
- Higher the mean temperature and lower the humidity, shorter is the flowering period.
- Pear blossoms open very quickly — weather is warm and dry.

PEACH

Botanical name: *Prunus persica*
Family: Rosaceae (sub Family: Prunoideae)

Peach is a tree native to China that bears juicy fruits. It is a small deciduous tree growing to 5–10 m tall. Peach trees grow very well in a fairly limited range, since they have a chilling requirement that subtropical areas cannot satisfy, and they are not very cold-hardy. The trees themselves can usually tolerate temperatures to around — 26 to -30°C, although the following season's flower buds are usually killed at these temperatures, leading to no crop that summer. Flower bud kill occurs at temperatures between -15°C and -25°C depending on timing of cold, with the buds becoming less cold tolerant in late winter. Certain cultivars are more tender and others can tolerate a few degrees more cold. In addition, a lot of summer heat is required to mature the crop, with mean temperatures of the hottest month between 20°C and 30°C.

Botany

A small tree with a spreading canopy, usually 2–3.5 m in cultivation. Trees are short-lived, generally living only 15–20 years, and even less in cultivation (average tree life expectancy in Georgia is 8 years). The deciduous trees, set in the orchard about 20 feet apart, are usually trimmed to 8–16 feet in height. Leaves linear with acute tips, folded slightly along the midrib, sickle-shaped in profile.

Flowers

- Light pink to carmine, to purplish; 1 inch in diameter.
- Single locule, single seed inside ovary, surrounded by hypanthium. Colour of inner surface of hypanthium is indicative of flesh color; whitish-green = white, gold = yellow.
- Flowers exhibit cleistogamy, pollinating themselves prior to opening.
- Petals can be large and showy, or small and curved on margins.
- Flowers are borne singly on short peduncles (almost sessile), from lateral buds on 1-year-old wood; usually 1–2 flower buds/node.
- Ornamental peaches contain fully double flowers, having many petals and a carnation-like appearance. Colours range from dark pink to white.
- Depending upon the variety the number of stamens varies from 33–59. The stamens are arranged in two to three whorls.
- The pistil with single style and clavate stigma with monocarpellay ovary having one ovule.

Fruits

- A drupe. The bony endocarp (pit) surrounds a single, large, ovate seed.
- The flesh is mesocarp, the skin the exocarp.
- Trees are very precocious; generally produce some fruits during second year in field.

Floral Biology

- Take 5 h to complete anthesis from the dom-shaped flower buds.
- Commence during 8–10 AM and continues up to 6 PM.
- Maximum number of flowers open between 10 AM and 12 noon.
- Anther dehiscence starts between 8 and 10 AM and continues up to 12 noon.
- Dehiscence of all anthers is completed with in a day and the remaining if any dehisce on the next day.
- The peak period of dehiscence is seen between 10 AM and 12 noon.
- Normally, the flowers are fully closed at 6 AM, but most of them are open by 10 AM and all are open by noon.
- Most cultivars produce pollen at the time the stigma is receptive.
- Appears as yellow powdery mass at the dehisced anther lobes.
- Fertility varies from 0–90%.
- Size ranges between 45–51 μ.
- Stigma receptivity starts 4 days before anthesis and they continue to be receptive till 3 days after anthesis.
- They do not close at night; they may stay open and the stigma may be receptive for 3 days.
- Maximum fruit set observed when the stigma were pollinated on the day of anthesis.

Pollination

- Peach showed a higher percentage of fruit set under open pollination than either self/natural cross pollination.
- Artificial cross-pollination increased the fruit setting significantly.
- Self-fertile, normally grown without a pollinizer in solid blocks.
- Pollination achieved in some cultivars before flowers open. A few old cultivars require pollinizers.
- There was slightly more fruit set on all cultivars when visited by bees.
- Most cultivars are self-fertile and a few are self-sterile.

PEACH

Variety	Year of release	Breeding method	Pedigree/ parentage	Important traits	Institutes/ universities
Pant Peach 1	1998	Selection	Seedling selection	Fruit medium-sized, have red pigmentation on surface, semi-cling stone	GBPUAT, Pantnagar

PLUM

Variety	Year of release	Breeding method	Pedigree/ parentage	Important traits	Institutes/ universities
Pant Plum 1	1993	Selection	Seedling selection	Dwarf, yellow coloured, subacidic fruit, good rootstock for other plum cultivars	GBPUAT, Pantnagar
Fla 12	1999	Selection	Seedling selection	Exotic type, much larger fruit than titron or Jamuni, more juicy, ripen about one week after Titron or jumuni	GBPUAT, Pantnagar

- Many self-sterile cultivars have been largely or completely eliminated from the market, regardless of their other good qualities, because interplanting of cultivars and insect pollination are necessary in their production.
- Unfortunately, references to self-sterility of such cultivars has tended to draw attention away from "self-fertile" cultivars and the possibility that they might not be capable of fertilizing themselves without the aid of an outside agency.

PLUM

Botanical name: *Prunus* sp.
Family: Rosaceae

Plum fruit is sweet and juicy and it can be eaten fresh or used in jam-making or other recipes. Plum juice can be fermented into plum wine; dried plums are known as prunes. Prunes are also sweet and juicy and contain several antioxidants. Plums and prunes are known for their laxative effect. This effect has been attributed to various compounds present in fruits, such as dietary fibre, sorbitol, and isatin. Prunes and prune juice are often used to help regulate the functioning of digestive system. These are two main sections within the *Prunophora* subgenus which contain cultivated plums, *Euprunus* (true plums) and *Prunocerasus* (plum – Chemies). All are diploid in nature. All the plums have a basic chromosome number of x = 8.

Botany

Shrub like or small tree like habit. Most of the species form suckers, more erect growing than peach. Flower perigynous, sepals 5, petals 5, petals shed off after anthesis. Stamens are numerous. Flowers are borne mostly in umbel like clusters of 2–3 individuals on short spurs and solitary or 2–3 in axils of 1-year-old wood. Pistill usually with only one style and single clavate stigma. Hercogamy in seen, *viz.* difference in length of stamen and style exist. Perfect flower 78%, intermediary flower — 16–28%, male flower 2–25%. Fruit A drupe, oval shaped or round in European types, round to conical or heart-shaped in Japanese types. Fruit size generally larger in Japanese types. Botanically called sulenta type, glabrous. Seeds are compressed, usually longer than broad and smooth or nearly so.

Floral Biology

- The maximum anthesis takes place between 12 noon and 2 PM and the increase of temperature hastened the process.

- Anther dehiscence takes more than one day to complete.
- Pollen fertility ranges from 56.2 to 83.2%.
- The germination of pollen-grains ranges from 4.3 to 70.2%.
- Generally dehiscence takes place on the day of flower opening and continuous up to next day also.
- Dehiscence is more between 11 AM and 10 PM depending upon the temperature on the day.
- The stigma becomes receptive when it is quite plump and shiny.
- Brownish and dull colour indicates the non-functional stigma.
- The stigma is fully receptive from one day before anthesis and remains so up to four days.
- A natural crop pollination is 3–16.6%.
- Honeybees and insects are major pollinators.
- In Japanese plums pollinizers are necessary for commercial production for about half the cultivars.
- The *domestica* types about half of the cultivars require pollinizers.
- Self-fruitful — it should exhibit an average of 30% fruit setting which is adequate for a commercial fruit production.
- The cultivars that are self-unfruitful set only 1–2% must be inter-planted with pollinizing cultivars.

ALMOND

Botanical name: *Prunus amygdalus*
Family: Rosaceae

Although almond is a native to the Mediterranean region, it has adapted to the climate of Kashmir. Big and shady, its trees grow up to a height of 15 m and have large, oval leaves tinged with red and fleshy fruits. These fruits turn bright red from green during the mature phase. The fruits are very aromatic and sweet. Almonds are considered healthiest nuts to eat because they have highest protein content. Almonds are also a rich source of calcium and many minerals like potassium and phosphorous. Almonds help in lowering cholesterol levels and contain oleic acid, which helps maintain a healthy heart.

There are two types of almonds: bitter and sweet. Almonds grown in Kashmir are slightly bitter but sweet ones are imported into India from Afghanistan in great quantities. Almond primarily a fruit species of temperate regions has two distinct features which have been exploited by breeders for cultivation in subtropical regions. First is,

ALMOND

Variety	Year of release	Breeding method	Pedigree/ parentage	Important traits	Institutes/ universities
Makhdoom	1995	Clonal selection	Seedlings of unknown variety	Medium nut size, semi-soft shelled medium-sized kernels, shelling 38–40%, mid-season crop, average yield, 5 q/ha	SKUAST(K), Srinagar
Parbat	1995	Clonal selection	Seedlings of unknown variety	Mid-bloomer cultivar, shelling 49%, average yield, 5 q/ha	SKUAST(K), Srinagar
Waris	1995	Clonal selection	Seedlings of unknown variety	Mid-bloomer cultivar, soft-shelled, shell yield is 6–7 q/ha	SKUAST(K), Srinagar
Shalimar	1995	Clonal selection	Seedlings of unknown variety	Mid-bloomer, high-shelling ability, attractive nuts, white in colour, plump average yield, 5 q/ha.	SKUAST(K), Srinagar

its low-chilling requirement (250–400 h at or below 7°C) is often easily met within subtropical regions and secondly, early spring frosts (almost non-existent in subtropical areas) are detrimental to blooms due to its early flowering habit in the spring.

The basic objectives in almond improvement are late-blooming behaviour (to escape spring frost damage to blossoms), cross-compatibility or self-fruitfulness, heavy-bearing and superior nut and kernel quality.

APRICOT

Botanical name: *Prunus armeniaca* Linn.
Family: Amygdalaceae

Apricot is an important fruit crop of midhill and dry temperate regions of the country. Cultivated apricot has its origin in North-Eastern China, whereas wild one, popularly known as *zardalu*, appears to be indigenous to India. It grows wild in hills of Shimla and Kinnaur districts in Himachal Pradesh. Fruit is delicious. It is rich in vitamin A and contains more carbohydrates, proteins, phosphorus and niacin than many other common fruits. Besides its use as dessert, it is also canned and dried. Fruit is processed into jam, nectar and squash. The kernel, which is either sweet or bitter depending upon the variety, is a valuable byproduct. Since sweet kernels taste like almond, they are used as its substitute in pastries and confectionery, while bitter ones are used for oil extraction.

Apricot is grown commercially in hills of Himachal Pradesh, Jammu and Kashmir, Uttar Pradesh and to a limited extent in north-eastern hills. Some drying type apricots are being grown in dry temperate areas of Kinnaur and Lahaul Spiti in Himachal Pradesh and Ladakh in Jammu and Kashmir. A deciduous, large and spreading tree with dark-brown to black bark; height, around 10 m, may go up to 15 m in some cases; girth, 82 cm; wood, hard and durable. Leaves, broad, cordate, dark green, petiolate (petiole, 24 cm long), alternate, having reticulate pinnate venation; length, 6.2 cm; breadth, 6.1 cm; margin, serrate. Flowers, simple, sessile, pentamerous, perigynous, actinomorphic, complete, hermaphrodite, light pink; 2–5 flowers per cluster; calyx, gamosepalous, companulate, with five sepals, light red, having 3 to 5 mm long lobes of sepals; corolla, polypetalous with 5 petals, rosaceous, valvate, actinomorphic, angular, imbricate, 1.5–2.0 cm long, white to light pink; androecium, polyandrous, with 35 stamens, dorsifixed; anther-lobes, yellow, episepalous, bithecus; gynoecium, monocarpellary, perigynous; ovary, unilocular, having basal placentation. Fruit, a drupe, velvety when young, but nearly

APRICOT

Variety	Year of release	Breeding method	Pedigree/ parentage	Important traits	Institutes/ universities
Chaubattia Alanker	1984	Induced mutation	Kaisha × Charmgaze Budwood irradiation with Gamma rays	Regular bearing, low chilling and very early maturity. Good keeping quality	Directorate of Horticulture and Food Processing, Chaubattia, Uttarakhand
Chaubattia Madhu	1984	Induced mutation	Turkey × Charmgaze Budwood irradiation with Gamma rays	Early-ripening, highly productive, regular in bearing	Directorate of Horticulture and Food Processing, Chaubattia, Uttarakhand
Chaubattia Kasri	1984	Induced mutation	St. Ambroise × Charmgaze Budwood irradiation with Gamma rays	Mid-season in ripening with good quality fruits	Directorate of Horticulture and Food Processing, Chaubattia, Uttarakhand

smooth at maturity, round to oblong; diameter, 2.5–2.6 cm; weight, 12.6 g; volume, 11.20 ml; fruit, externally yellow; pulp, deep yellow, less juicy than that of the cultivated apricots; endocarp, flat, smooth, stony and hard. Stone, 1.7–2 cm long, 1.6–1.7 cm broad, 2.5 g in weight, 1.3 ml in volume; 2 kernels, per seed; kernels, 1.5 cm long, 1.2 cm broad; weight, 879 mg; volume, 757 microlitres.

The flowering season continues from the first week to last week of March under Sanwara conditions. The fruits start ripening from mid-May to first week of June.

WALNUT

Botanical name: *Juglans regia* L.
Family: Juglandaceae

Walnut is most important temperate nut fruit of the country. It is grown in Jammu and Kashmir, Uttar Pradesh and Himachal Pradesh. There are no regular orchards of walnuts in the country because the existing plantations are generally of seedling origin. The seedling trees attain giant size and start bearing nuts of variable sizes and shapes after 10–15 years, whereas vegetatively-propagated plants are true-to-type and produce almost uniform-sized nuts after 4–5 years. They remain within manageable size. But the major constraint is low success in vegetative propagation. Limited availability of scion material from desired trees results in very few vegetatively propagated plants. Walnuts earn valuable foreign exchange.

Generally two types of walnut trees are found in Jammu & Kashmir, one that bears fruit and is noted for its wood, and another which is not fruit-bearing. The latter is known as 'zangul' in local parlance and is less strong and possesses no grains. Deciduous, monoecious trees, 12–15 m tall and rarely up to 60 m tall; bark brown or gray, smooth, fissured; leaf-scars without prominent pubescent band on upper edge; leaves alternate, foetid, pinnate, without stipules; leaflets to ovate-lanceolate, acuminate; margin irregularly serrate, glabrescent above, pubescent and glandular beneath; flowers developing from dormant bud of previous season's growth; staminate flowers in axillary, pendulous aments 5–15 cm long, developing 1–4 million pollen-grains each; flowers in axils of scales, with 2 bracteoles, perianth-segments 1–4, stamens 3–40; pistillate flowers in clusters of 3–9, developing as many nuts; in selected varieties not only terminal bud produces fruit, but all lateral buds on previous years growth also produce; perianth 4-lobed; fruit 3.5–5 cm in diameter, globose or slightly ridged, not splitting.

WALNUT

Variety	Year of release	Breeding method	Pedigree/ parentage	Important traits	Institutes/ universities
Hamdan	2001	Clonal selection	Seedling of unknown landrace	Nut weight is 14 g smooth in texture shelling ability is 54%	SKUAST(K), Srinagar
Suleiman	2001	Clonal selection	Seedling of unknown landrace	Precocious, nut weight is 20 g. Smooth shell surface, nut shape is roundish, shelling ability is 50%	SKUAST(K), Srinagar

The walnut wood is almost black, and grain here is much more pronounced than the wood of the trunk which is lighter in colour.

Black in colour, the walnut wood is used for making furniture and carvings. The branches have lightest colour, being almost blonde, and have no noticeable grain. The inherent worth of wood from each part of the tree differs. The root part is the most expensive and the branches having the lowest price.

The objectives of improvement in walnut have been to develop high quality, medium-shelled nuts for inshell purpose, and nuts with thin shells and high kernel recovery and extra light coloured kernels for shelling purpose.

SOLNACEOUS VEGETABLES

TOMATO

Botanical name: *Lycopersicon esculentum* Miller, 2n = 24
Family: Solanaceae

Tomato is most important and remunerative vegetable crop in India. A rich source of minerals, vitamins and organic acids, tomato fruit provides 3–4% total sugar, 4–7% total solids, 15–30 mg/100 g ascorbic acid, 7.5–10 mg/100 ml titratable acidity and 20–50 mg/100 g fruit weight of lycopene.Uttar Pradesh, Maharashtra, Karnataka, Bihar and Orissa, are major tomato-growing states in India. Three types are most common: fresh market tomatoes, small-sized "cherry" tomatoes, and processing tomatoes. Processing tomatoes have a bright red colour and high solids content that makes them suitable for making paste, catsup, or sauce etc.

Tomato has its origin in Peru, Ecuador and Bolivia on the basis of availability of its numerous wild and cultivated relatives.

Plant habit

Determinate and semi-determinate varieties produce stems that end with a flower cluster. Determinates are short and bushy while semi-determinate varieties grow slightly taller. Indeterminate varieties continually produce new leaves and flowers, and can grow very tall. Indeterminate varieties set fruits over a longer period. Indeterminate varieties should be staked and pruned. Most tomato plants have compound leaves, and are called regular leaf plants. But some cultivars have simple leaves known as potato leaf style because of their resemblance to potato. Of regular leaves, there are variations,

TOMATO

Variety	Year of release	Breeding method	Pedigree/ parentage	Important traits	Institutes/ universities
Sioux	1970	Introduction	American cultivar	Indeterminate, fruits uniform red, round, smooth, small, and less seeded. Yield 195–220 q/ha	IARI, Katrain
Marglobe	1970	Selection	Introduction from USA	Plants 70–80 cm tall, vigorous, erect and indeterminate in growth habit. Typically 4 or 5 flowers on truss, which bear 3 or 4 fruits. Fruits nearly round, deep scarlet in colour on ripening, 3–5 locules, juicy, mildly acidic, thick walled, smooth, solid and less seeded. Fruits have dark green shoulder, weight 100–150 g. Yield 200–250 q/ha	IARI, New Delhi
Pusa-120	1970	Selection	Selection from introduction	Semi-determinate, spreading, late maturing with dark green foliage. Fruits flatish-round, attractive, medium-large, uniform red, less acidic and less seeded; resistant to nematode; yield 300–320 q/ha	IARI, New Delhi
Pusa Ruby	1970	Selection	Sioux × Improved Meeruthi	Indeterminate, height 80-85 cm, spreading with few hardy branches. Fruits shape flat to round, slightly lobed (4–5 locules), size medium, colour red, yield 280–300 q/ha	IARI, New Delhi

TOMATO : *Contd.*

Variety	Year of release	Breeding method	Pedigree/ parentage	Important traits	Institutes/ universities
Best of All	1973	Selection	Introduction from USA	Indeterminate plant erect and height 100–200 cm. Fruits are medium in size, scarlet red in colour, round, juicy, small scar, thin skin, 3 loculed with green shoulder and borne in clusters of 4–5 fruits. Yield potential 300–315 q/ha	IARI, New Delhi
Pusa Early Dwarf	1973	Selection	Improved Meeruthi × Red Cloud	Plants are determinate, early maturing and dwarf with compact fruiting. Fruits are flatish-oblate, smooth, medium-large, red, furrow with 5–6 locules; suitable for rainy season cultivation in plains and March sowing in hills. Yield potential 300–325 q/ha	NBPGR, New Delhi
Pusa Gaurav	1988	Pedigree selection	Galmour × Watch	The plants are dwarf, bushy with moderate foliage cover. Fruits smooth, elliptical and borne in clusters. Whole fruit is suitable for processing and canning because it is without neck constriction and has higher TSS (6%) and better keeping quality at room temperature. Yield potential 330–350 q/ha	IARI, New Delhi
Pusa Hybrid 1	—	Heterosis	Pusa Sheetal × Chikoo	Determinate, compact with good foliage cover and prolific-bearing, fruits round,	IARI, New Delhi

TOMATO : ***Contd.***

Variety	Year of release	Breeding method	Pedigree/ parentage	Important traits	Institutes/ universities
				smooth and attractive, can set fruits up to a night temperature of 28°C	
Pusa Hybrid 2	1996	Heterosis	Pusa 120 × Pusa Gaurav	Plants are compact, semi-determinate with good foliage cover having prolific bearing. Fruits are round to flatish round, firm, smooth, attractive and develop a uniform red colour on ripening	IARI, New Delhi
Pusa Hybrid 4	1997	Heterosis	Pusa 120 × Chikoo	Plants are determinate, compact with dark green foliage, fruits are attractive, round, smooth, average weight 70–80 g; three locules, thick pericarp and uniform ripening; field resistance to root-knot nematode; suitable for semi-arid plateau and central high land. Yield potential 425–450 q/ha	IARI, New Delhi
Pusa Divya	1997	Heterosis	Long Style × Roma	Plants are indeterminate with profuse branching. Fruits are firm, thick, skinned, round to oval shaped, green with dark green stem end and turn dark red on ripening. Maturity period is 70–90 days from transplanting and 135–150 days from seed sowing in hills, however, this may be reduced in plain. Yield potential 350–450 q/ha	IARI, New Delhi

TOMATO : *Contd.*

Variety	Year of release	Breeding method	Pedigree/ parentage	Important traits	Institutes/ universities
Pusa Uphar	1999	Selection	—	Indeterminate, higher fruit setting under slightly adverse temperature with desirable quality attributes	IARI, New Delhi
Pusa Rohini	2004	Introduction and selection	—	Determinate growth habit, fruits round, medium-sized, thick pericarp, longer shelf-life, better market appeal, highly suitable for long distance transportation and processing. Average yield 412 q/ha. Suitable for growing during spring-summer season in northern Indian plains	IARI, New Delhi
Pusa Sheetal	—	Hybridization and selection	Balkan × Jemnoroshnese	Plants determinate; fruits set successfully at low night temperature up to 8°C, suitable for early spring season; fruits flattish round with yellow stem end, medium size, uniform ripening. Harvesting starts from early March in northern plains. Yield 350 q/ha	IARI, New Delhi
Pusa Sadabahar	2002	Introduction and selection	USSR	Plants determinate; suitable for growing under wide temperature of 8–30°C. Accommodate more number of plants per unit area, fruits oval to round, small. Yield 250–350 q/ha	IARI, New Delhi

TOMATO : *Contd.*

Variety	Year of release	Breeding method	Pedigree/ parentage	Important traits	Institutes/ universities
Punjab Selection No. 12	1975	Selection	—	Plants are dwarf, vigourous growth. Fruits are medium-sized and uiniform in colour. Yield potential 450–480 q/ha	PAU, Ludhiana
Punjab Chhuhara	1978	Pedgree selection	Introduced line from Israel and Shimla	Plants are determinate, dwarf and bushy with dense luxuriant foliage. Fruits are red, medium, pear shaped, firm, fleshy, bilocular, less sour, thin pericarp and less seeded; suitable for long distance transportation; highly susceptible to TLCV. Yield potential 230–250 q/ha	PAU, Ludhiana
Punjab Tropic	1978	Introduction	USA	Plant is 100 cm tall, indeterminate growth habit with luxuriant growth. Stem and foliage are dark green. Flowering starts 95 days after sowing. Fruits are large (100 g), round, red on maturity, borne in clusters (4–6 fruits) and remain totally covered with foliage. It has late fruit setting with slow development of fruits. This variety has yield potential of 225–240 q/ha in 141 days of crop duration	PAU, Ludhiana
Punjab Kesri	1978	—	—	Oval round fruits, medium in juice, good keeping quality, early-maturing, yield 225 q/ha	PAU, Ludhiana

TOMATO : *Contd.*

Variety	Year of release	Breeding method	Pedigree/ parentage	Important traits	Institutes/ universities
Punjab NR 7	1987	Pedigree selection	S 12 × NMR 1	It is a determinate variety, has bold, round flat fruits with an excellent blend of TSS and acidity. This variety also has less water requirement and fairly tolerant to sub-optimal temperatures on both lower and higher sides. Yield potential 380–400 q/ha	PAU, Ludhiana
TH 802	—	Heterosis	Healani × Castle Rock	Resistant to root knot nematode, plants determinate, fruits round, medium in size with thick pericarp, average yield 63 tonnes/ha	PAU, Ludhiana
TH 2312	—	Heterosis	VFN 8 × Punjab Chhuhara	Plants dwarf, spreading with dense foliage, fruits oval-round, medium sized, average yield 63 tonnes/ha, resistant to root knot nematode	PAU, Ludhiana
Solan Gola	1975	Selection	—	Plant are indeterminate having fruits of medium size, round, firm. Yield potential 375–390 q/ha	HPKV, Solan
Solan Sagun	1996	Heterosis	—	Plants are semi-indeterminate, erect with deep green foliage. First picking can be taken in 70 days. Fruits are slightly oval having deep red colour,	YSPUHF, Solan

TOMATO : *Contd.*

Variety	Year of release	Breeding method	Pedigree/ parentage	Important traits	Institutes/ universities
				average fruit weight 65 g and suitable for long distance transportation. Plants are tolerant to Alternaria blight and buckeye rot. Yield potential 490–500 q/ha	
Solan Vajr	1998	Selection	—	Plants are indeterminate with medium sized fruits, firm and heart shaped. Yield potential 450–475 q/ha	YSUHF, Solan
HS 101	1976	Pedigree method	Sel. 2–3 × Exotic cultivar	Plants are determinate, dwarf (55 cm) with dark green foliage and bear round, red fruits. Yield potential 310–325 q/ha	HAU, Hisar
HS 102 method	1976	Pedigree	S 12 × Pusa Early Dwarf	It is suitable for both winter and summer seasons and adapted to Haryana state. Plant is 45–50 cm tall and determinate in growth habit. Fruits borne in cluster (3–4 fruits/truss), small to medium in size, and 3–4 loculed and uniform red on ripening. It bears 40–50 fruits/plant. Yield potential 250–270 q/ha	HAU, Hisar
HS 110	—	Selection	Selection from an exotic line	Determinate, late, potato leaf type, fruits large, smooth	HAU, Hisar

TOMATO : *Contd.*

Variety	Year of release	Breeding method	Pedigree/ parentage	Important traits	Institutes/ universities
Hisar Lalit	1993	Pedgree selection	HS 101 × Resistant Bangalore	Plants are semi-determinate and early fruiting. Fruits are round and medium too large. It is resistant to root knot nematode and recommended for cultivation in root-knot nematode infested areas. Yield potential 300–325 q/ha	HAU, Hisar
Hisar Arun	1994	Pedgree selection	Pusa Early Dwarf × K 1	It is an early-maturing variety, plants determinate, dwarf, erect, with cut leave and synchronized clustered flowers, bear 15–20 fruits. Fruits are round, red, medium, average weight 65–70 g, 4–6 locules with deep red flesh. First picking starts 60–65 days after transplanting. Yield potential 170 q/ha and 280 q/ha	HAU, Hisar
Hisar Lalima	—	Pedgree selection	Pusa Early Dwarf × HS 101	Early, determinate, leaves deep cut, fruits round large, fleshy, high-yiedling	HAU, Hisar
Hisar Anmol	—	Pedgree selection	Hisar Arun × *L. hirsutum* f. *glabratum*	Field resistant to tomato leaf curl virus, plant determinate, fruits medium in size, round, red and fleshy	HAU, Hisar
CO 1	1969	Selection	It is a pureline selection isolated from American	Each plant bears on an average of 55 fruits. The fruits are nearly round and smooth. The fruits are pale green in	TNAU, Coimbatore

TOMATO : *Contd.*

Variety	Year of release	Breeding method	Pedigree/ parentage	Important traits	Institutes/ universities
			variety "Pearl Harbour"	colour when unripe and attractive capsicum red when ripe. The fruits are susceptible to cracking. The seed content in fruits is less and proportion of flesh to seed content is very high	
CO 2	1974	Selection	It is a selection from a Russian introduction	Plants are 70 cm tall, semi-dwarf, erect in growth with moderate branching, well-spread and needs no staking. Stem is light green and foliage is dark green in colour. Flowering occurs 65–70 days and first picking starts 100 days after *trans*-planting. Fruits are ovate, medium in size, smooth, green when unripe and bright red on ripening. Fruits are devoid of cracking. Plant contains 4–5 fruits/ truss and 20–25 fruits/plant. Yield potential 410–425 q/ha in 6–8 harvesting	TNAU, Coimbatore
CO 3 (Marutham)	1980	Mutation	CO 1	Plants are determinate, erect, compact, dwarf and bear 30–40 fruits. Fruits are round globular, smooth, parrot green (unripe), borne in cluster, medium (45–50 g), 25 ml/100 ml vitamins, 3.2% TSS; first picking starts 80–85 days after *trans*-planting. Yield potential 450–470 q/ha in 100–105 days of crop duration	TNAU, Coimbatore

TOMATO : *Contd.*

Variety	Year of release	Breeding method	Pedigree/ parentage	Important traits	Institutes/ universities
COTH 1	1998	Heterosis	IHR 709 × LE 812	The fruits are medium sized (50 g/fruit), bright red in color and borne in clusters of 4–5. The fruits are acidic (0.61%) with a TSS of 4.43° Brix. Yield 95.9 tonnes/ha	TNAU, Coimbatore
COLCRH 3	2006	Heterosis breeding	LCR 2 × CLN 2123 A	Resistant to leaf curl virus disease, high yielding, plants semi-determinate and suitable for high-density planting	TNAU, Coimbatore
PKM 1	1978	Mutant	Induced mutant from a local variety called Annanji	The plants are determinate. The fruits are flat round attractive with red colour with prominent green shoulders even after ripening. Fruits are uniform in shape, firm and ideally suitable for long distance transport	HC&RI, TNAU, Periyakulam
Paiyur 1	1988	Pedigree method	It is a hybrid derivative of a cross between Pusa Ruby and CO 3	The fruits are round, medium sized, slight ribbing at calyx end and with medium firmness ensuring keeping quality and suitable for distant transport. This variety has low incidence of fruit borer (*Heliothis armigera*) and diseases like leaf spot and leaf curl virus	TNAU, HRS, Paiyur
NTDR 1	1982	Pedgree selection	Bangalore × Kalyanpur Kuber	Fruits are small, average weight 40–50 g, flatish round, uniform red at maturity,	NDUAT, Faizabad

TOMATO : ***Contd.***

Variety	Year of release	Breeding method	Pedigree/ parentage	Important traits	Institutes/ universities
				suitable for processing, first picking starts at 60–65 days after transplanting. Plants are tolerant to root knot nematode and early blight. Yield 350–380 q/ha	
Narendra Tomato 2 (NDT 120)	1996	Selection	EC 119201	Plants are determinate with curly leaves. Fruits are round, pointed in back side, medium to large size smooth, solid, good keeping quality, single fruit weight 60–75 g. First harvested in 75–80 days. Yield potential 400–450 q/ha	NDUAT, Faizabad
Narendra Tomato 5 (NDT-96)	2006	Selection	—	Plants are indeterminate, leaves green, hairy. Fruits round, medium to large sized, smooth, juicy, single fruit weight 85–110 g. First fruit can be harvested in 75–80 days. Yield potential 400–450 q/ha	NDUAT, Faizabad
Narendra Tomato 6 (NDT-4)	2006	Pedgree selection	Pusa Ruby × Kalyanpur Kuber	Fruits are big, 80–90 g, uniform red at maturity, flat round with slightly grown with good keeping and processing attributes. Yield potential 350–375 q/ha	NDUAT, Faizabad
Kalyanpur Angoorlata	1983	Selection	From local germplasm	Plants are of creeping nature, 200–300 cm tall and indeterminate growth habit. Leaves are medium, light green with spaced leaflets. Fruits are small, oblong,	CSAUAT, Kanpur

TOMATO : *Contd.*

Variety	Year of release	Breeding method	Pedigree/ parentage	Important traits	Institutes/ universities
				8–10 fruits/trusses on average, skin firm and thick, less juicy and keeping quality 8–10 days at room temperature. Immature fruits are yellowish green and red on ripening. It is moderately resistant to leaf-curl mosaic and resistant to root-knot nematode. Yield potential 480–500 q/ha	
Azad T 2	1985	Pedigree selection	HB-5 × Kuber	Plants are determinate, compact with leaves of curved (inwards) margin. Fruits are small, round with 3–4 locules, slightly furrowed green at stem end, red flesh, 5.66% T.S.S, 3.68 mg/g lycopene, 15 mg/100 g of vitamin C; first picking starts 65–70 days after transplanting; moderately resistant to leaf curl mosaic virus; 75–80 days of active fruiting period. Yield potential 330–350 q/ha in 130–135 days	CSAUAT, Kanpur
Azad T 6	2006	Selection	—	Plants are determinate, fruits round, red, uniform, smooth high fruit number and firm. Yield potential 250–275 q/ha	CSAUAT, Kanpur
Junagarh Ruby	1984	Pedigree selection	Big Silari × Pusa Ruby	Plants are semi-spreading in growth habit with purple attractive growth and more foliage. Leaves are medium	GAU, Junagarh

TOMATO : *Contd.*

Variety	Year of release	Breeding method	Pedigree/ parentage	Important traits	Institutes/ universities
				in size and dark green. Fruits bore in cluster of (4–6 fruits/truss), round without grooves, 4 loculed, orange red on ripening with pinkish red flesh. Days to first picking are 85–90 days. On an average, it gives 20 pickings at short intervals. Yield potential 280–300 q/ha	
GT 2	2006	Pedigree selection	Angurlata × Punjab Chhuhara	Plant determinate, fruits heart shape, medium-sized, dark red at maturity and possess good shelf life. Plants are tolerant to leaf curl virus and leaf blight disease. Yield potential 330–350 q/ha	GAU, Anand
Pant Bahar	1985	Selection	AC 238	Plants are indeterminate, bushy (90 cm), profusely branched with light green foliage. Fruits are flattish-round, medium with 5–6 locules, slightly ridged; good storage and processing quality; first picking 75–80 days after transplanting. Yield potential 250–270 q/ha	GBPUAT, Pantnagar
Pant T 3	1988	Selection	Pureline selection	Plants are semi-determinate with thick round and hairy stem and dark green foliage. Fruits are round, smooth and	GBPUAT, Pantnagar

TOMATO : ***Contd.***

Variety	Year of release	Breeding method	Pedigree/ parentage	Important traits	Institutes/ universities
				uniform red; suitable for processing; more suitable for winter season cultivation; first picking starts 75–80 days after transplanting. Yield potential 325–350 q/ha	
Bhagya-shree	1992	—	—	Suitable for processing, high lycopene and low seed content	MPKV, Rahuri
Dhana-shree	1992	Pedigree method	407D P DBK × LA-124	Fruits medium small (60–70 g), round with orange colour, tolerant to spotted wilt and leaf curl, high yield potential	MPKV, Rahuri
Rajashree	1994	Heterosis	—	Fruits are red and medium sized. Fruit weight 60–80 g. Yield potential 750 q/ha, developed for distant market	MPKV, Rahuri
Vasundhara (Hybrid-28)	2001	Heterosis	—	Plants are determinate, Jasper red in colour, fruit weight is 70–75 g, good storability. Plants are resistant to virus and blight. Yield potential 500–525 q/ha	MPKV, Rahuri
Parbhani Yashshri (Sel. 14)	2001	Selection	—	Plants are semi-determinate, suitable for kharif and rabi season. Good in taste, tolerant to virus and blight. Yield potential 600–625 q/ha	MPKV, Rahuri

TOMATO : *Contd.*

Variety	Year of release	Breeding method	Pedigree/ parentage	Important traits	Institutes/ universities
Arka Vikas	1984	Selection	Pure line selection from an American introduction, Tip Top (Selection-22)	Plants are semi-determinate with narrow dark green foliage and good canopy. Fruits are medium-large (80–90 g), uniform deep red, oblate with light green shoulder; adapted to both rainfed and irrigated conditions; suitable for fresh market; tolerant to mosaic. Yield potential 400–425 q/ha in 140 days	IIHR, Bangalore
Arka Saurabh	1984	Selection	Pure line selection from a line V-685 introduced from Canada	Semi-determinate, fruits round, medium large, deep red, suitable for both fresh market and processing	IIHR, Bangalore
Arka Abha	1990	Selection	Pure line selection from VC-8-1-2-1 from AVRDC	Semi-determinate, fruits oblate, with light green shoulder, develop deep red colour on ripening, resistant to bacterial wilt, yield 43 tonnes/ha	IIHR, Bangalore
Arka Alok	1992	Selection	Pure line selection from IIHR 719-1-6 from AVRDC, Taiwan (BWR-5)	Determinate, resistant to bacterial wilt, suitable for fresh market, yield 46 tonnes/ha	IIHR, Bangalore
Arka Ashish	1990	Selection	UC-82-B line from USA	Determinate, dark green foliage, fruits oval, very firm, thick fleshed with two	IIHR, Bangalore

TOMATO : *Contd.*

Variety	Year of release	Breeding method	Pedigree/ parentage	Important traits	Institutes/ universities
				locules, tolerant to powdery mildew and fruit cracking, yield 38 tonnes/ha	
Arka Ahuti	1990	Selection	Pure line selection from Ottawa 60 from Canada	Semi-determinate, fruits oblong with 2–3 locules, suitable for processing, yield 42 tonnes/ha	IIHR, Bangalore
Arka Vishal	1990	Heterosis	IIHR 837 × IIHR 932	Indeterminate, fruits large, round with green shoulder, thick flesh, tolerant to fruit cracking, suitable for fresh market, yield 75 tonnes/ha	IIHR, Bangalore
Arka Vardan	1990	Heterosis	IIHR 550-3 × IIHR 932	Determinate, fruits large, round with green shoulder, tolerant to cracking with thick flesh, resistant to nematodes, yield 75 tonnes/ha	IIHR, Bangalore
Arka Meghali	1995	Pedigree selection	Arka Vikas × IIHR-554	Plants are semi-determinate. Fruits are superior to Pusa Ruby variety in size (63 g), flesh thickness and fruit firmness. Tolerant to moisture stress and heat. Yield potential 270–290 q/ha	IIHR, Bangalore
Arka Shreshta	1996	Heterosis	15 SBSB × IIHR 1614 (BHR-1)	Semi-determinate, fruits medium large, round with green shoulder, deep red, firm, resistant to bacterial wilt, yield 76 tonnes/ha	IIHR, Bangalore

TOMATO : *Contd.*

Variety	Year of release	Breeding method	Pedigree/ parentage	Important traits	Institutes/ universities
Arka Abhijit	1996	Heterosis	15 SBSB × IIHR 1334	Semi-determinate, fruits medium, round with green shoulder, deep red, firm, resistant to bacterial wilt, yield 65 tonnes/ha	IIHR, Bangalore
Utkal Pallavi	1992	Selection	—	Plants are determinate with curly leaf. Fruit borne in cluster consisting of 6–12, medium in size, pear-shaped, nipple at blossom end, firm, fleshy. Matures in 85–90 days after sowing. Resistant to bacterial wilt and nematode wilt. It also performs well in saline soil. Yield potential 375–380 q/ha	OUAT, Bhubanesh-war
Utkal Kumari (BT-10)	1998	Selection	Selection 22 × BT-2	Fruits are round, red medium-large with 5–6 days shelf life; first picking starts 90–95 days after sowing. Yield potential 400–415 q/ha in 90–100 days	OUAT, Bhubanesh-war
Sakthi	1993	Selection	LE 79	—	KAU, Vellanikkara
Mukthi	1998	Selection	Pure line selection from LE 79 (CL32D-0-1-19GS)	Plants are determinate with pubescent stem and normal leaf. Fruits are round, white and of medium size with average weigh of 58 g; first picking starts 75	KAU, Vellanikkara

TOMATO : *Contd.*

Variety	Year of release	Breeding method	Pedigree/ parentage	Important traits	Institutes/ universities
				days after sowing. Plants are resistant to bacterial wilt and fruit cracking. This variety gives an yield of 43.5 tonnes/ha in 95–100 days	
Vellayani Vijai	2006	Introduction	Introduction and selection from CLN1621F (AVRDC, Taiwan)	High-yielding (1.34 kg/plant), compact, early-maturing variety having bacterial wilt resistance and high temperature tolerance	KAU, Vellanikkara
Kashi Vishesh (H-86)	2006	Backcross pedigree selection	*L. hirsutum* f. *glabratum* B6013 donor parent	This variety is resistant to TLCV developed using *L. hirsutum* f. *glabratum* B6013. Plant determinate, foliage dark green, fruits colour red, shape spherical, size medium to large, fruit weight 80 g, first harvest can be taken at 70–75 days after transplanting. Yield potential 430–450 q/ha	IIVR, Varanasi
Kashi Amrit (DVRT-1)	2006	Backcross pedigree method	Sel-7 × *L. hirsutum* f. *glabratum* B6013	This is a determinate variety. Fruits round, size medium to large, colour attractive red, average weight 100 g, fleshy, suitable for cultivation under TLCV prone area. Yield potential 620–635 q/ha in 115–120 days	IIVR, Varanasi

TOMATO : *Contd.*

Variety	Year of release	Breeding method	Pedigree/ parentage	Important traits	Institutes/ universities
Kashi Anupam (DVRT-2)	2006	Backcross pedigree method	Sel-7 × *L. hirsutum* f. *glabratum* B6013	Plants are determinate, fruits large, flattish round (slightly indented at blossom end), attractive red with 5–6 locules, medium maturity (75–80 days after transplanting). Yield potential 570–600 q/ha	IIVR, Varanasi
Kashi Sharad (IIVR Sel-1)	2006	Pedigree selection	MTH-6 × Kalyani Eurish	Plants indeterminate in growth habit, leaves are broad, fruits are attractive red, slightly oval in shape and firm, thick pericarp suitable for longer shelf-life, numbers of fruits vary from 35 to 40, fruit weight 90–95 g. Yield potential 480–500 q/ha	IIVR, Varanasi
Kashi Hemant (IIVR Sel-2)	2006	Pedigree selection	Sel-18 × Flora Dade	Plants are determinate, concentrated fruit with attractive red colour and round shaped, it bears 30 fruits/plant, fruit weight varies from 80 to 85 g. Yield potential 400–425 q/ha	IIVR, Varanasi
VRTH 1	—	Heterosis	—	Determinate, early maturing, field tolerance to TLCV and early blight, average yield 1100 q/ha	IIVR, Varanasi
VRTH 2	—	Heterosis	—	Short duration hybrid, average yield 900 q/ha	IIVR, Varanasi

TOMATO : *Contd.*

Variety	Year of release	Breeding method	Pedigree/ parentage	Important traits	Institutes/ universities
Swarna Baibhav (CHTH 1)	2004	Selection	—	Fruits are round, deep red, firm, weight (140–150 g), contain high pulp, First harvest can be taken 55–60 days after planting. Plants are resistant to bacterial wilt. Sowing is recommended in August–September for Jharkhand and Bihar. Yield potential 700–725 q/ha	HARP, Ranchi
Swarna Naveen	2006	Pure line selection	—	Fruits are deep red oblong, weight (60–70 g) with TSS 5° Brix. Plants are resistant to bacterial wilt. First harvest can be taken 50–60 days after planting. Yield potential 580–600 q/ha	HARP, Ranchi
Swarna Lalima	2006	Pure line selection	—	Fruits are deep red, round, weight (120–125 g) with TSS 4° Brix. First harvesting can be taken 60–65 days after planting. Plants are resistant to bacterial wilt. Yield potential 500–550 q/ha	HARP, Ranchi
Swarna Sampada (HATH 3)	2006	Selection	—	Fruits are round, red, weight (120–130 g), firm, TSS 4.5–5.0° Brix, acidity 0.35–0.40% and contain high pulp. Plants are resistant to bacterial wilt and early blight disease. First harvesting can be taken 55–60 days after planting. Yield potential 600–650 q/ha	HARP, Ranchi

TOMATO : *Contd.*

Variety	Year of release	Breeding method	Pedigree/ parentage	Important traits	Institutes/ universities
Mani-khammu	2003	Pedigree method	EC 160193 × HS 101	Fruits attractive, medium sized with thick skin, suitable for fresh, processing and long distance transportation, resistant to fruit cracking, tolerant to leaf crul virus, tolerant to moisture stress	ICARRC, Barapani
Mani Leima	2006	Selection	—	Early-maturing, determinate and multi-branched variety. Fruits are medium sized, oval, nipple shaped and bilocular. It is suitable for processing and long distance transportation. It is resistance to fruit cracking and tolerance to leaf curl virus. It is also tolerant to moisture stress and suitable for spring summer season. Yield potential 630–650 q/ha	ICARRC, Barapani
VL Tamatar-1	2003	Pedigree	(Punjab Chuhara × A2) × Punjab Chuhara × Punjab Chuhara	Fruits red, round, smooth, skin thick, suitable for processing, Plant height 100 cm, maturity 65 DAT, resistant to fruit rot, susceptible to late blight, yield 200–250 q/ha	VPKAS, Almora

such as rugose leaves, which are deeply grooved, variegated, angora leaves, which have additional colours where a genetic mutation causes chlorophyll to be excluded from some portions of the leaves.

Flowers are borne in small forked racemose cyme. Anthesis starts at 6.30 AM and continues up to 11 AM. Anther dehiscence is longitudinal. It occurs 1–2 days after opening of corolla. The stamen sheds its pollen when the style grows up through anther tube, thus self-pollination occurs. When stigma protrudes above the anther tube, chances of cross-pollination through bees increases. The optimum temperature for pollination is around 21°C.

Tomato fruit is classified as a berry. As a true fruit, it develops from the ovary of the plant after fertilization, its flesh comprising the pericarp walls. The fruit contains hollow spaces full of seeds and moisture, called locular cavities. These vary among cultivated species, according to type. Some smaller varieties have two cavities, globe-shaped varieties typically have three to five, beefsteak tomatoes have a great number of smaller cavities, while paste tomatoes have very few, very small cavities.

BRINJAL

Botanical name: *Solanum melongena* L, 2n = 24
Family: Solanaceae

Brinjal is a popular vegetable crop grown in subtropical and tropical areas. It is called *brinjal* in India and *aubergine* in Europe. The name "eggplant" derives from shape of fruits of some varieties, which are white and shaped similarly to chicken eggs. It is a delicate perennial often cultivated as an annual. It grows 40–150 cm tall, with large coarsely-lobed leaves that are 10–20 cm long and 5–10 cm broad. (semi-) wild types can grow much larger, to 225 cm with large leaves over 30 cm long and 15 cm broad. The stem is often spiny. The flowers are white to purple, with a five-lobed corolla and yellow stamens. The fruit is fleshy, less than 3 cm in diameter on wild plants, but much larger in cultivated forms. The fruit is botanically classified as a berry, and contains numerous small and soft seeds.

Different varieties of eggplant produce fruits of different sizes, shapes and colours, especially purple, green, or white. A much wider range of shapes, sizes and colours is grown in India and elsewhere in Asia. Larger varieties weighing up to a kilogram grow between the Ganges and Yamuna rivers, while smaller ones are found elsewhere. Fruit colour varies from white to yellow or green as well as reddish-purple and dark purple. Some cultivars have a colour

BRINJAL

Variety	Year of release	Breeding method	Pedigree/ parentage	Important traits	Institutes/ universities
Pusa Purple Long	1970	Single plant selection	Batia cultivar of Punjab	Fruits are long (20–25 cm), purple, glossy and tender; suitable for summer and autumn seasons cultivation. Yield potential 280–310 q/ha	IARI, New Delhi
Pusa Purple Round	1970	Selection	Selection from local material	The plants are very tall with a thick greenish purple stem. The deep green leaves are highly serrated. On an average, each fruit weighs about 137 g with a bearing of only 6 fruits/plant. Resistant to shoot and fruit borer and little leaf disease.	IARI, New Delhi
Pusa Purple Cluster	1978	Single plant selection	Indigenous collection	Medium early fruits are borne in cluster of 4–9, 10–12 cm long and deep purple, moderately resistant to bacterial wilt. Yield potential 300–325 q/ha	IARI, New Delhi
Pusa Anmol	—	Heterosis	Pusa purple long × Hyderpur	Early and high-yielding	IARI, New Delhi
Pusa Kranti	1977	Pedigree method	(PPL × Hyderpur) × Wynad Giant	Plants medium tall, upright growth, purple pigmentation on young leaves and branches, fruits oblong, 15–20 cm long, green calyx, less seeded	IARI, New Delhi

BRINJAL : *Contd.*

Variety	Year of release	Breeding method	Pedigree/ parentage	Important traits	Institutes/ universities
Pusa Bhairav	—	Pedigree method	Progeny selection of PPL × 11a-12-2-1	Early, non-spiny, resistant to phomopsis fruit rot, dark purple, glossy, attractive, 12–15 cm long fruits	IARI, New Delhi
Pusa Anupam	1991	Heterosis	Pusa Kranti × Pusa Purple Cluster	The plants are medium in size and bushy, takes 135–145 days from sowing in first picking. Fruits are borne in cluster of 3–5. The fruits measure 14–18 cm length, 14–15 cm girth and 3.4–4.5 cm in diameter. The fruits are cylindrical, medium-long, tender, firm and purple in colour. Yield potential 300–400 q/ha with 30–35 fruits/plant	IARI, New Delhi
Pusa Hybrid 6	1993	Heterosis	Sel Br-112 × Sel 91-2	Plants are non-spiny, semi-erect, branched, vigorous, partial pigmentation on younger leaves and growing tips. Fruits are round, glossy, attractive, violet purple, partially pigmented peduncle; weight of 200–250 g; takes 85–90 days from sowing to first picking and further picking continue up to December; early bearing hybrid. Yield potential 450–470 q/ha	IARI, New Delhi

BRINJAL : *Contd.*

Variety	Year of release	Breeding method	Pedigree/ parentage	Important traits	Institutes/ universities
Pusa Hybrid 5	1994	Heterosis	Sel NDB-25 × Sel 129-5	Plants are vigorous, non-spiny with semi-erect branches. Fruits are long, glossy, attractive, dark purple with partially pigmented peduncle, average weight of 100 g; takes 80–85 days from sowing to first picking and further picking continue up to December in the central and northern Plains, whereas in Kerala, Tamil Nadu, Karnataka and Andhra Pradesh from November to May. Yield potential 520–540 q/ha	IARI, New Delhi
Pusa Hybrid 4	—	Heterosis	PPL × Hyderpur	—	
Pusa Hybrid 9	1997	Heterosis	Sel 91-2-1 × Sel 190-10-12	Plants non-spiny, strong erect branches, light pigmentation on young leaves, fruits oval, round, glossy, attractive, dark purple, peduncle partially pigmented, average weight 300 g, yield 67 tonnes/ha, recommended for Gujarat and Maharastra	IARI, New Delhi
Pusa Shyamla	—	Pedigree selection	Sel NDB-25-11 × Sel HE-12-2	Plants are non-spiny with erect branches, having light purple pigmentation partially on younger leaves. Its fruits are long, glossy, attractive, dark purple, each fruit	IARI, New Delhi

BRINJAL : *Contd.*

Variety	Year of release	Breeding method	Pedigree/ parentage	Important traits	Institutes/ universities
				weighing 80–90 g. The maturity period (days to first harvesting) is about 50–55 days from transplanting. Its average yield is 391.6 q/ha. It is recommended for growing during Autumn-winter season in northern Indian plains	
Pusa Uttam	1997	Pedigree selection	Progeny selection of GR × 91-2	Plants are semi-upright, vigorous and free from spines. Mature plants appear green with occasional light pigmentation on growing shoots. Leaf blade length and width average 15.5 cm and 8 cm with intermediate lobing on tip respectively. Flowers appear in clusters. Fruits are pendent, oval, medium, large-sized, and glossy with dark purple skin and green	IARI, New Delhi
Pusa Upkar	1997	Pedigree selection	Progeny selection of GR × 91-1	Plants are semi-upright (semi-erect stem), medium-tall (70 cm) with 5–6 branches and greenish-purple leaves; bear ~15 fruits. Fruits are round (9 cm diameter), dark purple, fruit weight 295 g. Yield potential 650-670 q/ha	IARI, New Delhi

BRINJAL : *Contd.*

Variety	Year of release	Breeding method	Pedigree/ parentage	Important traits	Institutes/ universities
Pusa Bindu	1997	Pedigree selection	Progeny selection of GR × 91-1	Plants are ~60 cm tall with green leaves. Fruits are borne in cluster of 4–5 fruits, having average fruit weight of 125 g each. Yield potential 250–300 q/ha	IARI, New Delhi
Pusa Ankur	1999	Pedigree selection	Progeny selection of GR × 91-1	Plants are semi-upright (semi-erect stem), medium-tall (65 cm) with 6–8 branches. Fruits are small, round and purple. Yield potential 560–580 q/ha	IARI, New Delhi
Black Beauty	1975	Introduction	—	Plants are 60–75 cm tall, erect and sturdy with dark green colour. Leaves are dark green in colour. Flowering starts 80–90 days after transplanting. Fruits are roundish oblong, 10–12 cm long of dark purplish black colour. Yield potential 250–270 q/ha	PAU, Ludhiana
Punjab Haryana Brinjal 4	1976	Pedigree selection	Hyderpur × Pusa Purple Long	Plants are bushy with pigmented stem. Fruits are long to medium, thin, dark purple in colour with light green flesh. Yield potential 225–250 q/ha	PAU, Ludhiana
Jamuni Gola	1978	Selection	Sel-16	Plants are medium in height, spreading, tornless and early maturing. Fruits are oval-round, plump and shining purple;	PAU, Ludhiana

BRINJAL : *Contd.*

Variety	Year of release	Breeding method	Pedigree/ parentage	Important traits	Institutes/ universities
				first picking starts after 65 days of *trans*-planting. Yield potential 250–270 q/ha	
Punjab Bahar	1978	Selection	—	Plants are 93 cm tall and non-spiny. Foliage is deep purple with green tinge colour. First picking starts 69 days after transplanting. Fruits are round, dark purple, shining, weighing about 200–300 g and less seeded. Fruits are suitable for Bhartha making. Yield potential 275–300 q/ha	PAU, Ludhiana
Punjab Chamkila	1978	Selection	—	This variety is suitable for summer and autumn seasons. Fruits are long, thin and dark purple. This variety has yield potential of 430–450 q/ha	PAU, Ludhiana
Punjab Barsati	1987	Pedigree selection	PPC × H-4	Fruits mediumlong, shining purple, suitable for rainy season, tolerant to fruit and shoot borer, average yield 140 q/ha	PAU, Ludhiana
Punjab Sada Bahar Baigan	1987	Pedigree selection	PPC × H-4	Fruits long, thin, deep purple, tolerant to fruit and shoot borer, average yield 125 q/ha	PAU, Ludhiana

BRINJAL : ***Contd.***

Variety	Year of release	Breeding method	Pedigree/ parentage	Important traits	Institutes/ universities
Punjab Neelam	1988	—	—	Fruits oval-round, medium, shining dark purple, suitable for February and August transplanting, average yield 140 q/ha	PAU, Ludhiana
Punjab Moti	1993	—	—	Small fruited, round shining, dark purple fruits, suitable for autumn and spring season, average yield 115 q/ha	PAU, Ludhiana
BH-2	—	Heterosis	Punjab Neelam × Punjab Barsati	Plants medium tall, erect, thornless, fruits oblong, shining, deep purple, large, avergae yield 442 q/ha	PAU, Ludhiana
Bhagya-mati	1982	Selection	—	Plants are 50–70 cm tall, purple shoots, which turn greenish with age. Emerging and mature leaves are purple, which turn yellow at senescence. Fruits are borne in cluster of 2–6 fruits at each node. Fruits oblong in shape of varying size from 6–8 cm in length at edible stage with deep shining purple colour that does not fade even at harvesting for 2–3 days. Yield potential 150–180 q/ha in 150–170 days	ANGRAU, Hyderabad
CO 1	1978	Selection	It is a pureline selection	The plants are erect, compact and bushy with green stem and leaves and greenish	TNAU, Coimbatore

BRINJAL : *Contd.*

Variety	Year of release	Breeding method	Pedigree/ parentage	Important traits	Institutes/ universities
				purple petiole. The fruits are oblong and medium-sized with pale green shade under white background. Each fruit on an average weight 50–60 g with good keeping quality. The seeds are soft even at full maturity. It is moderately resistant to root-knot and reniform nematodes. Yield potential 240–260 q/ha in 160 days	
CO 2	1988	Selection	It is a pureline selection from the local variety 'Varikkathiri' of Negamum, Coimbatore district	The fruits have dark purple stripes intermingled with light green color. They are oblong in shape with smooth calyx. Yield potential 35 tonnes/ha in 150 days	TNAU, Coimbatore
CoBH 1	2001	Heterosis	EP 45 × CO 2	The fruits are medium-sized, oblong in shape and dark violet in colour. The average single fruit weight is 60–65 g. It possesses higher ascorbic acid content (16.65 mg/100 g)	TNAU, Coimbatore
MDU 1	1982	Selection	Kallamp (a genotype of Madurai)	The plants are compact and medium spreading. The leaves are broad, green without pigments. Fruits are bright	TNAU, Coimbatore

BRINJAL : ***Contd.***

Variety	Year of release	Breeding method	Pedigree/ parentage	Important traits	Institutes/ universities
				purple in colour, round and each of 280 g in weight. As the fruit matures, the purple colour fades to pale pink. Yield potential 340–350 q/ha in 135–145 days of crop duration	
PKM 1	1984	Mutant	Mutant of a local type called 'Puzhuthi kathiri'	It yields on an average of 34.75 tonnes/ ha in 150–155 days. The fruits-measure 6–8 cm and 10–14 cm in length and girth respectively with a mean weight of 55 g. The fruits are small with green stripes. It is drought tolerant and can withstand long distance transport	HC&RI, TNAU, Periyakulam
PLR 1	1990	Selection	It is a reselection from a Nagpur ecotype	It possesses extended shelf-life (up to 10 days) and is suitable for growing in all seasons. It is cluster-bearing in nature and its fruits are medium sized, egg shaped, dark purple and glossy in appearance, which are preferred by the consumers	VRS, TNAU, Palur
KKM 1 (KSM-107)	1999	Selection	It is a pure line selection from Kulathur local near Tirunelveli	It is a pure line selection from Kulathur local near Tirnelveli. It is suitable for cultivation in both irrigated and rainfed conditions. Fruits are borne in cluster	TNAU, Coimbatore

BRINJAL : *Contd.*

Variety	Year of release	Breeding method	Pedigree/ parentage	Important traits	Institutes/ universities
				(2–4) with egg shaped and milky white in colour. The crop duration is 130–135 days. Yield potential 360-380 q/ha	
PPI (B) 1	2002	Selection	Selection from Karungal local type Vazhuthunangai	It is medium spreading with 30–35 lengthy pale green fruits, less seeded and bitter less. It is medium tolerant to brinjal fruit and shoot borers and wilt disease	HRS, TNAU, Pechiparai
Annamalai	1985	Selection	—	Plant height is 65.3 cm with a spread of 74.3 cm. Plant is characterized by disposition of young leaves and semi-erect. In winter, the fruits are deep purple, whereas, lighter in summer and suitable for ratooning. Fruits are somewhat club-shaped and slightly curved in the middle. This variety is resistant to aphids. Yield potential 310–325 q/ha	Annamalai Univerity, Annamalai Nagar
Kalyanpur T 3	1983	Selection	—	Fruits are round, light, purple and whitish green at stigmatic end. Yield potential 300–325 q/ha	CSAUAT, Kanpur

BRINJAL : *Contd.*

Variety	Year of release	Breeding method	Pedigree/ parentage	Important traits	Institutes/ universities
Azad Kranti	1985	Selection	—	Plants are medium-tall bear 12–15 fruits. Fruits are medium, thick-long, smooth, taper at distal end, dark purple with green calyx and white flesh, weight ranges from 100 to 110 g; first picking starts 60–65 days after transplanting; suitable for both late and early sown conditions; tolerant to fruit rot, little leaf, drought and frost; crop maturity duration is 175–180 days. Yield potential 310–330 q/ha in a fruiting duration of 140–150 days	CSAUAT, Kanpur
Junagarh Long	1984	Selection	—	Plant height is 120 cm, semi-spreading growth habit. Leaves are dark green with wavy margins. Plants are non-spiny. Fruits are medium in size, tender, pinkish purple in colour, long (~17 cm) with a girth of ~8 cm at the centre. Fruits are tapering at the distal end and 20–25 pickings can be taken at short intervals for edible purposes. This variety is free from collar rot disease under field condition. Yield potential 300–350 q/ha	GAU, Junagarh

BRINJAL : *Contd.*

Variety	Year of release	Breeding method	Pedigree/ parentage	Important traits	Institutes/ universities
GBH 1	1996	Heterosis	M-2 × M-35	The fruits are dark purple with single green styler spot at apex. Plants are tolerant to shoot-and-fruit borer as well as little leaf disease. Yield potential 560–580 q/ha	GAU, Anand
Pant Samrat	1985	Selection	—	Fruits are of attractive texture; resistant to phomopsis blight and bacterial wilt; less infestation of shoot-and-fruit borer and jassids; good for rainy season cultivation. Yield potential 390–415 q/ha	GBPUAT, Pantnagar
Pant Rituraj	1985	Selection	—	Plants are semi-erect with dark green foliage and prolific bearing. Fruits are round, purple, slightly tapering towards the bottom; first picking starts 60 days after transplanting. Yield potential 310–325 q/ha	GBPUAT, Pantnagar
Pant Brinjal Hybrid 1	1994	Heterosis	PB-129 × PB-225	This is a long fruited brinjal hybrid; plants are medium with purplish green stem. The leaves are dark green and there are purple tinge on the young leaves. The fruits are long, bright deep purple with soft flesh. The fruiting is	GBPUAT, Pantnagar

BRINJAL : *Contd.*

Variety	Year of release	Breeding method	Pedigree/ parentage	Important traits	Institutes/ universities
				in clusters. Plants take 70–75 days for first picking Plants are resistant to bacterial wilt and phomopsis blight under field condition. Yield potential 600–620 q/ha	
Arka Shirish	1986	Selection	Pure line selection from IIHR 194-1, a local collection from Karnataka	The fruits are very tender, extra-long, green, thick with attractive light green skin colour, good edible flesh and less seed. The flesh texture is good with better cooking quality. It is very high yielding and each plant bears 25 to 30 fruits. Yield potential 390–410 q/ha in 115 days after transplanting	IIHR, Bangalore
Arka Sheel	1986	Pure line selection	IIHR 192 (a local collection from Kodagu, Karnataka)	Plants are tall with green leaves, purple tinge at the base in young leaves and solitary bearing habit. Fruits are medium — long, deep shining purple skin with green calyx. Yield potential 380–400 q/ha in 150–160 days of crop duration	IIHR, Bangalore
Arka Kusumakar	1986	Seelction	Pure line selection from IIHR 193	Spreading plant habit with green stem and green leaves, fruits small borne in clusters, soft texture with good cooking quality, yield 40 tonnes/ha	IIHR, Bangalore

BRINJAL : *Contd.*

Variety	Year of release	Breeding method	Pedigree/ parentage	Important traits	Institutes/ universities
Arka Navneeth	1986	Heterosis breeding	F1 hybrid between IIHR 22-1 × Supreme	Plants are indeterminate; fruits are deep purple, black gossy, green calyx, weight of 350–400 g; white flesh, few slow maturing seeds; good cooking and keeping quality. Yield potential 700–710 q/ha in 120 days	IIHR, Bangalore
Arka Keshav	1990	Pedigree method	Dingrass Multiple Purple × Arka Sheel (IIHR-21)	Plants are tall and branched plants bearing long fruits in clusters. Red purple glossy fruit skin with green calyx. Fruits tender with slow seed maturity and free from bitter principles. Plants are resistant to bacterial wilt. Duration 150 days. Yield potential 450–470 q/ha. Resistant to bacterial wilt	IIHR, Bangalore
Arka Nidhi	1990	Pedigree method	Dingrass Multiple Purple × Arka Sheel (BWR-12)	Plants are tall and compact plants, bearing medium long fruits in clusters Dark green leaves with purple leaf base and purple veins, fruits tender with slow seed maturity and free from bitter principles Resistant to bacterial wilt. Yield potential 485–500 q/ha in 150 days	IIHR, Bangalore

BRINJAL : ***Contd.***

Variety	Year of release	Breeding method	Pedigree/ parentage	Important traits	Institutes/ universities
Arka Neelkanth	1990	Pedigree method	Dingrass Multiple Purple × Arka Sheel (BWR-54)	Tall and compact plants, bearing small fruits in clusters, violet purple glossy fruit skin with green purple calyx, fruits tender, slow seed maturity, resistant to bacterial wealth, yield 43 t/ha	
Aruna	1989	Selection	—	Plants are prolific bearing; fruits are small, roundish-oval, bright plum with red skin; first picking starts 95 days after transplanting. Yield potential 415–430 q/ha and 220–240 q/ha in rainy and summer seasons, respectively	PDKV, Akola
H 4	1976	Pedigree selection	Hyderpur × PPL	Fruit are medium to long, thin and dark purple in colour. Flesh is light green in colour. Yield potential 225–250 q/ha	HAU, Hisar
Hisar Shyamal (H-8)	1993	Selection	—	Plants non-spiny, semi-erect, branched, Fruits oblong glossy, attractive, violet purple, partially pigmented peduncle; weight of 225–280 g. Yield potential 400–425 q/ha	HAU, Hisar
BH 1	1996	Heterosis	Punjab Barsati × Jamuni Gola	Plants are medium in height, compact and thornless. Fruits are oblong, plumpy and shining purple. It is early	OUAT, Bhubaneshwar

BRINJAL : *Contd.*

Variety	Year of release	Breeding method	Pedigree/ parentage	Important traits	Institutes/ universities
				in fruiting and first picking starts after 52 days of transplanting. It is tolerant to fruit borer. This variety can be planted in February and August. Yield potential 510–530 q/ha	
BH 2	1996	Heterosis	Punjab Barsati × Punjab Neelam	Its leaves are green and purplish. Plants are medium, erect, spreading. Its fruits are oblong and deep purple. Its average fruit weight is 300 g. It is highly suitable for cooking as *Bartha*. Plants are tolerant to fruit borer. Yield potential 515–530 q/ha	OUAT, Bhubaneshwar
Narendra Baigan 1	1996	Selection	—	Plants are long, erect with greenish purple leaves. Fruits are dark purple colour, soft, less-seeded and oblong in shape. Fruits are ready to harvest in 70–75 days after planting. Yield potential 300–350 q/ha	NDUAT, Faizabad
Narendra Hybrid Brinjal 1 (NDBH-1)	1996	Heterosis	PBR-91-1 × K202-14-90-110	Plants are semi-erect, non-spiny, hardy, leaves are purple green; fruits are round, purple, attractive, shining, soft, fleshy, less seeded with good flavour, weight of 150–200 g; fruits are harvested 70–75	NDUAT, Faizabad

BRINJAL : ***Contd.***

Variety	Year of release	Breeding method	Pedigree/ parentage	Important traits	Institutes/ universities
				days after transplanting. Plants are moderately resistant to little leaf disease. Yield potential 450–500 q/ha	
Narendra Hyb Brinjal 3 (NDBH-3)	2006	Heterosis	—	Plants are spreading type, leaves light purple. Fruits round, large-sized, soft with shining purple colour. Fruits are ready for harvesting 75–80 days after planting. Yield potential 500–550 q/ha	NDUAT, Faizabad
NDBH 6	—	Heterosis	—	Plants semi-erect, non-spiny, fruits long, oval, attractive purple, soft, fleshy, moderately resistant to shoot-and-fruit borer, phomopsis blight and little leaf	NDUAT, Faizabad
Utkal Tarini (BB-7)	1992	Selection	—	Plants are medium in height. Fruits are oblong, medium-sized, deep purple, solitary bearing and sometimes two fruits per cluster. Picking starts 90–95 days after sowing. Plants are resistant to bacterial wilt. Yield potential 345–365 q/ha	OUAT, Bhubanesh-war
Utkal Madhuri (BB-44)	1998	Pedigree method	PBR 125-5 × Pipili-4	Plants are of medium stature with green stems and leaves, bear solitary fruits. Fruits are green (basal portion	OUAT, Bhubanesh-war

BRINJAL : *Contd.*

Variety	Year of release	Breeding method	Pedigree/ parentage	Important traits	Institutes/ universities
				white), long, medium with white flesh; 7–8 days shelf life; resistant to *Fusarium* wilt and drought; first picking starts 100–105 days after transplanting; prolonged fruiting period. Yield potential 320–340 q/ha in 160 days of crop duration	
Utkal Keshari (BB-26)	1998	Pedigree method	BB-11 × KJ-3-1	Plants are tall, erect with purple pigmentation on branches and leaves. Fruits are long, medium-large, uniform thick, basal portion slightly broad, deep purple, 5–6 days shelf life; tolerant to bacterial wilt and *Fusarium* wilt. Yield potential 325–340 q/ha in 130 to 140 days of crop duration	OUAT, Bhubaneswar
Utkal Jyoti (BB-13)	2004 method	Pedigree	KT-4 × BB-11	Plants are medium-tall, non-spinous and bear profuse fruits in cluster. Fruits are long, small-medium, purple with white flesh; good shelf life; tolerant to bacterial wilt. Yield potential 350–375 q/ha in 140–180 days of crop duration	OUAT, Bhubaneswar
JC 1	1998	Selection	—	Plant are tall, pedicel non-spiny, fruit elongated, medium sized, purple with pointed apex, no incidence of phomopsis	IGKV, Jabalpur

BRINJAL : *Contd.*

Variety	Year of release	Breeding method	Pedigree/ parentage	Important traits	Institutes/ universities
				blight and little leaf virus, wilt and borer infestation moderate. Yield potential 400–450 q/ha	
JC 2	1998	Selection	—	Plant are tall, pedicel spiny, fruit large, oblong purple with blunt apex, no incidence of little leaf and wilt phomopsis blight and fruit borer infestation. Yield potential 480–500 q/ha	IGKV, Jabalpur
Surya	1990	Selection	SM6-7 (SPS)	—	KAU, Vellanikkara
Sweta	1998	Selection	SM6-6	Swetha is a bacterial wilt resistant variety. Plants are bushy with purple tinge on leaf stalk, purple flower and tinge on leaf stalk. Fruits are medium-long, solitary (occasionally in cluster). This variety can be grown in May–August, September–December and January-March. Yield potential 300–325 q/ha in 135–150 days	KAU, Vellanikkara
Haritha	1998	Selection	SM-141 (SPS)	—	KAU, Vellanikkara

BRINJAL : *Contd.*

Variety	Year of release	Breeding method	Pedigree/ parentage	Important traits	Institutes/ universities
Swarna Mani (CHBR-1)	2004	Selection	—	Plants are prostrate (spreading stem), medium-tall (89.0 cm) with 6–9 branches and violet leaves, bear-15 fruits. Fruits are round (9.6 cm), dark violet, average weight of 320 g. Yield potential 750–770 q/ha	HARP, Ranchi
Swarna Pratibha (CH-309)	2004	Pure line selection	—	Fruit are medium-sized (250 g), and shiny purple colour. Plants are resistant to bacterial wilt. First harvesting can be taken 75–80 days after planting. Yield potential 470–500 q/ha	HARP, Ranchi
Kashi Prakash (IVBR-1)	2006	Selection	—	This is oblong-fruited variety. Plants are semi-upright, stem and leaves colour green. Fruits are attractive with light green spotted colour, calyx spiny, average weight 190 g. Yield potential 650–675 q/ha	IIVR, Varanasi
Kashi Taru (IVBL-9)	2006	Mass selection	—	Plant height 120–130 cm bearing 20–25 fruits/plant. The picking starts 75–80 days after transplanting. Fruits are long (31 cm), colour dark purple, glossy and tender; fruit weight 745–85 g, suitable for summer and autumn seasons cultivation. Yield potential 700–720 q/ha	IIVR, Varanasi

BRINJAL : *Contd.*

Variety	Year of release	Breeding method	Pedigree/ parentage	Important traits	Institutes/ universities
Kashi Sandesh (VRBHR-1)	2006	Heterosis	Pant Rituraj × BR-SPS-14	Plants semi-upright, stem green, leaves are purplish-green, flowers purple. Fruits are round light purple, medium-sized, average fruit weight is 225 g, First picking starts 75 days after *trans*-planting. Yield potential 800–820 q/ha	IIVR, Varanasi
Manjari Gota	1965	Selection	Local germplasm collected from Manjari, Pune	Plants medium tall, spreading, spines on leaf, mid-ribs and peduncle of fruits. Fruits round purple with white stripes, average yield 250 q/ha	MPKV, Rahuri
Vaishali	1985	Pedigree method	Manjari Gota × Arka Kusumakar	Plants medium tall and spreading, fruits oblong, purple with white stripes, fruits borne on clusters, average yield 350 q/ha	MPKV, Rahuri
Pragati	1988	Pedigree method	Vaishali × Arka Kusumakar	Plants hardy, resistant to lodging, fruits egg shaped, spiny, purple with white stripes, fruits borne on clusters, average yield 350 q/ha	MPKV, Rahuri
Krishna	1991	Hybrid	—	Plants tall, hardy, resistant to lodging, fruits egg shaped, spiny, fruits purple with white stripes, colour retention good, average yield 480 q/ha	MPKV, Rahuri

BRINJAL : *Contd.*

Variety	Year of release	Breeding method	Pedigree/ parentage	Important traits	Institutes/ universities
Phule Harit (RHRB-16)	2006	Selection	—	Plants are tolerant to fruit-borer and little leaf under field condition. Fruits are smooth and attractive green fruits with white strips at apical end, fruit bigger in size. This variety has yield potential of 350–370 q/ha in 180 days	MPKV, Rahuri
RHRBH 1	—	Heterosis	RHRB-1 × JB-16	Fruits non-spiny, purple with white stripes, average yield 576 q/ha	MPKV, Rahuri
RHRBH 2	—	Heterosis	RHRB-1 × RB-25	Plants short and spreading, fruits purple with white stripes, average yield 407 q/ha	MPKV, Rahuri
RHRBH 3	—	Heterosis	RHRB-1 × Vaishali	Plants tall and spreading, fruits purple with white stripes, spiny calyx, average yield 440 q/ha	MPKV, Rahuri
Swarna Shyamali	2006	Selection	—	Fruit are medium sized (250 g), round, attractive green colour with white stripes. Resistant to bacterial wilt. First harvesting can be taken 45-60 days after planting. Yield potential 500-525 q/ha	HARP, Ranchi
Swarna Shree	2006	Pure line selection	—	Fruits are creamy white, round to oval, soft and preferred for *bharta* preparation.	HARP, Ranchi

BRINJAL : *Contd.*

Variety	Year of release	Breeding method	Pedigree/ parentage	Important traits	Institutes/ universities
				Plants are moderately resistant to bacterial wilt. Good for July–August sowing. First harvesting can be taken 55–60 days after planting. Yield potential 550–600 q/ha	
Swarna Ajay (HABH-3)	2006	Pure line selection	—	Fruit are oblong, medium length (10–12 cm), weight (100–120 g) and attractive light purple. First harvesting can be taken 55–65 days after planting. Resistant to bacterial wilt and phomopsis blight. Yield potential 700–720 q/ha	HARP, Ranchi

gradient, from white at the stem to bright pink to deep purple or even black. Green or purple cultivars with white striping also exist.

Flowering commences 70–75 days after transplanting. Full bloom is observed 80 days after planting. The whole period of effective flowering lasts for 75 days. Anthesis starts at 5.35 AM and continues up to 7.35 AM with peak at 6.05 AM. No marked effect of atmospheric temperature and relative humidity is found. The dehiscence of anthers begins 30 min after anthesis. It commences at 6 AM and continues upto 8 AM with the maximum being at 6.35 AM. Anthers dehisce usually 15–20 min after flower buds open. The stigma is receptive from 2 days before anthesis and up to 8 days. The maximum receptivity is on the day of anthesis, which remains effective up to 4 days after anthesis. The pollen-grains remain viable from the day of anthesis to 10 days at 24.6°C and relative humidity of 82%. In long-styled flowers, fruit setting ranges from 70–80% while short styled flowers do not set fruits.

CHILLI

Botanical name: *Capsicum annum* Linn., 2n = 24
Family: Solanaceae

Chilli or pepper (*Capsicum*) is extensively cultivated throughout tropical Asia and equatorial America for its edible, pungent fruits. India is the largest producer and exporter of chilli. The genus *Capsicum* comprised all the varied forms of fleshy-fruited peppers grown as herbaceous annuals — red, green, and yellow pepper rich in vitamins A and C that are used in seasoning and as a vegetable food. It includes paprika, chili pepper, red pepper (cayenne), and bell pepper. The latter one is considered and eaten as a vegetable and is not covered in this section. The capsicums under each category vary tremendously and the species designation is not always possible. In general, paprika belongs to *C. annum* and the red peppers and chili peppers belong to the *C. frutescens*.

Its plant is bush-like and grows up to about 0.6 m, bearing white flowers that produce fruits in a variety of sizes, colours and shapes. The plants grow up to 1,800 m above mean sea level in the tropical areas. Their pungency is influenced by several factors, such as high night temperature and drought or over-watering.

Both self and cross-pollination occur, the later being about 16% by bees, ants and thrips. Flowering begins 1–2 months after planting and it takes another month for fruiting. Flowers open in morning between 5.00 to 6.00 AM. Anthers normally dehisce between 8.00 to

CHILLIES

Variety	Year of release	Breeding method	Pedigree/ parentage	Important traits	Institutes/ universities
Sindhur (CA 960)	1975	Selection	Hot Pourtgal	Fruits have deep red thick pericarp. The fruit measures 8.2 cm in length with capsaicin content of 0.81 mg/g of fruit. Pericarp is light green turning deep red on maturity with smooth surface. Top is blunt. Fruit has mild pungency. Seed content is 38%. Yield potential 12 q/ha (dry chilli) under rainfed and 60–80 q/ha under irrigated condition	RARS, Lam
Bhagya-lakshmi (G 4)	—	Selection	Selection from Thohian chillies from Sri Lanka	Fruits olive-green and turn dark red on ripening, fairly tolerant to diseases and insects	RARS, Lam
Andhra Jyoti (G 5)	1978	Pedigree selection	DG 2 × Bihar variety	Fruits are short and gundu type. Plants are tall and dense. The fruits are bright red measuring 5.1 cm in length and 6.3 cm in girth. The capsaicin content is 0.65 mg/g of fruit. Yield potential (dry) 20–30 q/ha	RARS, Lam
Sindhur	1978	Selection	CA 960	Plants are tall with green foliage. Fruits have deep red thick pericarp. The fruit measures 8.2 cm in length with capsaicin content of 0.81 mg/g of fruit. Pericarp of fruit is light green turning deep red on maturity with smooth surface. Top is	RARS, Lam

CHILLIES : *Contd.*

Variety	Year of release	Breeding method	Pedigree/ parentage	Important traits	Institutes/ universities
				blunt. Fruits are mild pungent. Seed content is 38%. Yield potential 12–14 q/ha (dry) under rainfed and 60 q/ha under irrigated condition	
Bhaskar	1986	Pedigree method	G 4 × Yellow Anther Mutant	Plants are compact, spreading with dark green leaves and yellow anthers. Fruits are thin, medium-long, thick red with red flesh and 45% seed content, pungent with high oleoresin; suitable for export; good shelf life and transport quality; first picking starts 130 days after sowing. Plants are tolerant to mite, thrips and aphids. Yield potential 15–17 q/ha (dry chilli)	RARS, Lam
Prasanth	2006	Selection	LCA 334	Plants are compact, spreading with dark green leaves and yellow anthers. Fruits are medium-long, green when immature and glossy red on ripening; suitable for both green and dry purposes, very good for export; good shelf life and transport quality; first picking starts about 130 days after sowing. Plants are tolerant to mites, thrips and aphids. Yield potential 150–170 q/ha	RARS, Lam

CHILLIES : *Contd.*

Variety	Year of release	Breeding method	Pedigree/ parentage	Important traits	Institutes/ universities
LCA 305	1995	Selection	Lam 305	Plants are bushy; fruits are light green with bright red colour after drying. Plants are tolerant to leaf curl virus. Yield potential (dry chilli) 20–25 q/ha	RARS, Lam
Pant C 1	1982	Selection	—	Plant is short stature with more primary branches. Fruits are erect, small size highly pungent, 5.5 cm long, light green and turning light red at maturity. First fruiting commences in two months and picking commences 100 days after transplanting. Plants are tolerant to mosaic and leaf curl virus. Yield potential 12–15 q/ha (dry)	GBPUAT, Pantnagar
CO 1	1982	Selection	Selection from local variety Sattur Samba	The plants are erect, medium tall and compact with medium branching. The fruits are green when unripe and bright shiny red on ripening. The fruits are 6–6.5 cm long with sharp tip and bulge shoulders. The seed content is 55%. The variety has high capsaicin content (0.72 mg/g). Yield potential 21 q/ha (dry chilli) in 210 days	TNAU, Coimbatore

CHILLIES : *Contd.*

Variety	Year of release	Breeding method	Pedigree/ parentage	Important traits	Institutes/ universities
CO 2	1984	Selection	Nambiyar local Gundu type [CA(P) 63]	The stem is angular, semi-dwarf and less-spreading. The fruits are oblong, thick and bright reds in colour. Seed content is high (60%). The capsaicin content of dry pod is 0.56%. It is suitable for harvest both as green pods and red ripe pods. Yield potential 110–125 q/ha (green chilli) in 210 days	TNAU, Coimbatore
CO 3	2001	Selection	Open pollinated progeny selection from Sri Lankan type	Plants are dwarf and compact. It is a samba type and suitable for very close spacing of 30 cm × 15 cm. It is also suitable for green chillies. It has very low stalk weight in comparison with pod weight. Unlike other cultivars, oleoresin and capsaicin contents are high and hence suitable for export. Yield potential 350–375 q/ha of green fruits	TNAU, Coimbatore
CO 4	2000	Selection	Single plant selection	Dwarf plants, suitable for vegetable purpose, 165 days duration	TNAU, Coimbatore
K 1	1985	Selection	Single plant selection from local variety Sattur Samba	The plants are tall and spreading. The crop duration is 210 days. The ripe fruits are red, long and contain high capsaicin. Yield potential 18–20 q/ha (dry chilli)	TNAU, RS, Kovilpatti

CHILLIES : ***Contd.***

Variety	Year of release	Breeding method	Pedigree/ parentage	Important traits	Institutes/ universities
K 2	1985	Selection	B 70 A (Assam type) × Sattur Samba	Plants are tall and semi-spreading. Fruits are long (7.3 cm) with 0.19 mm pericarp thickness, capsaicin content. Yield potential 9–12 q/ha (dry chilli)	TNAU, RS, Kovilpatti
MDU 1	1985	Mutation	K 1	It possesses broad and dark green leaves, compact plant type and determinate growth habit with less number of secondary branches. The fruits are borne in cluster of 4–9 at nodes as against single fruit borne at nodes as in K-1 or K-2 varieties. The fruits are long with dark shiny red colour. It has a total duration of 205–215 days. Yield potential 18–20 q/ha (dry chilli)	TNAU, Coimbatore
PKM 1	1991	Pedigree selection	AC.No.1797 × CO 1	Plants are dwarf (70 cm). The crop duration is 180 days. It is suitable for cultivation under irrigated conditions. It has very bold fruits, which are dark red in colour. Yield potential 30–40 q/ha	TNAU, Coimbatore
KKM (Ch) 1	2006	Pedigree selection	Hybrid derivative of Acc. 240 × CO 3	High-yielding, plants dwarf, compact, fruits are attractive red in colour, do not shrink much after drying, field tolerance to mosaic, moderate resistance to fruit rot	AC&RI, TNAU, Killikulam

CHILLIES : *Contd.*

Variety	Year of release	Breeding method	Pedigree/ parentage	Important traits	Institutes/ universities
PLR 1	1994	Selection	Single plant selection from Kanchengad local	'Gundu' type, suitable for green chilli, 210 days duration	TNAU, VRS, Palur
Chanchal	1985	Selection	—	Plants are medium tall with perennial habit. Foliage is dark green. Fruits are 25–30 cm long, 6–7 cm thick, erect, highly pungent, dark green when fresh and shining red on maturity. This variety is suitable for kitchen gardening. Plants are tolerant to leaf curl disease. Yield (dry chilli) potential 15–20 q/ha	CSAUAT, Kanpur
Punjab Lal	1987	Pedigree method	Perennial × Long Red	Plant is dwarf, bushy with dark green foliage. Fruit is erect medium size (4.25 cm × 0.79 cm) dark green and turning dark red at maturity. Fruits are rich in capsaicin (0.7 per cent), oleoresin, dry matter and red pigment. Plants are resistant to CMV and leaf curl virus and moderately resistant to fruit rot and die back. Yield (green chilli) potential 100–120 q/ha	PAU, Ludhiana
Punjab Guchhedar	1996	Selection	—	Fruit is small (4.68 cm), erect and borne in bunches. Fruit colour is deep red,	PAU, Ludhiana

CHILLIES : *Contd.*

Variety	Year of release	Breeding method	Pedigree/ parentage	Important traits	Institutes/ universities
				suitable for drying and processing. Yield potential 15–17 q/ha (dry chilli)	
Punjab Surkh	1996	Selection	ELS 2	Fruit is long (6.68 m) and deep red on maturity. It is very good for drying. Plants are tolerant to fruit rot and resistant to viral diseases. Yield potential (dry chilli) 20–25 q/ha	PAU, Ludhiana
CH 1	—	Hybrid	—	Male sterile based hybrid, fruits light green, tolerant to diseases and pests, average yield 95 q/ha	PAU, Ludhiana
CH 3	—	Hybrid	—	Male sterile based hybrid, plants medium dwarf, fruits pendent, dark green, suitable for processing, average yield 95 q/ha	PAU, Ludhiana
Jawahar Mirch 218	1989	Selection	J 218	Plants are dwarf (50–55 cm), spreading with light green foliage. Fruits are long, thin, bright green; good transport quality; tolerant to leaf curl and fruit rot; first picking starts 100–105 days after sowing. Yield potential 90–100 q/ha (fresh chillies) and 20–30 q/ha (dry chilli)	JNKVV, Jabalpur

CHILLIES : *Contd.*

Variety	Year of release	Breeding method	Pedigree/ parentage	Important traits	Institutes/ universities
Jawahar Mirch 283	1999	Selection	Selection from local genotype collected from adjoining areas of Jabalpur	Plants are dwarf (40–45 cm) with bushy habit. It produces 7–9 primary branches with dense foliage cover. The fruits are straight, firm, medium in length, thin (2.5–3.5 cm girth in the middle) and dark green in the colour. The red ripe fruits are bright red and pungent. Yield potential 80–100 q/ha (green chilli)	JNKVV, Jabalpur
Musalwadi	1991	Selection	Local collection	Plants are tall, dark green, spreading foliage with black patches. Fruits are smooth, medium-long, dark green and dark red at maturity; consumed as green as well as dry purposes. Plants are tolerant to die back and powdery mildew. Yield potential (green chilli) 110–125 q/ha in 180 days	MPKV, Rahuri
Jayanti	1996	Selection	AKC-86-39	Fruits are green while fresh and turn red on ripening. Plant height medium to tall (60–75 cm) with light green foliage. Yield potential 20–30 q/ha (dry chilli)	PDKV, Akola
Musalwadi	1987	Selection	Local collection from village Musalwadi, Ahmednagar	Palnts tall, spreading, foliage dark green, fruits smooth, medium long, tolerant to die back and wilt, average yield 19 q/ha (dry fruits)	MPKV, Rahuri

CHILLIES : *Contd.*

Variety	Year of release	Breeding method	Pedigree/ parentage	Important traits	Institutes/ universities
Agnirekha	1992	Pedigree method	Dondaicha × Jwala	Plants tall spreading, fruits long, bold, smooth and light green, average yield 250 q/ha	MPKV, Rahuri
Phule Jyothi	1995	Selection	—	Fruits dark green, dual purpose variety, tolerant to leaf curl and powdery mildew, resistant to lodging, suitable for *kharif* and summer planting	MPKV, Rahuri
Phule Suryamukhi	1996	Selection	—	Fruits green in colour, tolerant to leaf curl and powdery mildew, resistant to lodging, suitable for *kharif* and summer planting	MPKV, Rahuri
Phule Sai	—	Pedigree method	Pant C1 × Kamandalow	Plants tall, spreading, branching from ground level, fruits smooth medium long, cluster bearing, 5–6, erect, tolerant to thrips and wilt	MPKV, Rahuri
Kashmir Long 1	2001	Selection	—	Plants are medium tall, spreading, leaves ovate, green, stem round to angular with purple pigmentation on nodes, fruit bright red pendent long thick tapering with yellow seeds. Plants are moderately resistant to *Fusarium* wilt. Yield potential 160–180 q/ha of red ripe fruit yield	SKUA&T, Srinagar

CHILLIES : *Contd.*

Variety	Year of release	Breeding method	Pedigree/ parentage	Important traits	Institutes/ universities
Utkal Ava	2004	Selection	BC 14-2	Upward fruiting habit, suitable both for green and dry chilli purpose, deep red colour at ripening with good pungency. Fruits can be harvested in 125 to 130 days after sowing. Fruit length is 6.2 cm and girth is 3.7 cm. Plants are resistant to bacterial wilt. Yield potential 120–140 q/ha (green chilli) and 35–50 q/ha (dry chilli)	OUAT, Bhubaneswar
Gujarat Veg Chilli	2006	Selection	GVC-101	The fruits are medium pungent, long, compact, strait with green shining colour. The fruits have longer shelf-life and high ascorbic acid content. This variety has been recommended for cultivation Madhya Pradesh and Maharashtra. Yield potential 140–150 q/ha	AAU, Anand
DH-76 6	1982	Pedigree selection	Puri Red × Byadgi local	Plant height is 70–75 cm, bushy growth habit and highly branched. Stem is light green. Leaves are smaller than Byadgi variety. It has 25–30 fruits/plant. Fruits are thin (2 mm pericarp), pungent, long, light red with wrinkles. Seeds are light yellow with seed recovery of 60%. Seed yield is 1.2 q/ha. Yield potential 20 q/ha (dry chilli) in 180 days	IARI, New Delhi

CHILLIES : *Contd.*

Variety	Year of release	Breeding method	Pedigree/ parentage	Important traits	Institutes/ universities
Jwala (Pusa Jwala)	1983	Pedigree method	NP46A and Puri Red (Samba type)	Plant is dwarf, bushy, light green with broad leaves. Fruits are 9–10 cm long and thin, light green turning light red at maturity. Dry fruits are wrinkled and liable to break during packing and transport. The fruit contain capsaicin of 0.48 mg/g of fruit, thus highly pungency. Plants are fairly tolerant to thrips, mites and aphids. Yield potential (fresh) 75–90 q/ha	IARI, New Delhi
Pusa Sada Bahar	—	Pedigree selection	Pusa Jwala × IC 31339 (*C. frutescens*)	Fruit erect, 6–8 cm in cluster, 6–14 per cluster, resistant to CMV, TMV and leaf curl viruses	IARI, New Delhi
NP 46 A	—	Selection	HD Selection from Indigenous material	Plants dwarf, bushy; fruits 8–9 cm long, thin, light green, light red on ripening, tolerant to thrips. Yield 75 q/ha (green)	IARI, New Delhi
Kashi Anmol	2006	Selection	From introduced materials (K 2)	Plants are determinate, bushy and nodal pigmentation on stems. Fruit pendent, colour attractive green; suitable for green fruit production under chilli-wheat/chilli-potato cropping system; first picking takes 50 days after transplanting; crop duration 130-145 days. Yield potential 200–225 q/ha green fruit	IIVR, Varanasi

CHILLIES : *Contd.*

Variety	Year of release	Breeding method	Pedigree/ parentage	Important traits	Institutes/ universities
Kashi Surkh	2006	Heterosis	MS line; CCA4261 × Pusa Jwala	Plants are semi-determinate (1–1.2 m), erect and nodal pigmentation on stem. Fruits colour light green, straight, length 11–12 cm, suitable for green as well as red fruit production. The first harvesting starts after 55 days of transplanting. The plants are tolerant to thrips, mites and viruses. Yield potential 220–240 q/ha (green fruit) and 140–150 q/ha (dry chilli)	IIVR, Varanasi
Arka Lohit	1990	Selection	Selection from IIHR 324, a local collection	Fruits are dark green, smooth, straight; turn dark red at maturity, highly pungent (capsaicin 0.70%). This variety has yield potential of 15 tonnes/ha (dry chilli) in 180 days of crop duration	IIHR, Bangalore
Arka Suphal	2002	Pedigree method	Pant C1 × IIHR 517A	Plants tall and erect, fruits green, smooth, medium long (6–7 cm × 1 cm), pendent, turning deep red on maturity. Plant are tolerant to powdery mildew. Yield potential 230–250 q/ha (fresh), and 30–40 q/ha (dry chilli) in 180 days of crop duration	IIHR, Bangalore
Arka Meghna	2006	Selection	MSH-172	Dwarf type plant, branched early fruiting, fruit length medium, upward fruiting in bunches, 8–12 fruits per bunch,	IIHR, Bangalore

CHILLIES : *Contd.*

Variety	Year of release	Breeding method	Pedigree/ parentage	Important traits	Institutes/ universities
				fruit light green in colour while red at ripening. Yield potential 50–60 q/ha (dry chilli)	
Jwalasakhi	1990	Pedigree method	Vellanochi × Pusa Jwala	—	KAU, Vellanikkara
Ujwala	1996	Selection	CA 219-1-19-6	—	KAU, Vellanikkara
Vellayani Athulya	2006	Selection	Selection from local collection	High-yielding (650.33 g/plant), early maturing, shade tolerant, green chilli variety with light green, medium pungent fruits having excellent quality	KAU, Vellanikkara
Anguraha	2006	Selection	—	Plants are indeterminate, growth habit, solitary linear and pendulous, resistant to bacterial wilt. Fruits are light green at immature stage and red on ripening. Average fruit length 12.0 cm and average fruit weight is 3.6 g. Yield potential 270–300 q/ha	KAU, Vellanikkara
Gujarat Veg Chilli 111	2006	Selection	GVC-111	Fruits are attractive, medium-long, straight, pungent and slightly wrinkled surface with good shining. Yield potential 110–125 q/ha	GAU, Anand

CHILLIES : *Contd.*

Variety	Year of release	Breeding method	Pedigree/ parentage	Important traits	Institutes/ universities
Gujarat Veg Chilli 121	2006	Selection	ACS-97-2	Fruits are longer, thick, dark green with good shining and possess higher ascorbic acid content and better shelf-life. Yield potential 100–120 q/ha	GAU, Anand
RCH 1	2004	Selection	—	Plant height 95 cm, it takes 51 days to flower and bear 79 fruits/plant. Its takes 110 days to start fruiting after *trans*-planting and completly harvested in about 200 days. The fruits are long and bright red in colour. Fruit length is 13.1 cm and girth is 3.4 cm. The variety is well suited to arid and semi-arid condition of Rajasthan. Yield potential 20–30 q/ha of dry chilli	ARS, Durgapura
Paprika					
Arka Abir	—	Selection	Pure line selection from Dyavanur Dubba	Plants tall, fruits light green, wrinkled, turn deep red on maturity, high colour and low pungent, suitable for oleoresin extraction	IIHR, Bangalore
Kt Pl 19	—	Selection	Selection from exotic material from USA	Plants upright; fruits conical, flat 13–17 cm long, 2.2–3 cm in diameter, dark green, pendent, turn dark red at maturity. Suitable for powder; good for export	IARI, RS, Katrain

11.00 AM. Flower opening and anther dehiscence depend on weather conditions to a large extent. During cool as well as cloudy days, flower opening is delayed. Pollen is fertile on the day of anthesis and stigma becomes receptive 24 h after flower opening.

The pungency of red peppers and colour value for paprikas are the most important parameters. Hot peppers, used as relishes, pickled or ground into a fine powder for use as spices, derive their pungency from the compound capsaicin (8-methyl-*N*-vanillyl-6-enamide), a substance characterized by acrid and burning taste, that is located in internal partitions of fruits. First isolated in 1876, capsaicin stimulates gastric secretions and, if used in excess, causes inflammation. It is a tasteless, odourless white crystalline substance. Its level varies widely in capsicum peppers, from less than 0.05% in mildly pungent types to as high as 1.3% in hottest chilli. The pungency level is usually represented in Scoville heat values. Pungency levels vary in the same variety, by geographical region, and in maturity levels. Volatile oil content is low in all capscicums. The pigment responsible for colour in paprikas is capxanthin, a carotenoid. Other carotenoids present are capsorubin, zeaxanthin, lutein, kryptoxanthin and alpha and beta-carotene. The pigment content increases as fruit ripens and continues after maturity. The extractable colour of parika is usually expressed in ASTA colour value or in Colour Units, which are 40 times the ASTA colour.

SWEET PEPPER

Botanical name: *Capsicum annum* Linn., 2n = 24
Family: Solanaceae

Capsicum, also known as bell pepper or sweet pepper, is a popular vegetable in India. It is relatively a new entrant into our country. It is mainly cultivated in Himachal Pradesh, Uttar Pradesh, parts of Gujarat, Maharashtra, Karnataka, Ranchi region of Bihar, and hilly regions of Tamil Nadu. It grows well in summer season in hills and cooler season in the plains. Capsicum being a cool season crop is planted from February onwards to May/June (hills). It cannot withstand heavy rains during flowering or fruit setting. Very cold temperatures (<12°C) during night also hamper growth and fruit set. In plains, its crop is fitted in such a way that flowering and fruit setting can be under favourable conditions. Ideal temperatures for flowering is 26–28°C day and 16–18°C night. In northern plains, it is planted during February–June/July, in parts of Maharashtra in March–April and in Karnataka from June to December. With the advent of shade nets, it is possible to raise crops by lowering the temperature even during hotter months.

SWEET PEPPER

Variety	Year of release	Breeding method	Pedigree/ parentage	Important traits	Institutes/ universities
California Wonder	1975	Introduction	Introduction from USA	Plants are vigorous, upright and prolific. Fruits are smooth, 3–4 lobed with medium thick flesh. Yield potential 125–150 q/ha in 125–140 days	IARI, New Delhi
Chinese Giant	1975	Introduction	—	Plants are vigorous and prolific bearers. Fruits are smooth, 3–4 lobed, medium thick with dark green skin colour. Yield potential 130–150 q/ha in 125–150 days	IARI, New Delhi
Bull Nose	1978	Introduction	—	Plants are 50–60 cm tall, erect and sturdy. Foliage is dark green. Fruits are 8–10 cm long, thick walled 3 lobed, and deep green. Flesh is thick with non-pungent. Fruit becomes scarlet red on maturity. Flowering takes place 80–85 days after transplanting. Yield potential 100–120 q/ha	IARI, New Delhi
Pusa Deepti (KT 1)	1997	Heterosis	Yolo Wonder × Runion Yellow	Plants are upright, spreading with profused branching. Fruits are smooth, erect, conical, light green with thick flesh, 9–11 cm long and 3–5 cm in diameter. The fully ripe fruits are ready for picking 70–80 days after transplanting. This line gives an average	IARI, Katrain

SWEET PEPPER : *Contd.*

Variety	Year of release	Breeding method	Pedigree/ parentage	Important traits	Institutes/ universities
				of 55–59 fruits. Plants are tolerant to bacterial leaf spot and anthracnose under field conditions. Yield potential 250–350 q/ha	
Arka Basant	1986	Selection	Hungarian variety Soroksari	It is creamish white, conical, erect fruited variety with thick flesh and prolific-bearing. It has good aroma and crisp texture. The variety is suitable for both *kharif* and *rabi* seasons as well as under protected covers. It has got good export value of the fruits. Yield potential 150–200 q/ha in 125 to 140 days	IIHR, Bangalore
Arka Gaurav	1986	Selection	Selection from Golden Calwonder of USA	Plant is indeterminate with dark green fruiting. Fruits are blocky, 3–4 lobed, thick fleshed and medium sized (70–80 g). Fruits turn to orange-yellow on maturity. It is good for both *kharif* and *rabi* seasons. Foliage covers the fruits to avoid sunscald. Plants are tolerant to bacterial wilt disease. Yield potential 180–250 q/ha in 125 to 150 days	IIHR, Bangalore
Arka Mohini	1986	Selection	Selection from Titan of USA	Plants are determinate, with medium to large blocky fruits. Fruit is dark green	IIHR, Bangalore

SWEET PEPPER : *Contd.*

Variety	Year of release	Breeding method	Pedigree/ parentage	Important traits	Institutes/ universities
				surface, thick fleshed, 3–4 lobed and becomes red on ripening. This variety is suitable for both *kharif* and *rabi* seasons. Foliage covers the fruits avoiding the sunscald. Tipping of initial flowers is advocated to get higher yields. Yield potential 250–275 q/ha in 120 days	
Solan Hybrid 1	1996	Selection	—	Fruits are 3–4 lobed and bell shape. Yield potential 300–350 q/ha	YSPUHF, Solan
Nishat 1	2001	Selection	Capsicum Sel-2	Plants are indeterminate, medium (48 cm) with dark green foliage, green stem with purple strips. Fruits are dark green, turn yellow at physiological maturity, 3–4 lobed, non-pungent, bell shaped; fruit picking starts 45–50 days after transplanting. Yield potential 200–225 q/ha	SKUAS&T, Srinagar
DARL 202	2006	Hybridization	F1 hybrid	High yielding, adapted to hilly region, field resistance to *Fusarium* wilt, suitable for cultivation in open nad under protected conditions	DARL, Pithoragarh, Uttaranchal

Although a wide range of soils are suitable, red or sandy loam soil with a pH of 5.5–6.8 is ideal. Water stagnation is detrimental. High acid soils need to be reclaimed using lime. Liming can be done through use of calcium oxide, calcium bicarbonate or dolomite.

Capsicum has bidirectional root system and it is important that there is free unhampered root growth and root aeration for successful crop growth. While very heavy soils may pose problems in irrigation management and in keeping the surface loose, lighter soils with proper management strategy can result in good crop yields.

OKRA

Botanical name: *Abelmoschus esculentus* L. Moench
Family: Malvaceae

Okra or Bhendi is an important tropical vegetable. The genus is Asiatic in origin, the cultigen *A. esculentus* is variable. India, Ethiopia, West Africa and Tropical Asia are considered possible countries of origin. It is considered as a polyploid in nature. It is commonly grown through the warmer parts of temperate Asia, Southern Europe, northern Africa, USA, and in all parts of tropical regions. In India, okra is commercially grown in Gujarat, Maharashtra, Andhra Pradesh, Uttar Pradesh, Tamil Nadu, Karnataka, Haryana and Punjab.

Specific varieties are grown even in lower hills with moderate climate. Tender, green fruits are cooked in curry and soup. The roots and stems are used for clearing cane juice in preparation of '*gur*'. High iodine content of fruits helps control goitre while leaves are used in inflammation and dysentery. The fruits also help in renal colic, leucorrhoea and general weakness. In India, its crop has not adapted as leafy vegetable as in Far East countries. It has yet multiple uses. The dry seed contains 13–22% good edible oil and 20–24% protein. The oil is used in soap, cosmetic industry and as *vanaspati*, while protein is used for fortified feed preparations. Its crushed seeds are fed to cattle for more milk production and fibre is utilized in jute, textile and paper industry. It is an erect and herbaceous annual. Stem is green with or without reddish tinge. Leaves are alternate, 3–7 lobed palmate, hirsute and serrate. Flowers are solitary, axillary having epicalyx (up to 10). There are 5 yellow petals with crimson spot on claw. Staminal column consists of numerous stamens which are united to the base of petals. Stigma is 5–9 lobed. Fruit is capsule.

As the stem elongates, the lowermost flower buds open into flowers. There may be a period of 2, 3 or more days between the times of development of each flower, but never does more then one

OKRA

Variety	Year of release	Breeding method	Pedigree/ parentage	Important traits	Institutes/ universities
Pusa Makhmali	1955	Selection	Selection from local material collected from West Bengal	Early variety, smooth pods, straight, 5-edged, light green, slender, 15–20 cm long	IARI, New Delhi
Pusa Sawani	1958	Pedigree breeding	Derived from a cross of IC 1542 × Pusa Makhmali	Pods smooth, 5-edged, dark green, 18–20 cm long, distinguished by presence of a purple patch at the base of the yellow petal on both sides	IARI, New Delhi
Pusa A4	—	Pedigree breeding	HD Selection from [(*A. manihot* ssp. *manihot* Ghana × Pusa Sawani) × Prabhani Kranti]	—	IARI, New Delhi
Perkins Long Green		Introduction	Introduction from USA	—	IARI, New Delhi
CO 1	1976	Selection	Single plant selection from Red Wonder collected from Hyderabad	Pods long, slender, 5-ridged, glossy, smooth and scarlet red but colour disappears on cooking	TNAU, Coimbatore
CO 2	1987	Heterosis	Pusa Sawani × MDU 1	Fruits are fairly long with a mean length of 20.60 cm, girth of 6.8 cm and	TNAU, Coimbatore

OKRA : *Contd.*

Variety	Year of release	Breeding method	Pedigree/ parentage	Important traits	Institutes/ universities
				each weighs 33 g. Fruit surface is less hairy with a better consumer's appeal and market preference	
CO 3	1991	Heterosis	Parbhani Kranti × MDU 1	Suitable for growing during *kharif* as well as summer. The fruits are dark green medium in size and suits to fresh market	TNAU, Coimbatore
COBhH 1	2007	Heterosis	Vu selection × PA 4 (T)	—	TNAU, Coimbatore
MDU 1	1978	Mutation	Induced mutant, isolated from Pusa Sawani	Stem green with light purple pigmentation, fruits light green, about 20 cm long	TNAU, Coimbatore
Parbhani Kranti	1985	Pedigree breeding	Interspecific cross between *A. esculentus* cv. Pusa Sawani × *A. manihot*, an African species carrying ressitance to YVMV.	Leaves deeply lobed. Fruits dark green, smooth, tender, 5-ridged. YVMV resistant	Marathwada Agricultural University, Parbhani
Arka Anamika	1990	Pedigree breeding	Interspecific cross between *A.*	YVMV resistant, Fruits medium green, rough, 5-ridged, plants are medium tall	IIHR, Bangalore

OKRA : *Contd.*

Variety	Year of release	Breeding method	Pedigree/ parentage	Important traits	Institutes/ universities
			esculentus × *A. manihot* ssp. *tetraphyllus* followed by back cross	of about 100 cm with short internodal length and less branched. It is an early maturing and takes about 50 days for first flowering after sowing, 55 days for the first picking	
Arka Abhay	1992	Pedigree breeding	Interspecific cross between *A. esculentus* × *A. manihot* ssp. *tetraphyllus* followed by back cross	Tolerance to fruit-borer and resistant to YVMV, fruits lush green, tender and long	IIHR, Bangalore
Punjab Padmini	1982	Pedigree breeding	Cross between *A. esculentus* × *A. manihot* ssp. *manihot*	Pods dark green, shining, smooth, thin, 15–20 cm long, remain tender for 3–4 days, field resistance to YVMV, tolerance to jassids and cotton boll worm	PAU, Ludhiana
Punjab 7	1985	Pedigree breeding	—	Resistant to YVMV, fruits medium long, green, tender, 5-ridged, top of fruit is blunt and slightly furrowed	PAU, Ludhiana
Punjab 8	1995	—	—	Fruits thin, long, dark green, highly resistant to YVMV and tolerant to fruit borer, average yield 55 q/ha	PAU, Ludhiana

OKRA : *Contd.*

Variety	Year of release	Breeding method	Pedigree/ parentage	Important traits	Institutes/ universities
Gujarat Bhindi 1	1983	Selection	Pure line selection from unknown bulk sample received from IARI, New Delhi	Fruits 5-ridged, 14–15 cm long, 6–7 cm girth	GAU, Ahmedabad
AOH 262	—	Hybrid	—	Plants tall, average fruit weight 11 g, average yield 780 q/ha	GAU, Ahmedabad
AOH 263	—	Hybrid	—	Resistant to yellow vein mosaic virus, average yield 894 q/ha	GAU, Ahmedabad
Kiran	1990	Selection	Kilichundan local	—	KAU, Thrissur
Salkeerthi	1998	Selection	Reselection from NBPGR NO. 144	—	KAU, Thrissur
Aruna	1998	Selection	NBPGR No. 1343	—	KAU, Thrissur
Anjitha	2006	Mutation	Inter specific hybridization followed by mutation breeding and selection	High yielding (14.6 tonnes/ha), early maturing, YVM resistant variety	KAU, Thrissur

OKRA : ***Contd.***

Variety	Year of release	Breeding method	Pedigree/ parentage	Important traits	Institutes/ universities
Manjima	2006	Heterosis	Hybrid (Gowreesapattom local × NBPGR/ TCR-874)	High yielding (16 tonnes/ha), early maturing, YVM resistant variety	KAU, Thrissur
Kashi Vibhuti	2006	Pedigree selection	(NIC-9303 × IC111547) × IIVR-20	Plants height 50–70 cm. Flowering starts 38–40 days after sowing. Five ridges fruits 8–10 cm. Yield 17–18 tonnes/ha. Resistant to YVMV and OLCV	IIVR, Varanasi
Kashi Pragati	2006	Pedigree selection	(NIC-9303 × P.K.) × IIVR-20	Plants tall 130–175 cm, first flower appears at 36–38 days after sowing. Yield 18–19 tonnes/ha during rainy and 13–14 tonnes/ha during summer season, respectively. Resistant to YVMV and OLCV. Identified for the Zone IV and V	IIVR, Varanasi
Kashi Satdhari	2006	Pedigree selection	(P.K.X IC111547) × IIVR-20	Seven ridges. Plant height 80-130 cm. Flowering starts 42 days after sowing. Yield 11–14 tonnes/ha. Resistant against YVMV under field condition	IIVR, Varanasi
Kashi Bhairav	2006	Hybridization	IIVR-19 × IIVR-50	F_1 hybrid, plants are medium tall with 2–3 branches, fruit colour dark green,	IIVR, Varanasi

OKRA : *Contd.*

Variety	Year of release	Breeding method	Pedigree/ parentage	Important traits	Institutes/ universities
				length 10–12 cm at marketable stage and yield 16–18 tonnes/ha. Resistant to YVMV and OLCV under field condition. Identified for all zones of the country	
Kashi Mahima	2006	Hybridization	IIVR-21 × IIVR-15	Plants are tall (110–140 cm), flowers comes at 39–41 days of sowing, fruits five ridges, 11.3–12.6 cm long, yield 13–15 tonnes/ha. Resistant to OLCV. Identified for all zones of the country	IIVR, Varanasi
Shitla Jyoti	2001	Hybridization	PDIB-3 × PDIB-1	F_1 hybrid, plants are medium tall (110–150 cm), flowering starts at 30–40 days after sowing. Fruit colour green, length 12–14 cm. Yield 20–22 tonnes/ha. Resistant to YVMV and OLCV	IIVR, Varanasi
VL Bhindi 1	2003	Pedigree	Pusa Sawani × PS 1	Fruit green, long, 5 line, plant height 100–150 cm, maturity 55 days, tolerant to mosaic and spotted leaf disease, yield 145.0 q/ha	VPKAS, Almora
Varsha Uphar	—	—	—	Plants medium tall, suitable to rainy season, tolerant to yellow vein mosaic virus	CCSHAU, Hisar

OKRA : *Contd.*

Variety	Year of release	Breeding method	Pedigree/ parentage	Important traits	Institutes/ universities
Hisar Naveen	—	—	—	High-yielding, suitable to both summer and rainy season, tolerant to yellow vein mosaic virus	CCSHAU, Hisar
Hisar Unnat	—	—	—	High-yielding, most suitable to summer season	CCSHAU, Hisar
HBH 142	—	Hybridization	—	Suitable for rainy season, tolerant to yellow vein mosaic virus	CCSHAU, Hisar

flower appear on a single stem. A flower bud takes about 22–36 days from initiation to full bloom. Time of anthesis varies with cultivar, temperature and humidity. Usually the anthesis time is 8–10 AM. Anther dehiscence is transverse and occurs 15–20 min after anthesis. Dehiscence is completed in 5–10 min. Pollen fertility is maximum between an hour before and after opening of flowers. After pollination, pollen takes 2–6 h for fertilization. Flower remains open for a short time and they wither late in afternoon. Stigma is receptive at opening of flowers. Hence pollination at bud stage is not successful. There is no significant difference in fruit setting under open-pollination, self-pollinated by bagging alone and self-pollinated by hand-pollination of bagged flowers, indicating that it is potentially a self-pollinated crop. Though essentially self pollinated, because of its showy corolla, possibility of cross-pollination by insects cannot be ruled out. Consequently, cross-pollination of 4–19% with maximum of 42.2% has been reported. Hence, it can be classified as often cross-pollinated. The extent of cross-pollination in a particular place depends upon cultivar, competitive flora, insect population season, etc.

CUCURBITACEOUS VEGETABLES

BITTER GOURD

Botanical name: *Momordica charantia* L., 2n = 2x = 22
Family: Cucurbitaceae

Bitter gourd is a rich source of iron and vitamin C and an important curcurbit growth both as rainy season and spring-summer season crop in India. It is also extensively grown in China, Japan, South East Asia, tropical Africa and South America. The origin of bitter gourd remains unidentified. However, it is considered a native of tropical Asia, particularly eastern India and Southern China. Its plants are monoecious annuals with medium-sized vines. Staminate flowers are small, yellow and borne on long slender pedicles. The pistillate flowers are solitary, have small pedicel and are easily distinguishable by oblong to long distinct green colour ovary. Flowers are yellow. Leaves are segmented and have more length then width in the outline. There are five each calyx and corolla. Filaments are three, two are bilocular and one is unilocular. There are three short styles terminated by three bilobed or divided stigma.

Anthesis and dehiscence occur early in the morning. Therefore selfing and crossing should be attempted in forenoon preferably in early hours. For crossing purposes, female flower buds ready to

BITTER GOURD

Variety	Year of release	Breeding method	Pedigree/ parentage	Important traits	Institutes/ universities
CO 1	1978	Selection	Local type collected from Thudiyalur	The vines are medium tall (130 cm) and 5–7 branches/vine. First flowers open in 45–50 days and marketable fruit harvesting starts in 55–60 days after sowing and a total of 6–8 pickings can be done. The fruits are dark green, medium long (25–30 cm) and thick (6–8 cm) with characteristic warts. Each vine produces 20–22 fruits with average weight 100–120 g. Each fruit contains 24–30 seeds weighing 6–8 g. Its yield potential is 140–150 q/ha	TNAU, Coimbatore
CO BgoH 1	2001	Heterosis	F_1 hybrid developed by a crossing MC 84 × MDU 1	It recorded an average yield of 44.4 tonnes/ha. The average individual fruit weight is 300 g. Its yield potential is 51.29 tonnes/ha	TNAU, Coimbatore
MDU 1	1984	Mutation	Gamma irradiation of local cultivar MC 103	Flowers starts from 60 days of sowing and gives sex ratio of 1:20 of female and male flowers. The fruits are long with a average length of 40.34 cm and a girth of 17.54 cm, having average fruit weight of 410 g. Each vine bears on an average of 16.66 fruits. Its yield potential is 300–325 q/ha	AC&RI, TNAU, Madurai

BITTER GOURD : *Contd.*

Variety	Year of release	Breeding method	Pedigree/ parentage	Important traits	Institutes/ universities
Kalyanpur Baraha-masi	1983	Selection	—	Vines are vigorous, fruits long (30–50 cm) light green, thin and tapering. Plants are tolerant to fruitfly and mosaic virus. Recommended for *kharif* season cultivation in Uttar Pradesh. Its yield potential is 200–220 q/ha in 120 days	CSAUAT, Kanpur
Kalyanpur Sona	1996	Selection	Local material	Vine is medium in growth, fruits medium sized and green. Its yield potential is 120–130 q/ha	CSAUAT, Kanpur
Arka Harit	1986	Selection	Rajasthan collection	Fruits are spindles shaped attractive, glossy green with smooth regular ribs and thick flesh. Its yield potential is 125–135 q/ha in 120 days	IIHR, Bangalore
Pusa Do Mausami	1975	Selection	Selection from Indigenous material	Fruits dark green, 7–8 continuous ribs, each fruit weighs 100–120 g, yield potential 12–15 tonnes/ha in 120 days	IARI, New Delhi
Pusa Vishesh	1987	Selection	Selection from Indigenous material, Hapur, Uttar Pradesh	Fruit thick, medium long, glossy green; suitable for spring-summer season; vines short, hence more number of plants can be accommodated per unit area; first picking in 55–60 days. Yield 150 q/ha	IARI, New Delhi

BITTER GOURD : *Contd.*

Variety	Year of release	Breeding method	Pedigree/ parentage	Important traits	Institutes/ universities
Pusa Hybrid 1	1995	Heterosis	Pusa Domousmi × Pusa Vishesh	Vines medium in growth, fruits green, attractive, medium long, irregular smooth tubercles on surface, suitable for vegetable purpose and dehydration, averge yield 218 q/ha	IARI, New Delhi
Pusa Hybrid 2	2006	Hybridization	S 63 × Pusa Do Mausmi	Fruits dark green, medium-long, average fruit weight 90 g, early maturing, average yield 215 q/ha	IARI, New Delhi
Punjab 14	1987	Selection	Local material	Vines are comparatively short, fruit large, oblong and green in colour. It is suitable for spring summer and rainy seasons sowing. Its yield potential is 120–130 q/ha	PAU, Ludhiana
Konkan Tara (DPL-BG-2)	1996	Selection	—	Fruits are green, medium long (15–16 cm) and spindle-shaped with raised tubercles. Fruits have good keeping quality with a shelf life of 7–8 days at ambient temperature. Its yield potential is 230–250 q/ha	KKV, Konkan
Priya	1976	Selection	Kannur local	Fruits green, 40 cm length, 40–50 fruits/vine	KAU, Thrissur

BITTER GOURD : *Contd.*

Variety	Year of release	Breeding method	Pedigree/ parentage	Important traits	Institutes/ universities
Priyanka	1996	Selection	Local selection	—	KAU, Thrissur
Preethi	1998	Selection	MC 84	Plants have light green veins with average vine length of 6.8 m. Fruits are white, spiny, medium in size with 310 g of average weight. Its yield potential is 140–150 q/ha in 150 days	KAU, Thrissur
Hirkani	1991	Selection	Local germplasm	Fruits dark green, 15–20 cm long with prickles, average yield 138 q/ha	MPKV, Rahuri
Phule Green Gold (RHRBG 5)	2001	Pedigree selection	Green Long × Delhi Local	Fruits are dark-green, 25–30 cm long with raised tubercles. This is tolerant to downy mildew. Its yield potential is 220–235 q/ha in 220–235 days	MPKV, Rahuri
Phule Priyanka (RHRBG 1)	2001	Heterosis	RHR 5 × RHR 4	Fruits are medium long, dark green, and highly prickly. Plants are tolerant to downy mildew. Its yield potential is 200–250 q/ha in 180 days	MPKV, Rahuri
Pant Karela 2 (PBIG 1)	2006	Selection	PBIG 1	Fruits thin, 25 cm long, dark green with tapering ends. First marketing fruit picking at 50 days after seed sowing. Its yield potential is 150–160 q/ha	GBPUAT, Pantnagar

flower in next 1–2 days are covered by cotton pad wrapping/small paper bags. Likewise, male flower buds ready to open next morning are also covered. Pollination is carried out by rubbing staminate flower anthers from freshly opened flowers against the stigma of the protected pistillate flowers. The crossed flowers are covered as above for few more days to avoid contamination/pollination from other plants brought by insects. Tagging is simultaneously done by writing the percentage and date of pollination. In order to increase crossed fruit set, other open-pollinated fruits already set may be removed to avoid inhibitory effects of these fruits on further fruit setting.

SQUASH AND PUMPKIN

Botanical name: *Cucurbita* spp. 2n = 2x = 40
Family: Cucurbitaceae

Squash and pumpkin are the words quite often used interchangeably. However, the term 'squash' is more commonly used for *C. pepo* which is eaten as an immature fruit. The term 'pumpkin' is normally applied to edible fruit of a species of *Cucurbita* utilized when ripe as a table vegetable. The cultivated species of *Cucurbita* are widely distributed throughout tropical, subtropical and temperate regions in both the New and Old Worlds. In these areas, squash and pumpkin are important in agriculture, but considered as minor crops.

The genus *Cucurbita* is native to the Americas. The centre of diversity is in the tropics near Mexico-Guatemala border. The archaeological evidences indicate that these were widely cultivated in southwestern United States, Mexico, and northern South America in pre-Colombian times, *i.e.* prior to 1492 A.D. Four cultivated species of *Cucurbita* are herbaceous annuals with tiny growth and several runners. However, *C. pepo* has short internodes and bush type appearance. The vine (stem/runner) may be prickly or spiny, rounded or angled. There are often roots at the nodes. The tendrils are long and branched. Leaves are large, alternate, shallow to deeply-lobed and palmate. The flowers are large, showy with yellow or creamy corolla. The plant is monoecious. Flowers are unisexual and occur singly in axils of leaves. The staminate flowers are near the centre of the plant and have long, tender pedicles. The pistillate flowers have short, ridged pedicles and are distal to the staminate flowers. Tap root system is strong, efficient and goes substantially down in soil. The morphology of peduncle is distinct enough to be used in determining species limits. In *C. pepo*, peduncle is deeply furrowed and five to eight-ridged. In *C. moschate*, the peduncle is five ridged

SQUASH (*Cucurbita pepo*)

Variety	Year of release	Breeding method	Pedigree/ parentage	Important traits	Institutes/ universities
Early Yellow Prolific	1975	Introduction	—	Plants are bush type and early in flowering. Fruits medium-sized, warted and tapering towards stem end. Skin light yellow turns orange-yellow on maturity. Its yield potential is 250–275 q/ha	IARI, Katrain
Pusa Alankar	—	Heterosis	Australian Green × Yellow Prolific	Very early and high yielding, fruits dark green with light coloured stripes, 20–25 cm long, slightly tapering towards the stem end, tender with delicious flesh	IARI, Katrain
Punjab Chappan Kaddu 1	1983	Selection	—	Plants are bush type and have non-lobed leaves. Fruits are green, medium-sized, disc shaped This is an early maturing variety, ready to first harvesting 60 days after sowing. Single fruit weights 40–50 g. Plants are resistance to downy mildew, tolerance to CMV and pumpkin beetle in field condition. Yield potential 300–320 q/ha	PAU, Ludhiana
Patty Pan	1986	Introduction	Introduction from USA	Plants are erect with short internodes. Fruit are white, disc shaped with notches along the margin. Fruit weight 350–400 g. Its yield potential is 550–575 q/ha in 90–95 days	IIHR, Bangalore

PUMPKIN

Variety	Year of release	Breeding method	Pedigree/ parentage	Important traits	Institutes/ universities
CO 1	1971	Selection	Selection from local type	The fruits are flattened at the base measuring 34 cm in length and 26.0 cm girth. The flesh thickness is 4–5 cm. The seed content is less (1.2%) and each plant bears 6–7 fruits/plant. Each fruit weighs about 7.0 kg	TNAU, Coimbtore
CO 2	1978	Selection	Selection from local type	The vines are moderately vigorous, less-spreading and distance for high density planting (2.2 m × 1.9 m). Fruits are small, slightly ridged with bright orange flesh, 10.8% TSS, 3.4% total sugar, 1.9% reducing sugar, 9.2% starch, 10.0 mg/100 g ascorbic acid, 0.14% acidity. First picking starts 95–100 days after sowing. Its yield potential is 450–475 q/ha in 135 days	TNAU, Coimbtore
Ambili	1983	Selection	Local cultivar	Plants are spreading type, fruits are flat-round, medium (~6 kg weight), green with shallow furrows on surface. Its yield potential is 340–360 q/ha in 95–110 days	KAU, Vellanni-kaara
Suvarna	1998	Selection	Single plant selection, CM349	—	KAU, Vellanikkara

PUMPKIN : *Contd.*

Variety	Year of release	Breeding method	Pedigree/ parentage	Important traits	Institutes/ universities
Saras	2006	Selection	Local collection	Fruits elongated orange-fleshed and medium-sized, weighing 2.7 kg. Its yield potential is 390–410 q/ha	KAU, Vellanikkara
Sooraj	2003	Selection	—	Plants produce medium-sized globular fruits. Flesh is thick and orange coloured. Average fruit weight is ~3 kg. Its yield potential is 350–400 q/ha	KAU, Vellanikkara
Pusa Vishwas	1990	Selection	Local collection (Line SM-107)	Plants are of vigorous vegetative growth, dark green leaves with white spots including veins. Fruits are brown, spherical, average weight of 5 kg with thick and golden-yellow flesh. Its yield potential is 400–425 q/ha in 120 days	IARI, New Delhi
Pusa Vikas	1985	Selection	Selection from Indigenous material	Vines semi-dwarf to dwarf (2.0–2.5m long); leaves soft with light green or yellow spots; fruits small weighing 2 kg, flattish round; flesh yellow, rich in vitamin A. Yield 300 q/ha	IARI, New Delhi
Pusa Vishwas	1987	Selection	Selection from Indigenous material	Vigorous vegetative growth; leaves dark green with white spots including veins; fruits light brown, spherical with thick golden-yellow flesh, weigh 5 kg; maturity in 120 days. Yield 400 q/ha	IARI, New Delhi

PUMPKIN : *Contd.*

Variety	Year of release	Breeding method	Pedigree/ parentage	Important traits	Institutes/ universities
Pusa Hybrid 1	1996	Heterosis	Sel-107 × Sel-124	Fruits rounded-flat, medium-sized, average yield 520 q/ha. Suitable for Punjab, Kerala, Delhi and other parts of northern plains	IARI, New Delhi
Arka Chandan	1990	Selection	Local collection (IIHR-105-1-1)	Plants have vigorous vines, hairy stem, green broad leaves without any pattern. Fruits are round with flat blossom end, 2–3 kg weight, green with white patches on rinds (turn light brown at maturity); flesh is thick, firm, pleasant aroma, bright orange, sweet with 8–10% TSS, rich in carotene. Its yield potential is 325–350 q/ha in 120 days	IIHR, Bangalore
Arka Surya-mukhi	1984	Selection	Local collection from Dakshina Kannada	*Cucubita maxima*, fruits round with pressed ends, rind colour orange red with white stipes when mature, bright orange flesh colour, average fruit weight 1 kg, resistant to fruit fly	IIHR, Bangalore
Kashi Harit	2006	Pedigree selection	NDPK 24 × PKM	Vine short, leaves dark green with white spots including on veins. Fruits are green, spherical, weight 2.5–3.0 kg at green stage. Its yield potential is 300–350 q/ha in 65 days	IIVR, Varanasi

PUMPKIN : *Contd.*

Variety	Year of release	Breeding method	Pedigree/ parentage	Important traits	Institutes/ universities
Narendra Agrim	2006	Selection	—	It is a short vined, very early, small fruited and highly homogeneous open pollinated variety of pumpkin. Fruits round, dark green in colour. Plants bear female flowers four days earlier than male flowers and give its first picking within 55 days. Green fruits weigh 1.5–2.0 kg and mature fruits are of 3–4 kg weight. Its yield potential is 300–350 q/ha	NDUAT, Faizabad
Narendra Amrit	2006	Selection	Local collection	Fruits round, and pale-green-mottled in colour; become almond in colour at maturity. Single green fruits weigh 3 kg, while mature fruits are of 6 kg weight. The mature fruits have papery skin, thick flesh of about 7 cm with small seed cavity. Long storage life and can be kept safely for more than four months at ambient temperature. Its yield potential is 350–400 q/ha	NDUAT, Faizabad
Azad Pumpkin 1	2006	Selection	—	Plants are spreading type, leaves green hairy, early fruiting, close fruit bearing, green fruit with yellowish stripes, flesh thick, less number of seeds, fruit spherical flat. Its yield potential is 425–450 q/ha	CSAUAT, Kanpur

and flared at the point of fruit attachment. In *C. maxima*, peduncle is cylindrical or claviform but never prominently ridged. In *C. mixta*, peduncle is basically five-angled, rounded, but not at all or only slightly, enlarged at the fruit attachment. In *C. ficifolia*, peduncle is hard, smoothly angled, slightly flaring, but comparatively much smaller than in other species of *Cucurbita*.

Calyx and Corolla are campanulate. There are three anthers. Filaments are parly free. Pistil is oblong or discoid, unilocular with three to five placentae. Style is thick. The stigmas are three each two-lobed. Fruit is pepo. The pistillate and staminate flower buds are located a day before anthesis and are protected by tying the tip of e corolla tube. The following morning, as soon as the pollen sacs dehisce, staminate flowers are taken out and pollen-grains are applied on receptive stigma of pistillate flowers. This is easily done by rubbing the anthers against stigma. Pollination can be done from morning till noon.

BOTTLE GOURD

Botanical name: *Lagenaria siceraria,* 2n = 2x = 22
Family: Cucurbitaceae

Bottle gourd is one of the important cucurbits in India, both as a rainy season and summer vegetable. It is also known as white-flowered gourd. It is one of the most important vegetables of ancient China. It has been found in wild form in South Africa and India. However, it is indigenous to tropical Africa on the basis of variability in seeds and fruits. This species appears to have been domesticated independently in Asia, Africa and The New World. Bottle gourd is monoecious, annual vine with soft pubescence. The leaves are cordate-ovate to reinform-ovate, 15–30 cm across, not lobed. The flowers are white, solitary, showy and open at night. The flower has five petals. The staminate flowers are on long pedicels exceeding the foliage. The pistillate flowers are single with short peduncle and hairy ovary. Ovary has three placentae with numerous ovules.

The fruits are variable in size and shape. In shape, they are cylindrical, oval, egg-shaped, club-shaped, round etc. The fruits at maturity are hard-shelled, smooth-surfaced, green to whitish-green or tan in colour, and variously striped or mottled. There are three stamens, two as compound and one as single. Since flowers open at night, selfing/crossing must be done earliest in the morning. If possible, during night also, it could be done.

BOTTLE GOURD

Variety	Year of release	Breeding method	Pedigree/ parentage	Important traits	Institutes/ universities
Punjab Round	1978	Selection	—	Plants are vigorous and prolific-bearer. Fruits are round, flat, tender and shining. Its yield potential is 350–375 q/ha	PAU, Ludhiana
Punjab Komal	1988	—	—	Fruits oblong, medium-sized and light green, average yield 200 q/ha	PAU, Ludhiana
Punjab Long	1999	Selection	Selection from the locally grown material (LC-2-1)	Plants are vigorous. Fruits are long, tender and attractive. Its yield potential is 200–225 q/ha	PAU, Ludhiana
Kalyanpur Long Green	1983	Selection	—	Vine spreading type, fruit long, smooth, green, tender, blossom end semi round, homogenous fruit size, suitable for rainy season. Its yield potential is of 275–300 q/ha	CSAUAT, Kanpur
Lauki Azad Sankar 1	2006	Heterosis	—	Fruit long (30 cm), cylindrical, green, vines 3 to 3.5 m long, bears more number of fruits. Its yield potential is 580–600 q/ha	CSAUAT, Kanpur
CO 1	1981	Selection	Selection from a germplasm	Vine and leaves are light yellowish green. Fruits are round at the base with a prominent bottleneck at the top, medium-	TNAU, Coimbatore

BOTTLE GOURD : *Contd.*

Variety	Year of release	Breeding method	Pedigree/ parentage	Important traits	Institutes/ universities
				sized, attractive light green in colour with a mean weight of 2.02 kg. Yield potential 360–400 q/ha in 135 days	
Rajendra Chamatkar	1984	Selection	—	First picking starts after 68–70 days of seed sowing. Plants are prolific in bearing, long, cylindrical in shape, fruit weight 1.35 kg. Its yield potential is 175–200 q/ha	RAU, Pusa
Pant Shankar Lauki 1	1999	Heterosis	PBOG 22 × PBOG 40	The fruits are medium-sized long and somewhat cylindrical (about 35 cm long) and green. Vine length is about 5.5 m. The first picking is possible in about 60 days. It is suitable for planting in the plains as well as in the hills of northern India. Its yield potential is 400–425 q/ha	GBPUAT, Pantnagar
Pant Lauki 3 (PBOG 61)	2006	Selection	—	Fruits ~40 cm long, cylindrical, light green in colour. First fruit harvest in 60 days. Its yield potential is 350–370 q/ha	GBPUAT, Pantnagar
Pant Sankar Lauki 2	2006	Heterosis	PBOG 22 × PBOG 61	Fruits are about 40 cm long, club shaped with smooth green colour. The first green fruit harvest is in 65 days. Suitable for	GBPUAT, Pantnagar

BOTTLE GOURD : ***Contd.***

Variety	Year of release	Breeding method	Pedigree/ parentage	Important traits	Institutes/ universities
				plains and hills sowing. It can be sown from March to July in plains and April to May in hills. Yield potential of 400–425 q/ha	
Narendra Sankar Lauki 4 (NDBGH 4)	2006	Selection	—	Early in flowering, prolific-bearer with near cylindrical attractive fruits. It is suitable for cultivation throughout the year. The fruits remain soft for preparing palatable vegetable, even if they have grown out of normal size of picking. Yield potential 400–450 q/ha	NDUAT, Faizabad
Narendra Dharidar (NDBG-208-1)	2006	Selection	NDBG 208	Fruit striped green, bottle shaped with NDBG-I selected through pedigree method of selection. Its yield potential is 325–350 q/ha	NDUAT, Faizabad
Narendra Rashmi	2006	Selection	Local collection	Plants bear approximately 8 fruits per plant, moderately resistant, powdery mildew and downey mildew. Fruits are bottle-shaped, having shallow neck, white flash, average weight of 1.0 kg with 400–500 seeds. First harvesting starts 60 days after sowing. Its yield potential is 275–310 q/ha	NDUAT, Faizabad

BOTTLE GOURD : *Contd.*

Variety	Year of release	Breeding method	Pedigree/ parentage	Important traits	Institutes/ universities
Narendra Jyothi (NDBG 104)	2006	Selection	—	Fruits are slender, long, high-yielding, early-maturing. Its yield potential is 300–350 q/ha	NDUAT, Faizabad
Samrat	1992	Selection	Local germplasm collected from Dahanu, Thane, Maharashtra	Fruits 30–40 cm long, green in colour with dense pubescence and cylindrical in shape. Good keeping quality, average yield 400 q/ha	MPKV, Rahuri
Kashi Ganga (DVBG 1)	2006	Hybridization	Pedigree of IC 92465 × DVBG 151	Plant bears light green fruits of 30 cm length having 7cm diameter, fruit weight 800–900 g. Plants are tolerant to anthracnose and suitable for *kharif* and summer season cultivation. Its yield potential is 480–550 q/ha	IIVR, Varanasi
Kashi Bahar (VRH 1)	2006	Heterosis	—	Plants are vigorous and green. Single plant bears 12 light green straight fruits of 30–32 cm length and 780–850 g weight. It is suitable for both season cultivation in rainy, summer as well as river-bed areas. It is tolerant to anthracnose, downy mildew and *Cercospora* leaf spot under field condition. Its yield potential is 500–550 q/ha	IIVR, Varanasi

BOTTLE GOURD : *Contd.*

Variety	Year of release	Breeding method	Pedigree/ parentage	Important traits	Institutes/ universities
Arka Bahar	1986	Selection	Pure line selection from a local collection IIHR 20 from Karnataka	Fruits are straight, devoid of crookneck, medium-size and each weighing about 1 kg. Fruit rind is light green with shining flesh, tender with pleasant aroma. It is very good for cooking along with good keeping quality. Yield 400–500 q/ha in 120 days	IIHR, Bangalore
Thar Samridhi	2007	Hybridization	Banswara local 1 × Gujarat local 1	High yield potential	CIAH, Bikaner
Azad Nutan	2000	—	—	Ideal for both spring-summer and rainy season, fruits shining, light green and soft pubescent, neck-free	CSAUAT, Kanpur
Pusa Summer Prolific Long	1975	Selection	Selection from Indigenous material	Fruits pale green, 40–50 cm in length, Suitable for summer sowing	IARI, New Delhi
Pusa Summer Prolific Round	1975	Selection	Selection from Indigenous material	Fruits round, 15–18 cm in girth, green, high-yielding	IARI, New Delhi

BOTTLE GOUED : *Contd.*

Variety	Year of release	Breeding method	Pedigree/ parentage	Important traits	Institutes/ universities
Pusa Naveen	1992	Selection	Selection from Indigenous material	Fruits long (30–40 cm), green, straight, nearly cylindrical, suitable for packing for long distance transportation; first picking in 55 days. Suitable for both summer and *kharif* seasons. Yield 325 q/ha	IARI, New Delhi
Pusa Sandesh	1996	Selection	Selection from Indigenous material	Vines medium-long; fruits attractive green, round, deep oblate, medium-sized weighing 600 g; first picking in 55–60 days in *kharif* and 60–65 days in summer. Suitable for both summer and *kharif* seasons. Yield 320 q/ha	IARI, New Delhi
Pusa Samridhi	2005	Selection	Selection from Indigenous material	Its fruits are attractive green, smooth with normal pubescence, straight, slightly clubbed, without 'neck', 25–30 cm long and medium thick (6.5–7.5 cm) in diameter. The average fruit weight is 1 kg. Maturity (days to first fruit harvesting) — 50–55 days. Average fruit yield 310 q/ha in *kharif* and 275 q/ha in summer season. The fruits of this selection are superior in nutritional qualities (in vitamin C-8.07, phosphorus — 20.39, calcium-19.35 and iron-1.35 mg/100 g edible portion) to Pusa	IARI, New Delhi

BOTTLE GOURD : *Contd.*

Variety	Year of release	Breeding method	Pedigree/ parentage	Important traits	Institutes/ universities
				Naveen. The selection is recommended for commercial cultivation both in spring-summer and *kharif* seasons in northern plains	
Pusa Santushti	2008	Selection	Selection from Indigenous material	Its fruits attractive green, smooth with normal pubescence, pear shaped, fruit length 18.50 cm, fruit diameter 12.40 cm. The fruit sets under low temperature (10–12°C) as well as high temperature (35–40°C). The fruit weight is 0.8 to 1.0 kg. Maturity (days to first fruit harvesting 55–60 days). Average yield is 288.5 q/ha in *kharif* and 261.0 q/ha in summer season. The fruits are superior in nutritional qualities (Vitamin C – 7.12, phosphorus — 18.15, calcium 19.45 and iron — 1.10 mg/ 100 g edible portion) to Pusa Sandesh. This selection is suitable for commercial cultivation in spring-summer, *kharif* and also in autumn-winter seasons in northern plains	IARI, New Delhi
Pusa Meghdoot	—	Heterosis	PSPL × Selection 2	Early and gives 50% higher yield than PSPL. Fruits green, long and tender; suitable for summer and rainy seasons	IARI, New Delhi

BOTTLE GOURD : *Contd.*

Variety	Year of release	Breeding method	Pedigree/ parentage	Important traits	Institutes/ universities
Pusa Manjari	—	Heterosis	PSPR × Selection 11	Early and gives 48% higher yield than PSPR. Fruits round, green, tender; suitable for summer and rainy seasons	IARI, New Delhi
Pusa Hybrid 3	—	Heterosis	Pusa Naveen × Sel P 8	Fruits green, slightly club shaped without neck; suitable for easy packing and long distance transportation; first picking in 50–55 days. Yield 425 q/ha in summer and 470 q/ha in *kharif* season	IARI, New Delhi

MUSKMELON

Botanical name: *Cucumis melo* L., 2n = 2x = 24
Family: Cucurbitaceae

Muskmelon encompasses the netted, salmon-flesh cantaloupe, smooth-skinned green fleshed 'Honey Dew', the wrinkled-skinned, white-fleshed, 'Golden Beauty' and several other dessert melons in USA. Other forms with very different plant and fruit characters are seen in Orient and India. In addition several wild forms occur in Africa and India and all of these are interfertile. *Cucumis* comprises a genus of nearly 40 species including several of considerable economic importance such as cucumber, muskmelon and West Indian gherkin (*C. anguria* L.,). All species are indigenous to Africa which apparently were introduced to the West Indies from Africa and *C. sativus* and *C.hardwickii*, are natives of Asia. India, Persia, China and Southern Russia are considered secondary centres of diversity for muskmelon.

Three important species of the genus *Cucumis* are distinguishable on the basis of following key: fruits spiny, muricate or echinate; leaves deeply lobed, fruits small, prickly lemon — yellow when mature — *C. anguria*. Fruits glabrous or pubescent, smooth or netted, mostly oblong — *C. sativus*.

It is a polymorphic species where most cultivars are andromonoecious (staminate and perfect flowers) but other sex forms are also available. Stem is soft-hairy to glabrous, striate or angled, leaves orbicular to ovate to reniform, usually five-angled, sometimes shallowly three to seven — lobed, hairy or somewhat scabrous, 3–5 inches across. The staminate flowers are clustered. The pistillate flowers are solitary on short stout pedicels. There are various horticultural forms within *C. melo* based on fruit characteristics, namely, cantaloupes, nutmeg muskmelons, winter melons, white-skinned melons, snake melons, oriental pickling melons, mango melons, pomegranate melons which hybridize readily with each other and there is apparently very little sterility even among progenies from crosses involving variant types. Under open natural system, it is partly cross and partly self-pollinated by bees but there are reports of obtaining vigorous lines by inbreeding up to 25 generations.

Melons may be andromonoecious (hermaphrodite and staminate flowers) gynoecious (only pistillate flowers), or monoecious (pistillate and staminate flowers). Monoecious and andromonoecious are most common. Hand-pollination of andromonoecious types is a two step process. On the day prior to anthesis, the hermaphrodite flower is emasculated. Both pistillate and staminate flowers are covered to prevent insect contamination. Emasculation is not required on

MUSKMELON

Variety	Year of release	Breeding method	Pedigree/ parentage	Important traits	Institutes/ universities
Hara Madhu	1970	Selection	Local collection (Kutana type)	Fruits are round, tapering towards the stalk with open prominent green sutures, average weight 1 kg, do not slip from the stalk, thin rind, smooth and pale yellow at maturity; flesh is green, thick, very juicy with 12–15% TSS. Seeds are loosely packed in the seed cavity. Yield potential 130–150 q/ha	PAU, Ludhiana
Lucknow Safeda	1975	Selection	—	It is a very early variety with medium vine growth. Fruits are small, roundish, flat, yellow, smooth and weighing about 300 g. Flesh is creamy white, less juicy and sweet with good keeping quality. Yield potential 110–130 q/ha	PAU, Ludhiana
Punjab Sunehri	1978	Pedigree selection	Hara Madhu × Edisto	Fruits are round to elliptical and devoid of distinct sutures at maturity. Avarage fruit weight is 1 kg. It slips easily from the stalk. Its rind is thick, netted and dull yellow at maturity. Its flesh is thick, orange, fairly juicy and sweet (11% TSS). It has a long post-harvest life and has excellent transportability. Yield potential 160–180 q/ha	PAU, Ludhiana

MUSKMELON : *Contd.*

Variety	Year of release	Breeding method	Pedigree/ parentage	Important traits	Institutes/ universities
Punjab Hybrid 1	1987	Heterosis	MS 1 × Hara Madhu	Vines are vigorous, medium long and leaves are partly lobed. Fruits are globular with distinct sutures. Average fruit weight is ~800 g and slips from the stalk. Rind is quite thick, netted and creamy yellow. Flesh is thick, orange and juicy. It is sweet with 12% TSS. Yield potential 160–180 q/ha	PAU, Ludhiana
Punjab Rasila (MR 12)	1996	Selection	—	Fruits are nearly round. The rind is light yellow and green sutured with medium netting. The flesh is green, thick, juicy, 10% TSS and flavoursome. Fruit develops full-slip stage at maturity. Average fruit weight is 600 g. Yield potential 160–175 q/ha	PAU, Ludhiana
MH 10 (MHI 10)	1996	Heterosis	WI 998 × Punjab Sunehri	Fruits are round/oval, rind netted, non-sutured, thick, brown on maturity, average weight of 9.15 kg, flesh is thick, orange, medium juicy, TSS 9.10%, fruit developed 'full-slip' stage at maturity; good shelf-life and suitable for distant transportation. Yield potential 240–250 q/ha	PAU, Ludhiana

MUSKMELON : *Contd.*

Variety	Year of release	Breeding method	Pedigree/ parentage	Important traits	Institutes/ universities
Gujarat Musk-melon 1	1984	Selection	Collection from Sabarkantha district in Gujarat	Fruits are small, tasty, lemon yellow, smooth without suture and 5–6 fruits/ plant. Flesh is light green with 2.1 cm. The average weight of fruit is 0.982 kg with a diameter of 10 cm. Yield potential 140–170 q/ha	GAU, Anand
Gujarat Musk-melon 2	1984	Selection	Local germplasm collected from Nagpur area of Maharashtra	Flowering starts at 32 days after sowing. Fruits are medium-sized (diameter 11.5 cm), tasty, 4–5 fruits/ plant and average weight of fruit 1.3 kg. Skin is orange green with slight netting and no suture. Flesh is capucine orange with 2–6 cm thickness. This variety has yield potential of 120–150 q/ha	GAU, Anand
Gujarat Muskmelon 3 (GMM 3)	2006	Selection	Local collection	Fruits are medium-sized, round to slightly oval, yellowish brown spots on mature fruits with green strips. Flesh green, juicy, very sweet, TSS 12%, average fruit weight 800 g. Yield potential 110–125 q/ha	GAU, Anand
Pusa Sharabati	1978	Pedigree selection	Kutana × American Cantaloupe (PMR 6)	Fruits are round with nettled skin; flesh is salmon orange, firm, thick, 11–12% TSS with small seed cavity; good keeping	IARI, New Delhi

MUSKMELON : *Contd.*

Variety	Year of release	Breeding method	Pedigree/ parentage	Important traits	Institutes/ universities
				quality; suitable for riverbed cultivation. Yield potential 150–170 q/ha	
DMDR 2	—	—	Recurrent selection from FH 1 × Phoot	—	IARI, New Delhi
Pusa Rasraj	1993	Heterosis	M3 × DM	Fruits oval, skin smooth without netting, weighing 800 g to 1 kg, sweet (TSS 11%), maturity 75–80 days. Yield potential 225–240 q/ha	IARI, New Delhi
Pusa Madhuras	—	Selection	Rajasthan Collection	Fruits roundish flat, skin pale green, sparsely netted with dark green stripes and salmon orange flesh, TSS 12–14%	IARI, New Delhi
Arka Rajhans	1986	Selection	Single plant selection from a local collection (IIHR 107), Rajasthan	Plants are with densely hirsute stem, light green foliage with shallow lobes and bear 2–3 fruits. Fruits are round to slightly oval, medium-large, creamy white skin, weight varies from 1.25–2.0 kg, netted with shallow vein tracks at full slip stage; flesh is thick, firm and white with 11–14% of TSS, crisp texture, mild flavour; good transport and keeping quality. Plants are resistant to powdery mildew. Yield potential 285–300 q/ha in 85–90 days	IIHR, Bangalore

MUSKMELON : *Contd.*

Variety	Year of release	Breeding method	Pedigree/ parentage	Important traits	Institutes/ universities
Arka Jeet	1985	Selection	Single plant selection from Bati Strain, IIHR 103	Plants have dark green lobes leaves and angular stem. Fruits are small, round with smooth orange flat ends, 300–500 g of weight, dark green skin, thin golden orange at full step stage; flesh is white with medium soft texture, excellent flavour, TSS 15–17%; juicy and rich in vitamin C. Yield potential 150–175 q/ha in 90 days	IIHR, Bangalore
Kashi Madhu (IVMM 3)	2006	Selection	Mau melon	Fruits are round, with open prominent green sutures, weight 650–725 g, half slip in nature, thin rind, smooth and pale yellow at maturity, flesh salomon orange (mango colour), thick, very juicy with 13–14% TSS and seeds are loosely packed in the seed cavity. Post-harvest life is long with good transportability. Plants are tolerant to powdery and downy mildew. Yield potential 250–270 q/ha	IIVR, Varanasi
Durgapura Madhu	1974	Selection	—	Fruits are oblong with big seed cavity, average weight 500–600 g; flesh is light green dry texture with 14% TSS. Yield potential 130–150 q/ha	ARS, Durgapura

MUSKMELON : *Contd.*

Variety	Year of release	Breeding method	Pedigree/ parentage	Important traits	Institutes/ universities
RM 43	1997	Selection	Locally collected germplasm	Oblong pale green fruits with dark green 10–12 sutures, average weight 400–600 g, hard rind, small seed-cavity, green flesh, and TSS 10–12%. Fruits have good keeping quality and transportability. Yield potential 200–220 q/ha	RARS, Durgapura
MHY 3	1999	Pedigree method	Durgapura Madhu × Pusa Madhuras	Vines are vigorous and spreading about 2.5–3.0 m long, and fruits are round flat, and smooth surfaced with yellow rind and light green stripes. Flesh colour is light green, texture soft, and TSS 13–16%. Average fruit weight is 700–800 g. Yield potential 180–200 q/ha	RARS, Durgapura
RM 50	2004	Pedigree method	Durgapura Madhu × Sel-1	Vines are well-developed up to 2–2.5 m. Ovate-shaped fruits are pale-green in colour having ten green sutures, and TSS 14–16% with green flesh. Average fruit weight is 500 g. Yield potential 175–200 q/ha	RARS, Durgapura
MHY 5	2004	Pedigree method	Durgapura Madhu × Hara Madhu	Fruits with smooth rind and medium seed cavity are round flat, tapering at stem end and green yellow in colour. The flesh colour is light green, texture soft,	RARS, Durgapura

MUSKMELON : *Contd.*

Variety	Year of release	Breeding method	Pedigree/ parentage	Important traits	Institutes/ universities
				and TSS 13–16%. Average fruit weight is 700–800 g. Yield potential 180–200 q/ha	
Narendra Muskmelon 1 (NDM 2)	2006	Selection	Local collection	Fruits are round, large-sized with orange colour skin, green stripped at maturity. Flesh dark orange, thick with TSS — 9–11%. Single fruit weight 700–900 g. It is recommeneded for the cultivation in the states of Uttar Pradesh. Yield potential 125–150 q/ha	NDUAT, Faizabad

gynoecious and monoecious types. Hand-pollination is done at anthesis by gently rubbing pollen from anthers of staminate parent flower on the stigma of the pistillate parent flower. Pistillate flower after pollination is covered to prevent contamination by insects like honey bees and *Apis* spp. Emasculation and pollination can be done as one step procedure also in afternoon.

WATERMELON

Botanical name: *Cirtrullus lanatus,* 2n = 2x = 22
Family: Cucurbitaceae

Watermelon is most commonly grown in the Middle East, the USA, Africa, India, Japan and Europe. Plant growth is favoured by high temperature and adequate sunlight. The primary centre of diversity lies in Africa where wild forms are still found. Domestication is likely to have covered in Egypt and India. Plants are viny type, monoecious, with angular stem. Leaves are pinnately divided into 3 or 4 pairs of lobes. In a typical vine, lateral branches arise quite late and usually there is dominance of single runner stem. However, in some of the new dwarf types, multiple branching occurs simultaneously from the crown of the plant providing more potential bearing areas. Corolla is rotate, five-lobed, sulphur-yellow or chrome-yellow. The plant is classified as naturally cross-pollinated crop, but there is considerable amount of selfing and sibbing also.

Fruit size varies from 1.5 to 50 kg but 10–12 kg fruit size is most common. Fruit shape varies from long cylindrical to spherical with various intermediate shapes. Rind colour varies from white to shades of green with conspicuous stripes and mottled rind. Rind is hard but not durable. Fruit flesh colour may be white, yellow, orange, pink, red and texture may vary from firm to fibrous. The greater part of fruit flesh is mostly derived from placentae.

Flowers are less showy, mostly located as singly as separate pistillate and staminate flowers on the same plant (monoecious cultivars). The pistillate or hermaphrodite flowers normally occur in every seventh leaf axil while the intervening axils are occupied by the staminate flowers. Three stamens are at the base of the corolla.

In andromonoecious cultivars also, hermaphrodite flowers do not ensure self-pollination. They too need insect visit to effect pollination. Flowers open shortly after sun rise and remain open only one day. Usually, anthers have dehisced before the flower anthesis. Pollen is visually evident in sticky masses adhering to anther. The stigma is receptive throughout the day. Female bud to

WATERMELON

Variety	Year of release	Breeding method	Pedigree/ parentage	Important traits	Institutes/ universities
Durgapura Lal (RW 177-3)	2004	Heterosis breeding	Sugar Baby × KS-3566	Fruits are dark green with dark thin lining, round and juicy having hard rind, dark red crisp flesh, TSS 10–11% and average weight 4–5 kg. Picking starts 105–110 days after sowing. Plants are moderately resistant to blight and bud necrosis under field condition. Yield potential 430–450 q/ha	RARS, Durgapura
Durgapura Kesar	1978	Selection	—	Fruit weight is 4–5 kg, green skin with stripes. The flesh is yellow and moderately sweet. Seeds are large and yellow. Yield potential 400–450 q/ha	RARS, Durgapura
Durgapura Meetha	1978	Selection	—	Fruits are round with light green and thick rind, good keeping quality, weight varies from 7 to 8 kg; flesh is dark with 11% TSS, seeds with black tip and margin; gives yield of 40–50 in 125 days of crop duration. Yield potential 480–500 q/ha in 125 days	RARS, Durgapura
Durgapura Lal	2002	Pedigree method	Sugarbaby × K 3566 (Russian germplasm)	Leaves simple unlobed, fruits round, dark green, 4–5 kg, skin thin and hard, flesh dark red, TSS 10–11%	RARS, Durgapura

WATERMELON : *Contd.*

Variety	Year of release	Breeding method	Pedigree/ parentage	Important traits	Institutes/ universities
Shipper	1978	Introduction	—	Fruits are medium to large weighing 6–10 kg, round, with characteristic tripartite blossom end. Flesh is red with moderate sweetness (8–9% TSS). Rind is thick and dark green. Seeds are uniformly light brown. Yield potential 300–320 q/ha	PAU, Ludhiana
Arka Jyoti	1986	Heterosis breeding	IIHR 20 × Crimson Sweet	Fruits are round to oval, dark green with blue angular stripes, average weight of 5–6 kg; flesh is bright crimson, granular texture with 12–13% TSS; good keeping and transport quality; first picking starts after 90 days of sowing. Yield potential 700–720 q/ha	IIHR, Bangalore
Arka Manik	1990	Pedigree selection	IIHR 21 × Crimson Sweet	Fruits are round-oval with light green rind and dark green stripes, average weight of 6 kg; flesh is deep crimson, granular texture, pleasant aroma, high TSS (12–15%) with few small seeds arranged in row; good storage and transport quality; multiple resistance to powdery mildew, downy mildew and anthrancnose. Yield potential of 600–610 q/ha in 110–115 days	IIHR, Bangalore

WATERMELON : *Contd.*

Variety	Year of release	Breeding method	Pedigree/ parentage	Important traits	Institutes/ universities
Sugarbaby	1975	Introduction	—	Fruits are round with black rind, deep pink flesh with 11–13% TSS, brown small seeds weight varies from 2–5 kg. Yield potential 150–170 q/ha in 85 days	IARI, New Delhi
Asahi Yamato	—	Introduction	Introduction from Japan	Mid season; vines medium with deeply cut, lobed leaves; fruits medium-sized, 6–7 kg, round, pale green, smooth, non-striped skin; flesh deep red, sweet, TSS 9–11%; maturity 95–100 days. Yield 200 q/ha	IARI, New Delhi
New Hampshire Midget	—	Introduction	Introduction from USA	Fruits small, about 1.5–2 kg; skin light green with black stripes; flesh red, sweet	IARI, New Delhi
Pusa Bedana	—	Heterosis breeding	Tetra 2 × Pusa Rassal	Vine growth vigorous. Fruits have dark green skin with faint stripes, somewhat triangular shape, tough rind, red flesh with white remnants of false seeds, TSS 12–13%, 5–6 kg, produces 3–6 fruits per vine, very good keeping quality; maturity 115–120 days. Yield 200–250 q/ha	IARI, New Delhi

WATERMELON : *Contd.*

Variety	Year of release	Breeding method	Pedigree/ parentage	Important traits	Institutes/ universities
AHW 19	1998	Selection	Selection from local races	Early to mid maturity	CIAH, Bikaner
AHW 65	1998	Selection	Selection from local germplasm	Early maturing and high-yielding	CIAH, Bikaner
Thar Manak	2007	Selection	Selection from a cross between AHW 19 × Sugar Baby	Superior fruit quality, High-yielding than AHW 19	CIAH, Bikaner
PKM 1	1993	Selection	It is a selection from a local type	The skin is dark green with attractive pink flesh. Fruits are bigger in size with dark green skin and pinkish red flesh. Each fruit weighs 3–4 kg. It is suited to tropical plains of Tamil Nadu	HC&RI, TNAU, Periyakulam

be self-pollinated/cross-pollinated is protected by putting small screen cage over it or by wrapping it in unopened condition by cotton-rubber band. The staminate flowers are prepared by bending back the petals until they break followed by brushing the mass of sticky pollen against the stigmatic surface of the protected pistillate flower. For increasing fruit setting under controlled pollination, open-pollinated fruit should be removed as there is inhibitory effect produced by fruits already set.

ASH GOURD

Botanical name: *Benincasa hispida*
Family: Cucurbitaceae

Ash gourd is an annual vine trailing on the ground. It is grown in Andhra Pradesh, Tamil Nadu, Karnataka, Maharashtra and Kerala. It grows well in warm, humid tropical climate. The temperature of 22–35°C is ideal. Though a deep loamy soil is best suited, it can be grown on heavy clay as a rainfed crop. During summer, its crop can be grown in tank slopes. The optimum pH is 6.5–7.5. As the fruits develop, they become bigger in size and form an ashy coating on fruit surface. After full maturity the ashy bloom slowly drops off when the fruits become ready for harvesting 90–100 days after sowing. Fruits can be stored in a well-ventilated room for 4–5 months. The average yield is 250–300 q/ha.

CUCUMBER

Botanical name: *Cucumis sativus* L., 2n = 2x = 14
Family: Cucurbitaceae

Cucumber is one of the Asiatic species. It has 90 genera and 750 species. The African group has diploid chromosome number of 24. It is frost susceptible but grows well at temperature above 20°C. These are grown throughout the world to be consumed as fresh fruits, as slicing cucumber, and as pickles in immature stage. Cucumber is regarded the fourth most important vegetable crop after tomato, cabbage and onion. It probably originated in India from where it seems to have spread eastwards to China and westwards to Asia Minor, North Africa and Southern Europe and subsequently to entire Europe.

It is an annual plant species and is found to be day neutral. Under greenhouse 3 generations/year can be grown. Basically, it is monoecious, trailing or climbing vine with angled, hirsute or rough stems. Leaves are triangular-ovate, somewhat three lobed with mostly acute curves. The staminate flowers are in clusters with

ASH GOURD

Variety	Year of release	Breeding method	Pedigree/ parentage	Important traits	Institutes/ universities
CO 1	1971	Selection	Selection from a local type from Tamil Nadu	Fruits are globular, green, large oblong oval in shape, ashy coated, about 35 cm length and 22 cm in girth with a mean weight of about 6.8 kg. The flesh is white and thick and less seeded (0.8%)	TNAU, Coimbtore
CO 2	1984	Selection	Selection from Coimbatore local	Fruits are small, with average weight of 3 kg. Shape of the fruit is long spherical and flesh is green. Crop duration is about 120 days. Yield potential 230–250 q/ha	TNAU, Coimbtore
KAU Local	2001	Selection	BH 21	—	KAU, Vellannikaara
Indu	2001	Selection	AG 1	—	KAU, Vellannikaara
Pusa Ujwal	2004	Selection	Selection from Indigenous material DAG 1	Fruits oblong, ellipsoid, rind greenish white, flesh white, Average fruit weight 7.0 kg. Fruits are ideal for long distance transportation. Yield potential 480–500 q/ha (kharif season) and 410–425 q/ha (summer season)	IARI, New Delhi

ASH GOURD : *Contd.*

Variety	Year of release	Breeding method	Pedigree/ parentage	Important traits	Institutes/ universities
Kashi Ujjwal (IVAG 90)	2006	Selection	—	Fruits are globular, fruit weight is 10–12 kg, less-seeded, crop duration is 130–140 days. Yield potential 600–610 q/ha	IIVR, Varanasi
Kashi Dhawal (IVAG 502)	2006	Selection	—	Vine length 7.5–8 cm. Fruits are oblong with 11–12 kg weight. Flesh is white with 8.5–8.7 cm thickness and seed arrangement is linear, crop duration 120 days. Suitable for preparation of petha sweets. Yield potential 580–600 q/ha	IIVR, Varanasi

CUCUMBER

Variety	Year of release	Breeding method	Pedigree/ parentage	Important traits	Institutes/ universities
Japanese Long Green	1975	Introduction	—	Plants are early, prolific-bearing, first picking starts 45 days after sowing. Fruits are yellowish-green, 30–40 cm long, whitish-green with light green and crisp flesh. This variety essentially requires staking for straight fruits. Yield potential 150–275 q/ha	IARI, Katrain

CUCUMBER : *Contd.*

Variety	Year of release	Breeding method	Pedigree/ parentage	Important traits	Institutes/ universities
Straight eight	—	Introduction	Introduction from USA	Fruits medium in length, straight and cylindrical with round ends, skin medium green	IARI, New Delhi
Pusa Samyog	1972	Heterosis	Japanese Gyn. Line × Green Long Naples	Early-maturing and high-yielding, fruits long, cylindrical and dark green with yellow stripes, flesh crisp	IARI, New Delhi
Pusa Alankar	—	F_1 hybrid	Australian Green × Yellow Prolific	Very early and high yielding F_1 hybrid; fruits dark green with light coloured stripes, 20–25 cm long, slightly tapering towards the stem end, tender with delicious flesh	IARI, New Delhi
Pusa Uday	2005	Selection	Selection from Indigenous material	The plants (vines) are 1–1.5 m long. The flowering behaviour is monoecious. The fruits are medium in size (13–15 cm long), light green in colour with whitish green stripes (originating from blossom end and running up to one-third of the fruit length), straight, non-prickled and soft skinned. The flesh is very tender and sweet. The maturity period (days to first fruit harvesting) is about 48–52 days. The average yield is 15.6 tonnes/ha. It is suitable for cultivation in both spring-summer and rainy seasons in northern Indian plains.	IARI, New Delhi

CUCUMBER : *Contd.*

Variety	Year of release	Breeding method	Pedigree/ parentage	Important traits	Institutes/ universities
Solan Green	1975	Selection	—	Fruits are light green in colour, 11 to 15 cm long oblong shape yield 150–190 q/ha. Yield potential 180–200 q/ha	YSPUHF, Solan
Kalyanpur Green	1983	Selection	—	Fruits thick, green, tasty while brown at maturity. This variety is suitable for cultivation in rainy season. Yield potential 100–125 q/ha	CSAUAT, Kanpur
Sheetal	1989	Selection	—	Fruits are green, medium in length (25 cm), average weight 300–350 g and girth is 19.6 cm. The vine measures up to 2.7 m with 10 branches/vine. It bears 24 fruits/ plant and its duration is about 92 days. Sheetal can be raised from June to September, while the summer crop may be sown in January–February. Yield potential 380–400 q/ha in rainy season, and 200–230 q/ha in summer season	MPKV, Rahuri
Himangi	1992	Pedigree selection	Poinsett × Kalyanpur Ageti	Fruits white in colour, resistant to bronzing, average yield 190 q/ha	MPKV, Rahuri
Phule Shubhangi (Sel 75-1-10)	2001	Pedigree selection	—	Fruits green, colour of fruit remains after storage, surface smooth with trichomes. Plants are tolerant to pod-borer, anthracnose, leaf spot and leaf	MPKV, Rahuri

CUCUMBER : *Contd.*

Variety	Year of release	Breeding method	Pedigree/ parentage	Important traits	Institutes/ universities
				under field condition. Fruits are smooth and attractive green fruits with white stripes at apical end, fruit bigger in size. Yield potential 300–325 q/ha in 90 days	
PH 9	—	Hybrid	GYC-2 × Phule Shubhangi	Gynoecious line based hybrid, Fruits green, average weight 175 g, average yield 490 q/ha	MPKV, Rahuri
PH 18	—	Hybrid	GYC-4 × Poona Khira	Gynoecious line based hybrid, Fruits yellow, average weight 146 g	MPKV, Rahuri
Phule Priyanka	—	Hybrid	RHRB-5 × RHRBG-4	Fruits dark green, highly prickled, 20 cm long, suitable for rainy and summer seasons, tolerant to downy mildew, average yield 282.7 q/ha	MPKV, Rahuri
Gujarat Cucumber 1	2006	Selection	Local collection	Fruits are long, tender and attractive with light green stripes, soft and non-hairy with longer shelf-life. Yield potential 240–260 q/ha	GAU, Anand
Swarna Poorna (CH 20)	2004	Selection	Local collection	Fruits are cylindrical, long, medium-sized (300 g), light green without placental hollowness. Tolerant to powdery mildew. Fruit picking starts in 55–60 days	HARP, Ranchi

CUCUMBER : *Contd.*

Variety	Year of release	Breeding method	Pedigree/ parentage	Important traits	Institutes/ universities
				after sowing. Yield potential 310–335 q/ha in 100–120 days	
Swarna Ageti	2006	Selection	Local collection	This variety has been developed through mutation breeding at HARP, Ranchi. Fruits are cylindrical, long, medium-sized (200 g), green with no placental hollowness. Tolerant to powdery mildew. First picking starts 45–50 days after sowing. Yield potential 300–325 q/ha in 100–120 days	HARP, Ranchi
Swarna Sheetal	2006	Selection	Local collection	Fruits are cylindrical, long, medium in size (250 g), whitish green without placental hollowness. First picking starts in 60–65 days after sowing. Yield 400–450 q/ha in 100–120 days	HARP, Ranchi
Pant Khira 1	2006	Selection	Local collection	fruits are long (20 cm), cylindrical with light white stripes. The fruits attain first picking stage in 50–60 days. Recommended seed rate is 4 kg/ha. The optimum planting distance is 150 cm × 30 cm. This variety is suitable for planting from March–July in plains and April–May in hills. Yield 130–150 q/ha	GBPUAT, Pantnagar

CUCUMBER : *Contd.*

Variety	Year of release	Breeding method	Pedigree/ parentage	Important traits	Institutes/ universities
CO 1	1989	Selection	Selection from a local type of Kanyakumari District	The fruits are long (60–65 cm), slightly curved, tapering towards stalk end, fairly big and each weighs 2.0–3.0 kg at ripe stage. It has got a good consumer appeal and market preference	TNAU, Coimbatore

LONG MELON (*Cucumis melo* L. var *utilissimus*)

Variety	Year of release	Breeding method	Pedigree/ parentage	Important traits	Institutes/ universities
Arka Sheetal	1986	Selection	Pure line selection from a local collection, Lucknow	Fruits are medium long very tender, dark green, crisp, fleshy with crop duration of 100–110 days, average fruit weight is 90–100 g, free from bitter principle. Average fruit weight at marketable stage is 90–100 g. Yield potential 210–250 q/ha	IIHR, Bangalore
Punjab Long Melon 1 (H10)	1996	Selection	Local collection	Fruits long, thin and light green. Yield potential 200–220 q/ha	PAU, Ludhiana

short, slender pedicles. The pistillate flowers are usually solitary with stout, short pedicles. The calyx and corolla of staminate, pistillate and even hermaphrodite flowers are five-lobed. The staminate flowers have three stamens. Two stamens have two locules each and the third is unilocular. Filaments are free, but the stamens are more or less united by their anthers. The pistillate flowers are epigynous and hermaphrodite flowers are perigynous. The pistil consists of from one to five (but usually three) carpels which in turn, produce ovaries with a corresponding number of locules. The pistillate flowers contain up to five stigmas. The main stem of monoecious cucumber is usually characterized by three phases of sex expressions. Only staminate flowers are produced in the first phase followed by a phase of irregularly alternating female, male or mixed nodes and finally a phase of only pistillate flowers. Lateral shoots of monoecious cultivars usually have stronger female tendencies. Fruits from perigynous flowers are more rounded as opposed to elongated ones. The rounded fruits are horticulturally poor due to large and seed cavity.

For crossing purpose, pistillate flowers are closed with a rubber band/wrapped with cotton pad to protect against unwanted pollen about 1–2 days prior to opening. Anthesis takes place around 5.30–7.00 h. Dehiscence occurs around 4.30–5.00 h. Pollen fertility is up to 14 h. Similarly male flowers are also protected. After 1–2 days, pollination is carried out from pollen of protected staminate flower preferably in morning/forenoon. After about 4 weeks, its fruits are harvested and allowed to further ripening by about another week. For seed collection, fruits are cut-open and seeds are collected in a glass jar with water and left as such for two days to remove gelatinous mass. Then they are washed in a sieve and dried. Seeds should not be left in water for more than two days otherwise they may start germinating. Sixty days after pollination are usually preferred for optimum maturity of seeds. The success of controlled pollination may be enhanced by removal of any previously set fruit as first fertilized flower inhibits the development of subsequent fruits. Therefore, controlled pollination should be done as soon as possible after flowering begins.

POINTED GOURD

Botanical name: *Trichosanthes dioica*
Family: Cucurbitaceae

Pointed gourd or *parwal* is widely cultivated in Bihar, Uttar Pradesh, West Bengal, Orissa, Assam, Madhya Pradesh and Gujarat. In Andhra Pradesh and Tamil Nadu, it is grown in small pockets of hilly tracts.

POINTED GOURD

Variety	Year of release	Breeding method	Pedigree/ parentage	Important traits	Institutes/ universities
Rajendra Parwal 1	1994	Clonal selection	—	Fruits are long green with white strips and tapering at both ends. Average fruit weight is 40 g. Suitable for long distance transportation and tolerant to fruit fly. Yield potential 175–190 q/ha	RAU, Pusa
Rajendra Parwal 2	1994	Clonal selection	—	Fruits are drum-shaped, whitish-green with very light self strip and soft. Average fruit weight is 30 g. Tolerant to vine and fruit rot as well as fruit fly. Yield potential 160–180 q/ha	RAU, Pusa
Swarna Rekha	2006	Clonal selection	—	Recommended for transplant in February (rooted plant) and September (vine cutting). Fruits are elongated in shape, striped green, contain soft seed. Yield potential 175–200 q/ha	HARP, Ranchi
Swarna Alaukik	2006	Clonal selection	—	Fruit are elongated, light greenish, suitable for sweet preparation. Yield potential 230–250 q/ha	HARP, Ranchi

IVY GOURD (*Coccinia indica*)

Variety	Year of release	Breeding method	Pedigree/ parentage	Important traits	Institutes/ universities
Indira Kundru 35	2006	Clonal selection	—	Fruits are long and light green in colour having 6.0 cm fruit length and 2.43 cm fruit diameter. It is high-yielding variety, producing about 22.0 kg fruits/plant. Yield potential 410–450 q/ha	IGKV, Raipur
Indira Kundru 05	2006	Clonal selection	—	Fruits are light green and oval-shaped (4.30 cm fruit length and 2.63 cm fruit diameter). It is high-yielding variety, producing about 21.08 kg fruits/plant. Yield potential 400–420 q/ha	IGKV, Raipur
Sulabha	2006	Clonal selection	—	Fruits are long (9.25 cm), pale green, with average fruit weight is 18.48. The variety gives an average yield of 60 tonnes/ha. Is a clonal selection of CG-23. It takes 37 days from planting to first flowering. The first harvesting can be done 45–50 days after planting. Average fruit length is 9.25 cm, and fruit weight 18.48 g. Fruit shape is cylindrical. Fruit colour is light green with continuous striations. Yield potential 400–425 q/ha	KAU, Vellanikkara

Recently, it has been introduced in and around Hyderabad and Bangalore. *Parwal*, a semi-perennial creeper, remains dormant during winter. Fruits are globose, oblong, smooth, 5–12 cm long and 2–6 cm in breadth, striped, stripes are light green on young fruits and seeds are globose. Several cultivated *parwal* varieties differ in their size, shape and markings on fruits. Two more important forms are: one with large, oblong, deep green fruits, with longitudinal and somewhat obscure white bands and the other with shorter, thicker, pale-green fruits without markings. These are further classified into 4 groups depending on fruit size.

Its fruits in flushes are available from February to October. Vine starts bearing fruits 5 months after planting. Picking of fruits at weekly intervals gives more yield. Harvesting of fruits should be done 15–18 days after fruit setting. Delay in harvesting leads to development of hard seeds. Its yield varies from place-to-place depending on soil and climatic conditions. The fruit yield increases subsequently. It declines fourth year onwards.

RIDGED GOURD

Botanical name: *Luffa acutangula*
Family: Cucurbitaceae

Ridged gourd or ribbed gourd is also a monoecious viny vegetable. It is mainly cultivated in Andhra Pradesh, Tamil Nadu, Karnataka, Gujarat, Assam, West Bengal and Konkan region of Maharashtra. It grows very well in a warm hot climate, the optimum temperature being 25–30°C. Very high temperature especially in the early crop growth stage (more than 38°C) helps produce more male flowers, reducing the yield. Very low temperature also affects growth of vines. Sandy loam soil, rich in organic-matter content is most-suited for higher yield. Proper drainage is highly beneficial.

SPONGE GOURD

Botanical name: *Luffa cylindrica*
Family: Cucurbitaceae

Sponge gourd is a monoecious cucurbit vegetable. It is closely related to ridged gourd. Its tender fruits have smooth surface and contain vitamin A and C. Fully ripe, dried fruits having a large volume of fibrous portion are used for cleansing utensils, making shoe-soles and filters. Fibre is also used to manufacture table and bathroom mats. Sponge gourd can be grown from tropical to subtropical climates. Warm humid conditions favour its cultivation. Very low temperature is deleterious and frost kills its plants. Well-

RIDGE GOURD

Variety	Year of release	Breeding method	Pedigree/ parentage	Important traits	Institutes/ universities
CO 1	1976	Selection	Selection from the type collected from Aduthurai	Plants are vigorous, bear 10–12 fruits/ vine (3–4 kg). Fruits are 60–70 cm long with 3.0 cm diameter. Duration of crop is 125 days. Yield potential 210–150 q/ha	TNAU, Coimbatore
CO 2	1984	Selection	Selection from a germplasm	Each vine on an average produces 8–10 fruits weighing 9 to 10 kg. The fruits are green, very long (1 m) and fleshy. Yield potential 250–270 q/ha in 140–150 days	TNAU, Coimbatore
PKM 1	1980	Mutation breeding	H-160	The fruits are dark green 60–70 cm long with shallow grooves. They are characterized by very broad terminal than the basal end. The plants are tolerant to pumpkin beetle, fruit fly and leaf spot. Yield potential 250–280 q/ha in 160 days	TNAU, Coimbatore
Arka Sujat	1996	Pedigree selection	IIHR 54 × IIHR 18	Fruits are lush-green and tender, medium-long (35–45 cm) with an average fruit weight of 350 g. The variety has good transport and keeping quality. Yield potential 625–640 q/ha in 100–120 days	IIHR, Bangalore

RIDGE GOURD : *Contd.*

Variety	Year of release	Breeding method	Pedigree/ parentage	Important traits	Institutes/ universities
Arka Sumeet	2001	Pedigree selection	IIHR 54 × IIHR 18	Fruits are 55 cm long with 2.5 cm girth with prominent ridges and delicate aroma. Average fruit weight 380 g, crop duration 120 days. Fruits are lush-green, cylindrical with slow seed maturity. Yield potential 200–250 q/ha in 140–150 days	IIHR, Bangalore
Swarna Manjari	2006	Selection	—	Fruits are elongated, medium-sized, highly ridged, green and soft pulps contain less fiber. Plants are tolerant to powdery mildew. First picking starts 65–70 days after sowing. Yield potential 180–200 q/ha in 140–150 days	HARP, Ranchi
Swarna Uphar	2006	Selection	—	Fruits are elongate, medium-sized (200 g), weak ridge with soft pulp and less fiber. First picking starts 65–70 days after sowing. Yield potential 280–310 q/ha in 140–150 days	HARP, Ranchi
Haritha	2001	—	—	—	KAU, Vellanikkara
Deepthi	2006	Selection	—	Plants have green ovate intermediate sized leaves. First can be harvested in 63 days. Fruits are green, medium-sized	KAU, Vellanikkara

RIDGE GOURD : *Contd.*

Variety	Year of release	Breeding method	Pedigree/ parentage	Important traits	Institutes/ universities
				with finely wrinkled surface. Average fruit weight is 165 g. Plants are resistant to mosaic and downy mildew under field condition. Yield potential 300–325 q/ha	
Pusa Nasdar	—	Selection	Selection from Indigenous material	Mid season; flowering in 50 days; 15–20 fruits per vine, fruits ridged, light green. More suited for rainy season	IARI, New Delhi

SPONGE GOURD

Variety	Year of release	Breeding method	Pedigree/ parentage	Important traits	Institutes/ universities
Pusa Chikni	1975	Selection	Selection from Indigenous material	Fruits are smooth, dark green, and cylindrical. Single plant bears 15–17 fruits. Flowering starts 60 days after sowing, suitable for summer and rainy season cultivation. Yield potential 200–225 q/ha	IARI, New Delhi

SPONGE GOURD : *Contd.*

Variety	Year of release	Breeding method	Pedigree/ parentage	Important traits	Institutes/ universities
Pusa Supriya	1997	Selection	Selection from Indigenous material	Fruits are pale-green, smooth, 15–20 cm long, straight and slightly curved at the stem end, pointed distal-end, non-hairy, tender flesh, suitable for spring summer and *kharif* season. Fruits become ready for picking 50–55 days after sowing in spring-summer and 44–48 days after in *kharif* season. Yield potential 130–140 q/ha	IARI, New Delhi
Pusa Sneha	2004	Selection	Selection from Indigenous material	Fruits are attractive, dark green, 20–25 cm long, straight, peduncle long, smooth, non-hairy, flesh tender, skin hard suitable for long distance transportation. This variety is suitable for growing in both spring in both spring-summer and *kharif* seasons. Fruits become ready for first picking 50–55 days after sowing in spring summer and 40–45 days in *kharif* season. Yield potential 200–230 q/ha	IARI, New Delhi
Kalyanpur Hari Chikni	1983	Selection	—	It is an early variety, fruits long, smooth, dark-green, distal end pointed, suitable for rainy season. Yield potential 175 to 200 q/ha	CSAUAT, Kanpur

SPONGE GOURD : *Contd.*

Variety	Year of release	Breeding method	Pedigree/ parentage	Important traits	Institutes/ universities
Azad Toria Chikni 1	2006	Selection	—	It is an early-maturing variety. Fruits green, smooth, suitable for rainy season. Yield potential 180 to 200 q/ha.	CSAUAT, Kanpur
Phule Prajakta	2001	Selection	—	Fruits medium-sized, green, straight, slender, tapering at ends and flesh colour-pure white. Yield potential 130–150 q/ha in 145–150 days of crop duration	MPKV, Rahuri
Gujarat Sponge Gourd 1	2005	Selection	—	Fruits are medium-sized, light green in colour. Plants are moderately tolerant to downey mildew and leaf miner. Yield potential 120–140 q/ha	GAU, Anand
Swarna Prabha	2006	Selection	—	Fruit are medium in length (20–25 cm), weight (150–200 g), light greenish, pulp soft and contain less fibre. First harvesting can be taken 70–75 days after planting. Yield potential 200–250 q/ha	HARP, Ranchi

ROUND GOURD (TINDA) *Praecitrullus fistulosus*

Variety	Year of release	Breeding method	Pedigree/ parentage	Important traits	Institutes/ universities
Punjab Tinda (S 48)	1978	Selection	Local cultivar	Fruits are medium-size, and the fruit surface is shining and pubescent. Flesh is white, less-seeded, tender and has good cooking quality. It takes about 60 days from sowing to marketable maturity. Yield potential 45–50 q/ha	PAU, Ludhiana
Arka Tinda	1986	Pedigree selection	T-3-4 × T-8-2	It is an early summer season variety. Fruits are round, light green with soft hair and tender flesh at marketable stage. Plants are tolerant to fruit fly. Yield potential 95–100 q/ha in 90–100 days	IIHR, Bangalore
Hisar Tinda	2006	Selection	From the local cultivar	It is an early-maturing and high-yielding variety. Fruits round, medium-sized and tender. Yield potential 75–100 q/ha	HAU, Hisar

drained loamy soils are ideally-suited though it can be grown on a wide variety of soils. Its fruits should be harvested at tender stage. If allowed to develop fully, they become spongy or fibrous which are unfit for vegetable purpose. On an average, its yield is 150–200 q/ha.

SNAKE GOURD

Botanical name: *Trichosanthes cucumerina*
Family: Cucurbitaceae

Snake gourd is an annual, climbing vine-providing both long and short fruits. India is its native home. It prefers tropical, warm and humid climate. It can be grown successfully up to 1,500 m above mean sea-level. It cannot tolerate frost and grows very well at 25–30°C. Heavy rains during flowering affect its yield. A loamy soil with a pH of 6.0–7.0 is ideal for its cultivation. Its fruits are ready for picking 70–80 days after sowing. Fully-developed and tender fruits are harvested once in 5–7 days for vegetable purpose. If allowed to mature fully, fruits ripen within 2–3 days of harvesting. Its average yield is 180–200 q/ha.

CRUCIFERS AND TEMPERATE VEGETABLES

CABBAGE

Botanical name: *Brassica oleracea* subsp. *capitata*, 2n = 2x = 18
Family: Brassicaceae

Cabbage is the most important vegetable grown throughout the world. Modern hard-head cabbage cultivars originated from wild non-heading brassicas somewhere in the eastern Mediterranean and Asia Minor region. A cabbage flower has four sepals, six stamens in tertadynamous condition (two short and four long stamens) and a bicarpellary ovary which is superior and has a false septum. Ovules are attached on both the sides of septum. Two active nectarines are located between the bases of short stamens and ovary. The buds open under pressure of rapidly growing petals and become fully expanded in about 12 h. Flowers are slightly protogynous and cabbage is naturally cross-pollinated due to sporophytic self-incompatibility. Pollination is brought about by the bees and flies. Bud pollination is effective to achieve selfing. For cross-pollination flower buds expected to open within 1–2 days are emasculated and are pollinated immediately with desired pollen using a brush/flower stamens.

SNAKE GOURD

Variety	Year of release	Breeding method	Pedigree/ parentage	Important traits	Institutes/ universities
CO 1	1976	Pure line selection	Local cultivar collected from Alangulam, Tirunelveli district	Fruits are long (160–180 cm), dark green with white strips and flesh whitish green. It is an early-maturing, vines moderately vigorous and medium spreading, weighing 500–750 g on an average. The flesh is smooth, light green and with good cooking quality. Yield potential 180–190 q/ha in 135 days	TNAU, Coimbatore
CO 2	1986	Selection	Pureline selection from a local type of Coimbatore district	The vines are less spreading and hence seeds can be sown at close spacing of 1.5 m × 1.5 m. The fruits are short and stout, light greenish white weighs 400–600 g. Yield potential 350–370 q/ha in 105 days	TNAU, Coimbatore
MDU 1	1981	Heterosis	F_1 hybrid between Panripudal and Selection	It is an easily flowering type (84 days) with a sex ratio of 1:38. It produces 13 fruits per vine weighing 7.15 kg with an average yield of 31.75 tonnes/ha in 145 days. The fruits are medium long (66.94 cm) with white stripes under green background. Each fruit on an average weighs 55 g	AC&RI, TNAU, Madurai
PKM 1	1979	Mutation breeding	Induced mutant from H.375	Fruit is dark green in colour with white stripes on outer side and light green on inside, with a mean fruit weight of 700 g. The fruits are extra	TNAU, Coimbatore

SNAKE GOURD : *Contd.*

Variety	Year of release	Breeding method	Pedigree/ parentage	Important traits	Institutes/ universities
				long (180–200 cm). Fruits are of small (45.65 g) with green stripes. Yield potential 255–270 q/ha in 145 days	
Konkan Shweta	1994	Selection	—	Fruits are medium-long (90–100 cm) and white in colour. Fruits have good flesh if harvested timely, otherwise it becomes hollow. Yield potential 150–200 q/ha in 120–130 days	KKV, Dapoli
Kaumudi	1996	Selection	Local selection	—	KAU, Vellanikkara
Baby	2006	Selection	—	Fruits are small, attractive white, suitable for distant marketing. Average fruit weight is 475 g with length and width of 36.5 and 22 cm respectively. This is recommended for the cultivation in Kerala. Yield potential 390–420 q/ha	KAU, Vellanikkara
Manushree	2006	Selection	—	Fruits are ready for harvesting in eighth week. Fruits are white attractive, uniform medium long (65–70 cm), green markings at pedicel end. Average fruit weight is 750 g with 67 cm length. Yield potential 575–620 q/ha	KAU, Vellanikkara

SNAP MELON (*Cucumis melo* var. *momordica*)

Variety	Year of release	Breeding method	Pedigree/ parentage	Important traits	Institutes/ universities
Pusa Shandar	2004	Selection	Selection from Indigenous material	Fruits are oblong, cylindrical, average weight 700 g, skin creamy white with light pink flesh. Its yield potential is 400–425 q/ha in summer, and 370–380 q/ha in rainy season	IARI, New Delhi
AHS 10	1998	Selection	Selection from local land races of arid region	High-yielding, 225–230 q/ha	CIAH, Bikaner
AHS 82	1998	Selection	Selection from local germplasm	High-yielding, 245–250 q/ha	CIAH, Bikaner

KACHARI (*Cucumis callosus*)

Variety	Year of release	Breeding method	Pedigree/ parentage	Important traits	Institutes/ universities
AHK 119	1998	Selection	Selection from land races	Early, high-yielding, 95–100 q/ha	CIAH, Bikaner
AHK 200	1998	Selection	Selection from land races of arid region	Early, high-yielding, 115–120 q/ha	CIAH, Bikaner

KAKDI (*Cucumis* spp.)

Variety	Year of release	Breeding method	Pedigree/ parentage	Important traits	Institutes/ universities
AHC 2	1998	Selection	Selection from local material known as 'Arya'	Yield 175–202 q/ha	CIAH, Bikaner
AHC 13	1998	Selection	Selection from local germplasm	Yield 85–125 q/ha	CIAH, Bikaner

CABBAGE

Variety	Year of release	Breeding method	Pedigree/ parentage	Important traits	Institutes/ universities
Golden Acre	1970	Selection	Copenhagen Market	It takes about 60–65 days from *trans*-planting to head formation. It has fewer outer leaves, which are cup-shaped and arranged in two whorls. The heads are solid, short core and weighing 1.0–1.5 kg. The heads should be harvested immediately after head formation otherwise cracking may take place. The variety is highly suitable for spring and summer growing in hills. Its yield potential is 175–200 q/ha	IARI, RS, Katrain
Pusa Drum Head	1970	Selection	Drum Head Group	Plants have wider frame with 20–25 light green outer leaves with prominent midribs and venation. Heads are solid and flat, weighing 3–4 kg. It takes about 80–90 days from transplanting to head formation. Plants are field resistant to black leg. Its yield potential is 220–210 q/ha	IARI, RS, Katrain
Early Drum Head	1978	Introduction	—	Tit is an early variety having light green foliage. Leaves are medium to large, curving inward and enclosing the head loosely. Leaves forming the head fold over each other at the centre. The	IARI, RS, Katrain

CABBAGE : *Contd.*

Variety	Year of release	Breeding method	Pedigree/ parentage	Important traits	Institutes/ universities
				heads are flat. Its yield potential is 180–200 q/ha	
Copenhagen (Market)	1978	Introduction	—	This is still being grown but on a limited area. This is an introduction, which has been replaced by Golden Acre and 'Pride of India' because of their earliness and smaller head size. Its yield potential is 175–200 q/ha	IARI, RS, Katrain
Pusa Mukta	1989	Pedigree selection	EC 24855 × EC 10109	Heads are compact, slightly flatish-round with loose wrapper leaf at the top, 1.5–2.0 kg in weight. Average head weight is 1.5–2.0 kg. It is resistant to black rot (*Xanthomonas campestris*). This variety is recommended for cultivation in black rot prone areas. Its yield potential is 200–225 q/ha	IARI, RS, Katrain
Pusa Ageti	2000	Selection	Recurrent selection from TKCBH-28	Heat tolerant variety which can produce seeds in subtropical climate	IARI, New Delhi
Pride of India	1973	Introduction	—	The plant type of this variety is similar to 'Golden Acre' but it is about a week later in maturity. It has bigger sized heads weighing about 1.5 to 2.0 kg. Yield potential 215–240 q/ha	YSPUHF, Solan

CABBAGE : *Contd.*

Variety	Year of release	Breeding method	Pedigree/ parentage	Important traits	Institutes/ universities
PUCH 3 (Palampur Green)	1996	Mass selection	—	Leaves are broad, Smooth and dark green with prominent white and tender petiole. Its yield potential is 380–400 q/ha	HPKV, Palampur
Kinner Red (Red Cabbage)	1999	Mass selection	—	Head round at top and oval at the base of purple colour with red pigmentation leaves upright clasping round the head. Its yield potential is 250–275 q/ha	HPKV, Palampur

CAULIFLOWER

Botanical name: *Brassica oleracea* L. var *botrytis*, 2n = 2x = 18
Family: Brassicaceae

Cauliflower is grown both in hills and plains and from 11°N to 35°N. Important cauliflower-growing states are Karnataka, West Bengal, Punjab and Bihar. It is also commonly grown in northern Himalayas and in the Niligiri hills in south. Cauliflower is harvested from August or early-September to late-February or early-March in north Indian plains and from March to November in hills. Cauliflower has been domesticated in the Mediterranean region since there is greatest variability in wild types of *B. oleracea*. It originated in Island of Cyprus from where it moved to Syria, Turkey Egypt, Italy, Spain and north-western Europe. The cole crops, including cauliflower and cabbage, have descended from a common kale like ancestor, the wild cabbage (*B. oleracea* L. var. *sylenstris* L.) which is still found in Western and Southern Europe and North Africa.

The stem of vegetative cauliflower plant is rather short, and it thickens to about the same extent as that of cabbage. The leaves are large, generally oblong, the younger ones being nearly always sessile. In contrast to situation in other cole crops, buds do not usually arise in the leaf axils. During transition to generative phase, which, in contrast to the situation in most other cole crops, is accomplished in many types of cauliflower during first year, peduncles in axils of bracts formed by the main growing point branch repeatedly, so that branches even of the fifth order can arise. At first, numerous peduncles do not grow lengthwise but become thick and fleshy. Thus a colourless, roughly spherical, terminal compact head arises of which the upper surface consists of vast numbers of naked apical meristems. The peduncles are composed of thin-walled parenchyma and vascular bundles which do not become woody. The young 'curd' is at first entirely covered by the foliage, on becoming visible it already has a diameter of over 5 cm. After sometime the flower stems elongate, and a number of the apices develop into normal flowers. Soon after floral primordia appear some of the second order branches bearing them elongate leaving other normal non-elongated second order branches behind. These later bear lateral floral primordial which abort. Later several third order branches with floral primordial around their apex elongate leaving behind rest of the third order branches with abortive floral primordia. These processes of elongation and abortion may be repeated till floral primordial develop into normal flowers. There are 4 sepals, 4 petals, 6 stamens, and two carpels. The carpels form a superior ovary with false septum and two rows of campylotropous ovules. The androecium is tetradynamous *i.e.* there

CAULIFLOWER

Variety	Year of release	Breeding method	Pedigree/ parentage	Important traits	Institutes/ universities
Pusa Snow-ball 16	1970	Mass selection	Exotic collection	Late, curd white, solid self blanched. Its yield potential is 150–200 q/ha	IARI, RS, Katrain
Pusa Katki	1970	Mass selection	—	This is one of the earliest variety which gets matured in October–November and having medium plants, bluish green, waxy leaves. It is suitable for sowing during mid-May. Late planting beyond middle of August does not give good size curds. Its yield potential is 130–140 q/ha	IARI, New Delhi
Pusa Deepali	1978	Mass selection	Open pollinated progeny selection from Indigenous collection	Leaves are erect, short, green and waxy. Curds are compact, self-blanching, white and medium in size. Uniform curds are well-protected by leaves and resulting no riceyness. Curds are ready in late-October when the average temperature is around 20–25°C. Since it is recommended for sowing from May end to early June. This is recommended for general cultivation in states of Delhi and Punjab. Its yield potential is 150–175 q/ha	IARI, New Delhi
Pusa Snowball 1	1978	Mass selection	EC 12013 × EC 12012	It is a late variety suitable for cool season. The optimum average	IARI, RS, Katrain

CAULIFLOWER : ***Contd.***

Variety	Year of release	Breeding method	Pedigree/ parentage	Important traits	Institutes/ universities
				temperatures for curd initiation and development are 10–16°C. Its sowing time in north India is from mid-September to October end. Leaves are straight, upright and inner leaves tightly cover the curd. Curds are very compact, medium in size and snow white. Its yield potential is 140–160 q/ha	
Pusa Snowball 2	1978	Mass selection	Selection from EC 12012	Its outer leaves are upright while inner leaves cover the curd initially. The curd remains white even on exposure. Curds become ready by the end of January end or early-February (10–16°C). This variety could not become popular because of its poor seeding ability. Its yield potential is 170–190 q/ha	IARI, RS, Katrain
Pusa Shubra	1985	Mass selection	MGS2-3	Its plants are erect with long stalk and bluish green leaves. Curd compact and white, curd weight 700–800 g. Plants are highly tolerant to riceyness. It takes 125–130 days for 50% curd and best temperature for curding is 12–16°C. Its yield potential is 250–300 q/ha	IARI, New Delhi

CAULIFLOWER : *Contd.*

Variety	Year of release	Breeding method	Pedigree/ parentage	Important traits	Institutes/ universities
Pusa Snowball K 1	1988	Mass selection	Selection from EC 12012	Plants are tolerant to black rot *Xanthomonas campestris*. Plants bear best quality curds which are snow white in colour and retain it even if the harvesting is delayed. Leaves are puckered, serrated and light green, harvesting; tolerant to black rot. Its yield potential is 200–225 q/ha	IARI, RS, Katrain
Pusa Early Synthetic	1993	Hybridization (synthetic)	Synthesis from six inbreds	Plants are erect with bluish green leaves. Curds are small-medium, flat, creamy white and compact; suitable for early planting; resistant to *riceyness*. Its yield potential is 120–150 q/ha	IARI, New Delhi
Pusa Hybrid 2	1994	Heterosis	CC × 18-19	Plants are semi-erect with bluish-green upright leaves, curds are creamy-white, highly compact, average weight 907 g; matures from mid-November to mid-December in north Indian plains. Plants are resistant to downy mildew. Its yield potential is 250–270 q/ha	IARI, New Delhi

CAULIFLOWER : *Contd.*

Variety	Year of release	Breeding method	Pedigree/ parentage	Important traits	Institutes/ universities
Pusa Snowball K 25	2004	Pedigree method	Mass pedigree method of selection from cross of heading broccoli (EC 10376) and Pusa Snowball	High-yielding, resistant to black rot and tolerant to *Sclerotia* rot, late maturing.	IARI, RS, Katrain
Pusa Sharad	2004	Mass selection	OP (Selection from Indigenous collection) (Sel-309-1-2)	Plants are semi-erect and open type with small stalk. Leaves are glabrous, bluish-green, long petiolate, oblong with narrow apex, wavy margin, prominent mid-rib and ear-like lobes at the base of lamina. Its marketable curd becomes ready 85 days after transplanting. Curds are white, knobby, very compact, semi-dome shaped and about 750–1000 g in weight. Its yield potential is 260–280 q/ha	IARI, New Delhi
Pusa Meghna	2004	Mass selection	Open pollinated (Selection from Indigenous collection) DC/98-2	Curds are retentive white, compact and medium size (400 g). Curds become ready for harvesting after 95 days of seed sowing. Its yield potential is 125–150 q/ha	IARI, New Delhi
Pusa Himjyoti	1986	Recurrent selection	Local selection from Ganganagar	Plants with erect bluish green leaves having waxy coating, self blanched; retentive white curds, 500–600g in	IARI, Katrain

CAULIFLOWER : *Contd.*

Variety	Year of release	Breeding method	Pedigree/ parentage	Important traits	Institutes/ universities
				weight; takes 60–75 days to harvesting. Only variety suitable from April to July sowings in the hills above 1,000 m height. Yield, 160 q/ha	
Pant Gobhi 4	1995	Mass selection	235-S	This is a November maturity group variety derived through simple recurrent selection from a local variety (Aghani). Plants have medium-long stem and sparse semi-erect leaves with 15 cm of stalk. Curds are semi-spherical, creamish white, medium, compact and non-ricey. Its takes about 115–120 days for curd maturity. Its yield potential is 150–170 q/ha	GBPUAT, Pantnagar
Early Kuwari	1973	Mass selection	—	Leaves are bluish green, and wavy with waxy bloom. Curds are semi-spherical with even surface; curd availability period is mid-September to mid-October from May sown crop. It is ready for harvesting 100–200 days after sowing. This variety has been recommended for cultivation in states of Punjab, Haryana, Himachal Pradesh and Delhi. Its yield potential is 160–180 q/ha	PAU, Ludhiana

CAULIFLOWER : ***Contd.***

Variety	Year of release	Breeding method	Pedigree/ parentage	Important traits	Institutes/ universities
Punjab Giant 26	1978	—	—	Suitable for main season, curd large, average yield 110 q/ha	PAU, Ludhiana
Hisar 1	1976	Mass selection	—	It bears medium to large-sized white heads. It is suitable for mid-late season under Haryana conditions. Its yield potential is 250–275 q/ha	HAU, Hisar
Kashi Kuwanri (IVCE 2)	2006	Mass selection	—	This variety is identified under early maturity group. Its sowing time is June-end to July-under Indian plain condition. It can tolerate high rainfall during its vegetative growth. Curds are semi-dome type, white compact, texture fine and curd weight about 300–450 g	IIVR, Varanasi
Ooty 1	1998	Selection	It is a selection from open pollinated progenies of local types	The crop can be raised throughout the year. The curds are attractive creamy white in colour, uniform and big-sized. Each curd weighs 2.6 kg and is free from riceyness. Its yield potential is 46.4 tonnes/ha in 120 days	HRS, TNAU, Ooty

are two short and four long stamens. The pollen-grains are 30–40 μ in diameter and have three germination pores. The bright yellow petals become 15–25 mm long and about 10 mm wide. The sepals are erect.

The buds open under the pressure of rapidly growing petals. This process starts in the afternoon, and usually the flowers become fully expanded during the following morning. The anthers open a few hours later, flowers being slightly protogynous. The flowers are pollinated by insects, particularly bees, which collect pollen and nectar. The nectar is secreted by two nectarines situated between the basis of the short stamens and the ovary. Situated outside the basis of the pairs of long stamens are also two nectarines, but these are not active. Flowers are borne in racemes on the main stem and its branches. The inflorescence may attain a length of 1–2 m, but the slender pedicles are only 1.5–2 cm long. The fruits are glabrous siliques, 4–5 mm wide and sometimes over 10 cm long, with two rows of seeds lying along the edges of the replum (false septum, an outgrowth of the placentae). A silique contains from ten to thirty seeds. Three to four weeks after the opening of the flower from which it is formed, the silique reaches its maximum length. When it is ripe, dehiscence takes place through the two valves breaking away from below upwards, leaving the seeds attached to the placentas.

BROCCOLI

Botanical name: *Brassica oleracea* var. *Italica*
Family: Brassicaceae

Broccoli is of 2 types, heading and purple or green sprouting. Sprouting broccoli is more popular in India. Heading broccoli forms curds like cauliflower, while sprouting broccoli contains a group of green, immature buds and thick fleshy flower stalk forming a head. In India, its cultivation is negligible but now it is becoming increasingly popular in hotels in Mumbai, Kolkata, Delhi and Chennai. It is mostly cultivated in hilly areas of Himachal Pradesh, Uttar Pradesh, Jammu & Kashmir, Nilgiri hills and northern plains.

It is a cool season crop resistant to mild frost. The temperature of 20–25°C is optimum for its proper growth, while 15–20°C for heading stage. The heads become loose with rise in temperature. Broccoli can be grown in a wide variety of soils but deep loamy soil is best-suited. Soil should be well-drained and sufficiently fertilized. Broccoli requires moist soil for fast and proper growth. The shoots become more fibrous under dry soil. The pH of 5.0–6.5 is optimum.

BROCCOLI

Variety	Year of release	Breeding method	Pedigree/ parentage	Important traits	Institutes/ universities
Pusa Broccoli Kt 1	1996	Recurrent selection	Selection from exotic material	Compact head, light green with small buds weighing 250–400 g, maturity 85–95 days, yield 125 q/ha	IARI, New Delhi
Palam Samridhi	1997	Mass selection	—	Green sprouting broccoli is tender, fresh and full of flavour. It can be consumed raw as salad, in cooked form and also be pickled. Terminal head weight 300–400 g	HPKV, Palampur
Palam Haritia	2005	Mass selection	—	It is a sprouting broccoli with dark green upright leaves having purple reddish tinge. The heads attain marketable stage in about 145–150 days. Its yield potential is 230–250q/ha	HPKV, Palampur
Palam Vichitra	2005	Mass selection	—	It is a heading broccoli, medium-sized, open dark green leaves. Purple tinge is prominent on stem at seedling stage and on leaf margins at full growing stage. The head is purple and compact. The variety is rich in nutrients especially vitamins and minerals with high cosmetic appeal. The heads attain marketable size in 115–120 days. Its yield potential is 230–260 q/ha	HPKV, Palampur

BROCCOLI : *Contd.*

Variety	Year of release	Breeding method	Pedigree/ parentage	Important traits	Institutes/ universities
Palam Kanchan	2005	Mass selection	—	It is a heading broccoli with long, broad, bluish green upright leaves having prominent white midrib and veins. The head is large in size, compact, attractive yellowish-green in colour. It is rich in vitamin A. The heads attain marketable size in 140–145 days. Its yield potential is 250–275 q/ha	HPKV, Palampur
Punjab Broccoli 1	1998	—	—	Sprouts compact, attractive and succulent, suitable for salad and cooking, average yield 70 q/ha	PAU, Ludhiana

KNOL KHOL

Botanical name: *Brassica oleracea* var. *gongylodes*, 2n = 2x = 18
Family: Brassicaceae

Knol khol, kohlrabi is a herbaceous vegetable crop grown on a limited scale. Its breeding system is similar to that of other *B. oleracea* but self-incompatibility level is intermediate between that of sprouting broccoli and Brussels' sprouts. The spherical root is formed from a swollen stem to which leaf bases are attached. It has been described by the writes in ancient Rome. It developed in northern Europe in the 15th century.

The swollen edible portion is stem. Leaves are attached on this bulb like swollen structure. The conducting vessels which supply leaves often form tough inedible tissue. In young kohl rabi seedling the stem above the epicotyl shows primary growth in thickness, while growth in length is inhibited. This results in the development of a aerial tuber, borne on a short stem which consists of hypocotyl and some unthickened internodes. The secondary growth in thickness is restricted mainly to unthickened basal parts of stem. Usually petiolate and rather small leaves are arranged in a compressed spiral on bulbous part. The tubers are green or violet and generally round to flat round. There are also types with more elongated tubers, which are transitional to marrow-stem kale. It is biennial and plants bolt at low temperature. Therefore, breeding work is possible only under temperate conditions, for example, IARI, Regional Station, Katrain in Himachal Pradesh. Flower structure and pollination details are essentially the same as for other Brassicas.

CARROT

Botanical name: *Daucus carota* L., 2n = 2x = 18
Family: Umbelliferae

Carrot is widely grown both for fresh consumption and processing. It is a rich source of high carotene content (provitamin A) and has anticarcinogenic effect due to carotene. The genus *Daucus* has many wild forms that grow mostly in the Mediterranean region and south-west Asia. Afghanistan is believed to be the primary centre of genetic diversity. There are evidences that purple carrot together with a yellow variant spread from Afghanistan to Mediterranean region as early as the tenth or eleventh century. White and orange carrots are probably mutations of yellow form. The domestic carrot readily crosses with widely adapted wild carrot known as Queen Anne's Lace.

KNOL KHOL

Variety	Year of release	Breeding method	Pedigree/ parentage	Important traits	Institutes/ universities
Large Green	1975	Introduction	—	The knobs are early with small tops tender, and flavoured. Its yield potential is 225–250 q/ha	IARI, RS, Katrain
White Vienna	1975	Introduction	—	The knobs are white, medium sized, tender, oval-shaped with distinct small tops. Its yield potential is 180–200 q/ha	IARI, RS, Katrain
Purple Vienna	1975	Introduction	—	Leaves purplish blue, knobs globular round, large, bluish purple, flesh light green	IARI, RS, Katrain
Palam Tenderknob	2005	Mass selection	—	It is characterized by small green foliage. The knobs are round, flat, stringless and fleshy. These attain marketable size about a week earlier than White Vienna. Its yield potential is 250–275 q/ha	HPKV, Palampur

CARROT

Variety	Year of release	Breeding method	Pedigree/ parentage	Important traits	Institutes/ universities
Tropical/Asiatic varieties					
Pusa Kesar	1963	Selection	Selection from a cross of Local Red and Nantes	Short tops. Roots deep red, tapering, small, rich in β-carotene (7753 IU/100 g edible portion), self-coloured core. Roots stay longer in field without bolting. Sets seeds freely in north Indian plains. Yield 25 tonnes/ha in 100–120 days	IARI, New Delhi
Pusa Meghali	1985	Selection	Selection from a cross of Pusa Kesar and Nantes	Short tops. Roots are smooth with orange flesh, self-coloured core and rich in β-carotene (11571 IU/100 g edible portion). Sets seeds freely in north Indian plains. Yield 25–26 tonnes/ ha in 100–120 days	IARI, New Delhi
Hisar Gairic	1993	Mass selection	—	Leaves light green. Roots long (18.1 cm), 3.9 cm in diameter, attractive, tapering, light brick red, thin self-core, less fibre and forking. 96.2 mg carotene/100 g fresh weight. Matures in 120 days. Average yield 28.8 tonnes/ha	CCSHAU, Hisar
Sel No 29	—	Selection	Selection from local material	Roots are long, tapering, thin and light-red in colour	PAU, Ludhiana

CARROT : ***Contd.***

Variety	Year of release	Breeding method	Pedigree/ parentage	Important traits	Institutes/ universities
Sel No 233	1978	Pedigree method	Derivative of a cross of Nantes and Sel. No 29	Tops small. Leaves 30–35 cm in size. Roots are 15 cm long, smooth, semi-cylindrical and orange. Core is light orange in colour. Root formation in 90–100 days. Less cracking and forking.	PAU, Ludhiana
Temperate/European varieties					
Nantes	1955	Introduction	Introduction from USA	Temperate type; roots small, orange, perfectly cylindrical, abruptly ending in small thin tail, sweet taste, self-coloured core with orange flesh. Suitable for sowing from mid-October to early-December; maturity 90–100 days. Yield 120 q/ha	IARI, Regional Station, Katrain
Pusa Yamdagni	1986	Pedigree method	Selection from a cross of EC 9981 and Nantes	Roots are 15–16 cm long, orange with self coloured core, rich in carotene, slightly tapered and semi-stumpy with medium tops. Yield 10 tonnes/ha in 90–100 days	IARI Regional Station, Katrain
Chaman	—	—	—	Foliage dark green and semi erect. Roots are long, cylindrical, semi-blunt, tolerant to cracking and forcing. Flesh is orange, sweet, fine textured. Yield 25–27 tonnes/ha	SKUAT, Srinagar

CARROT : *Contd.*

Variety	Year of release	Breeding method	Pedigree/ parentage	Important traits	Institutes/ universities
Ooty 1	1997	Selection	Selection from half-sib progeny of local type DC 3	Roots long, slightly tapering, deep orange with self coloured core. Central core is thin, fleshy and palatable. Free from premature bolting and resistant to powdery mildew, leaf spot. Medium maturing and suitable for growing throughout the year. Yield 49 tonnes/ha	Horticultural Research Station, TNAU, Ooty
Arka Suraj	2007	Pedigree method	Nantes and IIHR 253	Smooth root surface; Conical shape, Root length 15–18 cm, root diameter 3–4 cm; TSS 8–10%, carotene content 11.27 mg%, tolerant to powdery mildew and nematode. It flowers and sets seeds under tropical condition	IIHR Bangalore

BEETROOT

Variety	Year of release	Breeding method	Pedigree/ parentage	Important traits	Institutes/ universities
Ooty 1	1992	Selection	Selection from local material	It yields on an average of 31.45 tonnes/ha of roots. The roots are blood red in colour with thin skin and good quality. It contains 1.52% protein, 10.25% carbohydrate and 6 mg of vitamin C /100 g of pulp. It can be used as a salad.	HRS, TNAU, Ooty

The inflorescence of carrot is a compound umbel. A primary umbel can have over 1,000 flowers at maturity, whereas secondary, tertiary and quaternary umbels bear fewer flowers. Floral development is centripetal, *i.e.* flowers to dehisce first are on the outer edges of outer umbellets. Carrot is protandrous. After straightening of filament, the pollen is shed and stamens quickly fall. After this, petals open fully and the style elongates. The style is divided into two parts. The petals of petaloid plants are persistent unlike those of brown-anther, male sterile plants. Flowers are epigynous. There are five each small sepals, petals, and stamens, and two carpels. Emasculation is laborious and time consuming. As soon as the first bud in an umbel opens, the whole umbel of the female parent is bagged in a muslin/cloth bag. The flowers are removed daily until peak flowering reaches. Anthers are removed from the early opening outer flowers in the outer whorl of umbellets until sufficient flowers are emasculated. Unopened central florets in the emasculated umbellets and all late-flowering umbellets are removed. Thus, only the emasculated flowers are left on female inflorescence inside the bag. A pollen bearing umbel from previously protected male plant is inserted into the bag of female parent along with some house-flies to ensure pollination. Daily for a few days in morning, male umbel is gently rubbed against the emasculated umbel to enhance artificial cross-pollination. Sometimes, 1–2 flowering umbels of both the parents are enclosed in the same cloth along with some house-flies. Seeds from each parent is sown in adjacent rows. The hybrids and the parents could be identified (not always) and necessary rouging done to remove the selfed plants.

BEET ROOT

Botanical name: *Beta vulgaris* ssp *vulgaris*, 2n = 2x =18

Family: Chenopodiaceae

Beet root is an important crop in eastern and central Europe. It is of lesser importance in Western Europe and the USA where it is known as garden beet. The swollen roots are consumed as vegetable and salad. The earliest form of domesticated beet was leaf beet. The leaves were eaten and the roots were used for medical purposes only. This species is believed to have originated form Beta maritime known as sea beat which is indigenous to Southern Europe.

The beet root is closely related to sugarbeet, with which it is cross-compatible. It has the almost unique characteristic (in vegetable kingdom) of being wind-pollinated like its related species spinach, and is, therefore, a very prolific pollen producer. Breeding work with this species should be carried out under pollen-proof conditions employing filtered air and this may be one of the reasons why this crop has been largely neglected by vegetable breeders. Garden beet is a biennial producing enlarged hypocotyls (roots) and a rosette of leaves in first year and flowers and seed in second year. Enlargement of hypocotyl is due to growth of several concentric vascular combia which comprise the rings of beet. It requires cold temperature (4–10°C) treatment for 2 weeks or longer for flower induction. The inflorescence is a large spike. The flowers are small, inconspicuous without corolla, but with green calyx which becomes thicker and covers the seed completely. This forms what is called the beet seed or multigerm seed which, botanically is a fruit containing usually 2–6 seeds. The true seeds are small, kidney-shaped and brown.

RADISH

Botanical name: *Raphanus sativus* L., 2n = 2x = 18
Family: Brassicaceae

Radish, a root crop, is an important vegetable in many countries. Radish is grown widely in India, Egypt, Japan, Europe and China. Radish probably originated in Europe and Asia. It had been under extensive cultivation in Egypt since long. It was introduced in England and France in the beginning of 16th century. In 1806, it was introduced in America. Radish does not exist in wild state, it is believed to have originated from *R. raphanistrum* which is widely distributed as weed in Europe.

The edible portion of radish develops from the primary root and hypocotyl. The inflorescence is a typical terminal raceme. The flowers are small, usually white in colour and resemble those in cabbage and cauliflower. Sepals (four) are erect and petals (four) are clawed. Radish is cross-pollinated due to sporophytic system of self-incompatibility. It shows considerable inbreeding depression on selfing. It is entomophilous. It is pollinated mainly by wild honeybees and wild-flower flies. Stigma receptivity is maintained up to four days after anthesis. Selfing can be accomplished by bud-pollination. The flower buds are pollinated two days prior to opening by their own pollen by applying fresh pollen from previously bagged flowers of the same plant. Emasculation is not necessary in bud-pollination. After pollination, buds are to be protected from foreign pollen by enclosing the particular branch bearing those buds in a muslin cloth bag. In crossing the same technique is used as in bud-pollination except that in the crossing, buds of female parent are emasculated a day prior to opening and are pollinated by pollen collected from flowers of male parent which were also bagged before opening. The artificial pollination is done by hand by shaking the pollen over the stigma directly from the freshly opened but previously bagged buds of male

RADISH

Variety	Year of release	Breeding method	Pedigree/ parentage	Important traits	Institutes/ universities
Pusa Desi	1965	Selection	Selection from Indigenous material	Roots white with green shoulders, 30–35 cm long, medium thick, tapering, pungent; tops medium; leaves dark green. Suitable for sowing from mid-August to September in northern plains, maturity 50–55 days, yield 300 q/ha	IARI, New Delhi
Pusa Reshmi	—	Selection	Selection from Indigenous material	Asiatic type, roots 30–45 cm long, white with green tinge on top, maturity 50–60 days	IARI, New Delhi
Pusa Chetki	1996	Selection	Selfing and massing of seeds collected from Denmark	Tolerant to high temperature and humid weather conditions, roots are white, smooth, medium long and almost stumpy	IARI, New Delhi
Japanese White	—	Introduction	Introduction from Japan	25–30 cm long roots, 5 cm diameter, cylindrical and blunt at tip, skin is pure white, smooth, flesh is snow-white, smooth, crisp, solid and mild pungent	IARI, RS, Katrain
Pusa Himani	—	Hybridization	Black × Japanese White	Roots are 30–35 cm long, semi-stumpy, pure white with whitish-green shoulder, mildly pungent, crispy and sweet flavoured, tops are short with green cut semierect leaves	IARI, RS, Katrain

RADISH : *Contd.*

Variety	Year of release	Breeding method	Pedigree/ parentage	Important traits	Institutes/ universities
Arka Nishant	1991	Selection	Mass selection from IIHR 72	Roots long, marble white in colour with crisp texture and mild pungency, resistant to pithiness, premature bolting, root branching and forking	IIHR, Bangalore
CO 1	1981	Selection	It is a selection from germplasm type (RS 44)	Its crop duration is 40–45 days and hence suitable for cropping systems. It yields 20–25 tonnes/ha. The roots are milky white, long (22 cm) thick 12.5 cm girth) and each weighs 220 g	HRS, TNAU, Ooty
Palam Hriday	2003	Selection	—	Pink fleshed, rich in vitamin C, early-maturing	HPAU, Palampur
Punjab Safed	1975	—	—	Roots white and long with semiblunt tip, average yield 160 q/ha	PAU, Ludhiana
Punjab Ageti	1986	—	—	Suitable for April–August sowing, roots long, thin with lower half white and upper red, average yield 150 q/ha	PAU, Ludhiana
Punjab Pasand	1997	—	—	Suitable for main season, roots long, white, semi-stumped, average yield 214 q/ha	PAU, Ludhiana

parent. When a large quantity of crossed seed is required, the roots of radish female and male parents are planted in alternate rows, spaced 60 cm apart. Later about 3–4 days before opening of buds, the plants are covered under an insect-proof wire net or plastic cage of 22–24 mesh. Usually 2 plants, 1 female and 1 male are covered under small cage, or sometimes a cage is used to cover 4 plants, 2 female and 2 male plants. A small honeybee colony is placed inside the cage, 3–4 days before opening of buds. This method is followed when it is possible to rogue out the selfed or sibmated plants in the seedling or root stage with the help of a dominant marker gene or when the male and female lines are homozygous for self-incompatibility alleles but are cross compatible. This procedure can also be used to produce sibmated seeds to maintain a variety under insect proof cages. However, this is necessary to place about 20–30 plants under a cage to avoid inbreeding depression. A wire net or plastic net cage of 3 m × 3 m × 2.5 m (height) with a small door on one side is convenient for this purpose.

LEGUMINOUS VEGETABLES

FRENCH BEAN

Botanical name: *Phaseolus vulgaris* L., 2n = 2x = 22
Family: Leguminosae

Frenchbean is an important legume crop to be used as green pod vegetable (known by various names as snap bean, string bean, garden bean, fresh bean) or dry seeds (known as dry bean). In Western World (USA, Western Europe), the fresh pod/processed pod consumption is quite substantial. In India, green pod as well as dry seed consumption is conspicuous but figures on area of production are not available. In India, it is primarily grown in Jammu and Kashmir, Himachal Pradesh and hills of Uttar Pradesh. Production is slightly visible in plains also.

Its tap roots are with poor nodule formation. Stem could be dwarf bush type or pole type depending upon growth habit. Stem growth habit is of two types, *viz.* determinate vs determinate type. Leaves are trifoliate. Flowers are papilionaceous. Keel of corolla and style are coiled trough 360° (1–5 turns). Androecium is in diadelphous condition (9 + 1). Vexillary stamens are free from base upward. For emasculation an unopened flower bud is selected. The standard is detached from below with a pair of forceps and bent backward. The keel is pulled in pieces with the forceps. Care should be taken to turn in the same direction as the spiral winding, otherwise the style

FRENCH BEAN

Variety	Year of release	Breeding method	Pedigree/ parentage	Important traits	Institutes/ universities
Arka Komal	1990	Pure line selection	IIHR-60 (collection from Australia)	Plants erect and bushy. Flowers light pink. Pods straight, flat, tender, green and suitable for transport. Seeds light brown, oblong and large. Green pod yield 18.0 tonnes/ha in 70 days	IIHR, Bangalore
Arka Suvidha	2006	Pedigree selection	Blue Crop × Contender	Plants bushy and photo-insensitive. Flowers light pink. Pods straight, oval, light green, fleshy and stringless. Green pod yield 18.0 tonnes/ha in 70 days.	IIHR, Bangalore
Arka Bold	1998	Pure line selection	IIHR 220 from Hungary	Plants bushy and Photo-insensitive. Flowers white. Pods flat and stringless, fleshy, crisp, extra large (1.6 cm) and medium long. Resistant to rust. Green pod yield 15 tonnes/ha in 70 days	IIHR, Bangalore
Arka Anoop	2007	Pedigree selection	Arka Bold × Arka Komal	Plants bushy and Photo-insensitive. Flowers white. Pods flat straight, fleshy, crisp and long (17–18 cm). Resistant to rust and tolerant to bacterial blight. Green pod yield 20 tonnes/ha in 70 days	IIHR, Bangalore
Contender	1960	Introduction	Introduction from USA	Plants bushy, flowers pink. Pods are round green about 15 cm long, non stringy, meaty and slightly curved.	IARI New Delhi

FRENCH BEAN : *Contd.*

Variety	Year of release	Breeding method	Pedigree/ parentage	Important traits	Institutes/ universities
				Green pod yield 8 tonnes /ha in 70 days. Susceptible to rust and MYMV	
Pusa Parvathi	1972	Mutation breeding	Yellow "wax pod" EC 1906 from USA	Plants bushy, flowers pink, pods are round green about 15–18 cm long, straight flattish round, stringless and green in colour. Green pod yield 8 tonnes/ha in 70 days. Highly susceptible to rust and MYMV	IARI, Regional Station, Katrain
Pusa Himalata	—	Introduction	Exotic introduction EC 38913	It is a pole type variety (2.5 m long). Each node bears 3–4 flowers. Pods are medium-sized (14.0 cm) round meaty stringless and light green. First picking starts 60 days after sowing. Pod yield 20 tonnes/ha	IARI, Regional Station, Katrain
Phule Surekha	2000	Pure line selection	Jampa improved variety	Semi-determinate, pods straight, suitable for *kharif*, *rabi* and summer (photo-insensitive), resistant to anthracnose, leaf crinkle, bean yellow-mosaic virus and wilt (stem fly)	MPKV, Rahuri
Phule Suyash	2004	Pedigree selection	Waghya × Shahapur local	Pods round to semi-round, straight to slightly curved pods and are long, smooth and more fleshy. Pod yield 16 tonnes/ha in 75 days	MPKV, Rahuri

FRENCH BEAN : *Contd.*

Variety	Year of release	Breeding method	Pedigree/ parentage	Important traits	Institutes/ universities
VL Lata Bean 12	1973	Introduction	Introduced	—	VPKAS Almora, Uttaranchal
VL Lata Bean 17	1973	Introduction	Introduced from USA as Kentuki wonder	—	VPKAS Almora, Uttaranchal
VL-Bauni Bean 1	1985	Pedigree selection	Composite of snapbush and rival	Flowers are white and purple tinged. Pods are medium long (12.0 cm), round fleshy non-stringy, with pale green colour. Pod yield 8 t/ha in 75 days	VPKAS Almora, Uttarakhand
VL Bean 2	2008	Pedigree selection	Interspecific hybridization followed by Pedigree selection, Contender *(P. vulgaris)* P.B.L 257 *(P. multiflorus)*	Plants are viny and photo-insensitive. Pods are green round, stringless, 13–14 cm long. Pod yield 12.5 tonnes/ha	VPKAS Almora, Uttaranchal
Pant Anupama (UPF 191)	1984	Selection	—	Plants are dwarf with dark green foliage. Pods tender, smooth, non stringy, medium long, round, straight, transcluscent and green pods concentrated at mid height. It is resistant to angular leaf spot and	GBPUAT, Pantnagar

FRENCH BEAN : *Contd.*

Variety	Year of release	Breeding method	Pedigree/ parentage	Important traits	Institutes/ universities
				moderately resistant to common mosaic virus. Pod yield 9 tonnes/ha	
Pant Bean 2 (UPF 626)	1996	Pedigree selection	Turkish Brown × Contender	Plants are bushy and sturdy. Pods are flattish round straight, and non stringy at edible stage. The seeds are brown with mottling and are bold. It is resistant to rust and moderately resistant to common mosaic virus. Pod yield 9 tonnes/ha	GBPUAT, Pantnagar
Laxmi	—	Pedigree selection	Contender × Local variety (pole type)	Plants are viny and photo-insensitive. Pods are formed in clusters of three, 13–14 cm long, stringless, green round. Green pod yield 12–14 tonnes/ha	Dr. Y.S. Parmar University of Horticulture and Forestry, Solan, HP
TKD 1	1998	Selection	Selected from germplasm population	Pods are less fibrous. The crop duration is 90 days. It yields 5.6 tonnes of green pods/ha. Dry seed yield is 2.78 tonnes/ha. Seeds turn white in maturity used for cooking. Suitable for growing in the hill ranges of Tamil Nadu, *viz.* Shevroys, Palani hills and Nilgiris	HRS, TNAU, Thadiyan-kudisai

FRENCH BEAN : *Contd.*

Variety	Year of release	Breeding method	Pedigree/ parentage	Important traits	Institutes/ universities
YCD 1	1994	Selection	It is a pure line selection from a local indigenous type collected from Shevroy hill ranges of Tamil Nadu	The variety is a dual purpose one, useful as green pod and grain. It is adaptable to the rainfed situations of Shevroys. It is suitable for cultivation in *kharif* (June–July) season and possesses a higher yield potential of 9.75 tonnes/ha of green pods and grain yield of 6.3 tonnes/ha	HRS, TNAU, Yercaud
Ooty 1	1999	Selection	Pure line selection from accession PV 26	It is a pole type adapted to temperate zone from 1,800 to 2,500 m above MSL. It yields 33.68 tonnes/ha	HRS, TNAU, Ooty
Ooty 2	2002	Selection	Single line selection from PV 15 1	It is dwarf and dense plant, bears round, fleshy fiberless, pale green pods. It contains high protein (22.38%), calcium (1.22%) and magnesium (0.34%) and highly suitable for cooking. It is medium tolerant to pest and diseases. It has recorded a yield of 14.30 tonnes/ha of green fruits in 90 days	HRS, TNAU, Ooty
Kashi Param	2006	Pure line selection	—	Plants are determinate (70 cm long). Pod is fleshy, round, dark green colour 14.7 cm length. Seed colour dark brown. Green pod yield 12–14 tonnes/ha	IIVR, Varanasi

may break. After pulling off keel, the stamens are removed. For the supply of pollen, freshly opened flowers are collected. The thickly pollinated stigmas emerge as soon as the wings are pressed downward. This stigma is rubbed against the stigma of the emasculated bud. Sometimes, thickly pollinated stigma of open flower to be used as male is hooked into the stigma of flower bud to be pollinated. Pollination without emasculation has also been suggested. In this method the standard is detached and unfolded. By pressing the left-hand wing downward, the unpollinated stigma emerges. Pollen is rubbed into this stigma. The emasculation and pollinations are done simultaneously in the forenoons.

VEGETABLE COWPEA

Botanical name: *Vigna unguiculata*, 2n = 2x = 22
Family: Leguminosae

Cowpea also called as southernpea and blackeyed pea, is a well-adapted to the tropical areas. The major cowpea-growing countries are Nigeria, Niger, Burkina Faso, Ghana, Kenya, Uganda, Malawi, Tanzania (all in Africa), India, Sri Lanka, Burma, Bangladesh, Philippines, Indonesia and Thailand. It is primarily used in the form of dry seeds, fodder, green pod, green manure and cover crops. Growth habit ranges from erect, determinate, non-branching type to prostrate or climbing, indeterminate, with profuse branching. It has strong tap root system with several lateral roots. Stems are cylindrical and slightly ribbed, twisting, sometimes hollow and glabrous. Stems may be green or pigmented (purple). Leaves are alternate, trifoliate, with one symmetrical terminal leaflet and two asymmetrical leaflets. Petioles are 3–25 cm long with a swollen pulvinus at the base. Inflorescence is an unbranched axillary raceme bearing several flowers at the terminal end of peduncles. The peduncles vary from 5 to 60 cm in length and are slightly twisted and ribbed. Calyx is longitudinally ribbed, tubular with 2–15 mm long subequal lobes.

The corolla is papilionaceous with an erect standard petal spreading at anthesis. The pigmentation pattern of corolla varies from white to solid mauve with yellow spots near the base of the standard petal. The wings are adherent to the boat-shaped keel, enclosing the androecium and gynoecium. The stamens are diadelphous (9 + 1). Anthers are bright yellow. Ovary is monocarpellary, unilocular with many ovules. Pods are pendent or vertically attached to the raceme axis. They are mostly linear, although curved ends coiled shapes are also found. The length of pods may vary from the less than 11 to more than 100 cm.

VEGETABLE COWPEA

Variety	Year of release	Breeding method	Pedigree/ parentage	Important traits	Institutes/ universities
Arka Garima	1990	Interspecific hybridization	Interspecific hybridization followed by back cross and selection, IITA TUV 762 × *V. unguiculata* subsp *sesquipedalis*	Plants tall, photo-insensitive. Pods light green, long, thick, round, fleshy and stringless. Suitable for vegetable purpose. Tolerant to heat and low moisture stress. Pod yield 18 tonnes/ha	IIHR, Bangalore
Arka Suman	2006	Pedigree selection	Arka Garima × Pusa Komal	Very early cultivar. Plants erect, bushy, photoinsensitive, early, with pods borne above the canopy. Pods green thin, medium long, tender, fleshy without parchment. Duration 70–75 days	IIHR, Bangalore
Arka Samrudhi	2006	Pedigree selection	Arka Garima × Pusa Komal	Very early, plants erect, bushy, photo-insensitive, early, with pods borne above the canopy. Pods green, medium thick, medium long, tender, fleshy without parchment. Duration 70–75 days	IIHR, Bangalore
KMV 1	1996	Selection	Manjeri Red Plain	—	KAU, Vellanikkara
Malika	1992	Selection	Single plant selection from Thiruvananthapuram	—	KAU, Vellanikkara

VEGETABLE COWPEA : *Contd.*

Variety	Year of release	Breeding method	Pedigree/ parentage	Important traits	Institutes/ universities
Sharika	1993	Selection	Valiyavila local (SPS)	—	KAU, Vellanikkara
Lola	2001	Pure line selection	Kerala Local	Plants viny, pods green smooth, long (30.35 cm) and tender. Yield 20 tonnes/ha	KAU, Vellanikkara
Vyjayanthi	1998	Pure line selection	Perumpadavam local (VS 21-1)	Plants viny, pods smooth, long (30.35 cm) and tender. Yield 20 tonnes/ha	KAU, Vellanikkara
Vellayani Jyothika	2006	Selection	Sreekaryam local	Pole type, High yielding (19.33 tonnes/ ha) with long light green pods	KAU, Vellanikkara
Pusa Phalguni	—	Selection	Selection from exotic material from Phillipines	Plants are dwarf with bushy habit. It gives two flushes of dark green erect pods of 10–12 cm lengthwithsmall cylindrical white seeds. Pods are ready for harvest in about 60days. Green pod yield 7.5 tonnes/ha	IARI, New Delhi
Pusa Barsati	—	Selection	Selection from exotic material from Canada Dolique Du Tonkin from Canada (EC 4281)	It is an early cultivar (45 days to first green pod picking) suitable for growing during rainy season. It gives 2–3 flushes of light green pendent pods. The pods are 25–28 cm long and contain large green seeds. Green pod yield 7.5 tonnes/ha	IARI, New Delhi

VEGETABLE COWPEA : *Contd.*

Variety	Year of release	Breeding method	Pedigree/ parentage	Important traits	Institutes/ universities
Pusa Dofasli	—	Pedigree method	Pusa Phalguni × Phillippines introduction	It flowers in 35–40 days after sowing. Pods are light green erect about 18 cm long. Plants bushy, photo-insensitive and suitable for growing in spring-summer and rainy seasons.Pod yield 8.0 tonnes/ha	IARI, New Delhi
Pusa Komal	1983	Hybridization and Selection	Pusa Dofasli, EC 26410, and P 426 (bacterial blight resistant line)	Plants bushy 55–60 days duration. Pods light green non fibrous, resistant to bacterial blight. Pod yield 10 tonnes/ha	IARI, New Delhi
Narendra Lobia 1	1993	Pedigree method	Pusa Komal × Varanasi Local	Plants determinate, grows to a height of 40–45 cm. Foliage is green with large leaves. Pods green 28–32 cm long with purple terminal end. Edible pod maturity 45–48 days. Pod yield 9.0 tonnes/ha	NDUAT, Faizabad
Sel 263	1992	Selection	A germplasm collection from IIHR	A dual season and dwarf variety. Pods are green, thick meaty, tender, meaty, tender, medium long (20 cm) and contain 3.57% crude protein on fresh weight basis. It has resistance to mosaic and yields 10–12 tonnes/ha	PAU Ludhiana

VEGETABLE COWPEA : ***Contd.***

Variety	Year of release	Breeding method	Pedigree/ parentage	Important traits	Institutes/ universities
Bidhan Barbati 1	2001	Pedigree breeding	EC 243954 × EC 305827 (inter-culti-group hybridization followed by modified back cross-pedigree method)	Plants compact, determinate with dark green foliage, yield 13.4 t/ha, resistant to cowpea mosaic and cowpea golden mosaic virus	BCKV, Nadia
Bidhan Barbati 2	2001	Pedigree breeding	V-70 × Sel Tm 3 (inter-cultigroup hybridization followed by modified back cross-pedigree method)	Semi-determinate, yield 15.9 t/ha, very low incidence of cowpea mosaic and cowpea golden mosaic virus	BCKV, Nadia
Kashi Shyamal	2006	Selection from local collection	—	Dwarf and bushy variety (70–75 cm height), suitable for both *kharif* and *zaid* season, 3–4 branches, early flowering (40 days after sowing), first harvesting after 48 days of sowing, produces 35–40 pods per plant with pod length of 25–30 cm. Green pod yield 7–8 tonnes/ha. Tolerant to *Golden mosaic* virus	IIVR, Varanasi

VEGETABLE COWPEA : *Contd.*

Variety	Year of release	Breeding method	Pedigree/ parentage	Important traits	Institutes/ universities
Kashi Gauri	2006	Hybridization	IIHR Sel. 16 × Sel. 2-1	Bush type, dwarf, photo-insensitive and early variety suitable for sowing in both spring summer and rainy season. Flowers in 35–38 days and pods are ready for harvest in 45–48 days. Pods are 25–30 cm long, light green, soft, fleshy and free from parchment layer. Resistant to golden mosaic virus and *Pseudocercospora cruenta*. Green pod yield 10–12 tonnes/ha	IIVR, Varanasi

VEGETABLE DOLICHOS BEAN or LAB LAB (*Dolichos lablab*)

Variety	Year of release	Breeding method	Pedigree/ parentage	Important traits	Institutes/ universities
Arka Jay	1990	Back cross and pedigree selection	Hebbal Avare × IIHR 93	Plants dwarf, bushy, erect and photo insensitive. Flowers purple. Pods long, light green slightly curved, without parchment. Vegetable type with excellent cooking qualities. Tolerant to low moisture stress. Pod yield 12 tonnes/ha	IIHR, Bangalore

VEGETABLE DOLICHOS BEAN or LAB LAB : *Contd.*

Variety	Year of release	Breeding method	Pedigree/ parentage	Important traits	Institutes/ universities
Arka Vijay	1990	Back cross and pedigree selection	PEP × Hebbal Avare-3	Plants dwarf, bushy, erect and photo insensitive. Flowers white colored. Pods short dark green. Pods with characteristic aroma, without parchment. Vegetable type with excellent cooking qualities. Tolerant to low moisture stress. Pod yield 12 tonnes/ha	IIHR, Bangalore
Konkan Bhushan	1991	Pedigree selection	Hebbal Avare 3 × Wal 1	It is a bush type plant, 60–75 cm tall and matures in 55–60 days after sowing. The pods are tender and stringless. A plant produces 125–180 pods. Green pod yield 13 tonnes/ha in 90 days	Konkan Krishi Vidyapeeth, Akola
Hima	2006	Selection	Selection from local collection	High yielding (13.34 kg/plant), hardy, medium maturing, pole type with white flowers, light green broad and straight pods and coffee brown seeds	KAU, Vellanikkara
Grace	2006	Selection	Selection from local collection	High yielding (13.6 kg/plant), hardy, early maturing, pole type with purple stem, lilac flowers and slightly curved greenish purple pods and black seeds	KAU, Vellanikkara
Pusa Early Prolific	Before 1970	Selection	Selection from Indigenous material	Early, climbing type; pods light green, long, thin borne in bunches. Suitable for both spring and autumn	IARI, New Delhi

VEGETABLE DOLICHOS BEAN or LAB LAB : *Contd.*

Variety	Year of release	Breeding method	Pedigree/ parentage	Important traits	Institutes/ universities
Pusa Sem 2	1990	Selection	Selection from Indigenous material	Pods dark green, very tender, stringless, semi round, 16–17 cm long with 5–6 seeds per pod, 11–12 pods/spike; excellent organoleptic quality. Highly tolerant to viruses, anthracnose disease, insects like aphid, jassid, pod borer and frost; first picking in 120–125 days. Yield 150 q/ha	IARI, New Delhi
Pusa Sem 3	1990	Selection	Selection from Indigenous material	Pods green, meaty, very tender, stringless, flat, 15 cm long with 5–6 seeds/pod, 9–10 pods/cluster; excellent organoleptic quality. Tolerant to viruses, anthracnose disease and insects like aphid, jassid and pod borer; first picking in 115–120 days. Yield 170 q/ha	IARI, New Delhi

Cowpea flowers are large and showy. Flowers open only once between 7 and 9 AM. On cloudy days, flowers may open even in the afternoon. Though the flowers open late in the morning, the dehiscence of anthers is much earlier. It may vary from 10 PM to 0.45 AM. The dehiscence is influenced by presence of moonlight, a clear sky and a dry warm atmosphere. During dark nights, dehiscence tends to be delayed. Due to dehiscence taking place before opening of flowers, cowpea is strictly self-pollinated. Since dehiscence of anthers is much in advance of blooming, emasculation needs to be carried out in mature flower buds in preceding evening. The flower buds likely to bloom the next day (recognized by large size, yellowish colour of back if standard petal) is selected for emasculation. The bud is held between the thumb and the forefinger with the keel side uppermost. A needle is run along the ridge where the two edges of the standard unite.

One side of the standard is brought down and secured in position with thumb. Same thing is done with one of the wings. After this the exposed keel is slit on the exposed side, about 1/16 inch from the stigma. A section of keel is also brought down and secured in position under the end of thumb. Now 10 stamens are seen. They are removed with pointed forceps. Afterwards, the disturbed parts of standard, wing and keel are brought in original position as far as possible. To prevent drying out of the emasculated bud, a leaflet may be folded and pinned around the bud. A tissue paper can be used to cover and protect the bud. Pollination is done next morning from a freshly opened flower. The standard and wings of male flower are removed. By slight depression of the keel, stigma covered with pollen grains protrudes out. This itself can be used as a brush for pollination. Cowpea flowers are highly sensitive and drop off early with slight mechanical disturbance or injury. Therefore, much labour and time are devoted to get enough crossed seed.

GARDEN PEA

Botanical name: *Pisum sativum* L., 2n = 2x = 14
Family: Leguminosae

Pea is consumed as fresh vegetables or dry seeds throughout the world. In India it is grown as winter vegetable in plains and as summer vegetable in hills. The geographical region comprising of Central Asia, the Near East, Abyssinia, and the Mediterranean is considered as its centre of origin based on genetic diversity. It is an annual herbaceous plant. It has a tap root system. Stems are slender, usually single, and upright in growth. Leaves are pinnately compound with two to several leaflets. The rachis terminates in a simple or

GARDEN PEA

Variety	Year of release	Breeding method	Pedigree/ parentage	Important traits	Institutes/ universities
Arka Ajit	1993	Back cross and pedigree selection	Bonneville, Freezer 656 Erygel and IIHR209	A mid season variety. Pods 8–9 cm long; seeds bold green, sweet, shelling percentage 55, resistant to both powdery mildew and rust. Duration 90 days and Yield 10 tonnes/ha. It is recommended for Uttar Pradesh, Rajasthan and Karnataka	IIHR, Bangalore
Arka Karthik	2007	Pedigree selection	Arka Ajit × IIHR 554	Plants bushy, erect and resistant to both powdery mildew and rust. Pods are long (11–12 cm) with 8–10 green. Sweet seeds. Crop duration 90 days, pod yield 11 tonnes/ha	IIHR, Bangalore
Arka Sampoorna	2007	Back cross and pedigree selection	Bonneville, Freezer 656, Manoa Sugar and IIHR209	This is whole pod edible pea. The plants are bushy erect. Pods are flat, parchment-less, crisp, sweet, and resistant to powdery mildew and rust. Duration 80 days and yield 8 tonnes/ha	IIHR, Bangalore
Bonneville	1975	Introduction from USA	Bonneville of USA	Mid season var. Light green pods. Wrinkle seeded variety. 8 cm long pods with 7–8, green, sweet, and bold seeds/ pod. Maturity 55–60 days. Yield 10–12 tonnes/ha	IARI, New Delhi

GARDEN PEA : *Contd.*

Variety	Year of release	Breeding method	Pedigree/ parentage	Important traits	Institutes/ universities
Lincoln	1987	Introduction	Introduction from USA	Mid season, plants medium long, double poded. pods about 9 cm long, dark green and curved at the top, 8–9 seeds/pod. Suitable for sowing up to 15th December. More suitable for hills. Maturity 90–95 days. Yield 11–13 tonnes/ha	IARI, RS, Katrain
Arkel	1975	Introduction from England	Arkel from England	Plant dwarf. 40–60 cm tall, pods dark green about 8 cm long, slightly curved and well filled. Grains very sweet and wrinkled when dry. Suitable for sowing in mid October. Maturity 55–60 days. Yield 100–125 q/ha	IARI, New Delhi
Early Badger	1975	Introduction	Perfection horse ford × Notts Excelsior	Early, dwarf, wrinkle seeded variety, pods pale green, 7.5 cm long pods. Suitable for early October sowing. Maturity 60–65 days. Yield 50–60 q/ha	IARI, New Delhi
Meteor	1972	Introduction	Introduction from UK	Plant dwarf. 50-60 cm tall, round seeded variety suitable for early-October sowing. Maturity 55–60 days. Yield 90–100 q/ha	IARI, New Delhi
Pusa Pragati	2001	Introduction	Introduction from Germany and selection	Pods long (10 cm), green with 9 seeds per pod; first picking 60-65 days; resistant to powdery mildew. Yield 70 q/ha	IARI, New Delhi

GARDEN PEA : *Contd.*

Variety	Year of release	Breeding method	Pedigree/ parentage	Important traits	Institutes/ universities
Jawahar Matar 1	—	Pedigree method	Type 9 × Greater Progress	Mid season variety, plant height 82 cm. leaves normal shaped, green in colour. Pod straight with bead like outgrowth at the lower end. Average size of the pod is 8.8 cm × 1.4 cm, 8–9 seeds/pod and mature seeds green in colour and wrinkle. Maturity 85–90 days. Yield 100–110 q/ha	JNKVV, Jabalpur
Jawahar Matar 4	—	Pedigree selection	Type 19 × Little marble	Plant height 70 cm, green pods, stem and foliage. Pods well filled. Mature seeds wrinkled, green. Maturity 65–70 days. Yield 60–70 q/ha	JNKVV, Jabalpur
Kashi Nandini	2006	Pedigree selection	P-1542 × VT-2-1	Early-maturing. Plant height 47–51 cm and 50% plants bears flowers at 34 days after sowing. Plants are erect, dark green, height 60 cm with 7–8 pods/plant. Pods are 8–9 cm long, attractive well-filled with 8–9 seeds, shelling 47–48% and yield 11–12 tonnes/ha. Tolerant to leaf miner and pod borer	IIVR Varanasi
Kashi Udai	2006	Pedigree selection	Arkel × FC 1	Early maturing. Plant height 58–62 cm and 50% plants bear flowers at 35–37 days after sowing. Dark green foliage and short internodes with 8–10 pods/ plant.	IIVR Varanasi

GARDEN PEA : *Contd.*

Variety	Year of release	Breeding method	Pedigree/ parentage	Important traits	Institutes/ universities
				Pods are 9–10 cm long, filled with 8–9 bold seed, shelling 48%, yield 10–11 tonnes/ha	
Kashi Mukti	2006	Pedigree selection	No 7 × PM 5	Early maturing, powdery mildew resistant. Plant height 50–53 cm and 50% plants bear flowers at 35–36 days after sowing. Pods are 8.5–9 cm long, filled with 8–9 bold, soft textured seeds, shelling 48–49%, yield 11–12 tonnes/ha	IIVR Varanasi
Kashi Shakhi	2006	Pedigree selection	Hara Bona × NDVP 8	Medium maturing. Plant height 90–98 cm, 50% plants bear flowers at 54–56 days after sowing. Dark green foliage with 11–12 pods per plant. Pods 10–10.5 cm long, filled with 7.5–8.5 bold seed, shelling 48–49%, yield 14–16 tonnes/ha	IIVR Varanasi
VL Matar 3	1989	Pedigree method	Old sugar × Early wrinkled Dwarf	Average plant height 67 cm, light green foliage, and light green pod. Average pod length 6.8 cm with 5 seeds/pod. Resistant to powdery mildew. Maturity 105 days. Yield 10–11 tonnes/ha	VPKAS Almora, Uttarakhand
VL Ageti Matar 7	1984	Pedigree method	Pant Uphar × Arkel	Pods light green, bold, plant height 65–70 cm, maturity 120–125 days	VPKAS Almora,

GARDEN PEA : *Contd.*

Variety	Year of release	Breeding method	Pedigree/ parentage	Important traits	Institutes/ universities
				(seeding to first picking) during November sowing in mid-hills, escapes powdery mildew due to early maturity, it has longer shelf-life	Uttarakhand
Vivek Matar 6	1997	Pedigree method	IP 3/VL 3	—	VPKAS Almora, Uttarakhand
Vivek Matar 8	1997	Pedigree method	Perfection × Bonneville	Plant height 65–70 cm, maturity 135–140 days (seeding to first picking) during November sowing in mid hills, pods light green, tolerant to powdery mildew and white rot, pod yield 11–12 tonnes/ha	VPKAS Almora, Uttarakhand
Vivek Matar 9	2005	Pedigree method	VP 8591 × P 185	Pod dark green, long, plant height 60–70 cm, maturity 130–140 days (seeding to first picking) during November growing in mid hills, tolerant to powdery mildew and white rot, yield 90–110 q/ha green pod	VPKAS Almora, Uttarakhand
Vivek Matar 10	2007	Pedigree method	Azad Matar 1 × PMR 3	Pods green, long and curved, seeds light green, shriveled, plant height 55 cm,	VPKAS Almora,

GARDEN PEA : *Contd.*

Variety	Year of release	Breeding method	Pedigree/ parentage	Important traits	Institutes/ universities
				maturity early, 125 duration for first harvesting, yield 95 q/ha of green pods	Uttarakhand
Pant Uphar	1985	—	—	Plant height 75–80 cm. relatively thin small leaflets, foliage light, green, and length of pod 7–8 cm. Susceptible to powdery mildew. Maturity 75–80 days. Yield 95–100 q/ha	GBPAU&T, Panthnagar
Ooty 1	2000	Selection	Pureline selection (PS-33-1) among 40 germplasm types	The plants are dwarf with dark green, long and smooth pods with 9 seeds. The pods have good cooking quality. The crop yields 11.1 tonnes/ha under rainfed conditions and 12.9 tonnes /ha under irrigation. The seeds contain high protein (19.94%)	HRS, TNAU, Ooty
Palam Priya	2001	Pedigree selection	Bonneville × P 388	Medium growth habit, self-stalking, double podding, high yield, resistance to powdery mildew, tolerant to leaf miner	HPKV, Palampur

CLUSTERBEAN (*Cyamopsis tetragonolobus*)

Variety	Year of release	Breeding method	Pedigree/ parentage	Important traits	Institutes/ universities
Pusa Mausami	1960	Selection	Selection from indigenous material	Densely branched, pods smooth, bright green, 10–13 cm, late variety, suitable for rainy season	IARI, New Delhi
Pusa Sadabahar	1960	Selection	Selection from indigenous material	High-yielding variety, pods green, 12–13 cm length, tender, suitable for summer and rainy season	IARI, New Delhi
Pusa Naubahar	1965	Pedigree method	Pusa Mausami × Pusa Sadabahar	Suitable for summer and rainy season	IARI, New Delhi
Goma Manjri	1997-98	Selection	Selection from local germplasm	Erect single stem, yield potential 88–103 q/ha	CHES, Godhra

branched tendril. There are large stipules at the base of the leaf. The plant may be single-stemmed or many axillary stems may originate at the cotyledonary node or any superior node, especially if apical growing point is destroyed, leaflets of a pair are opposite or slightly alternate. The lower leaflets are larger than the upper leaflets. The margins of leaflets and stipules may be entire or serrated.

The inflorescence is raceme arising from the axil of a leaf. The lowest node at which flower initiation occurs is normally constant under a given set of conditions and is used in classifying the varieties into early and late types. Most early cultivars produce the first flower from modes 5–11 and late cultivars start flowering at about nodes 13–15. Early cultivars are often single flowered or bear some single and some double flowers. Late cultivars are usually double/triple flowered. The flowers are typical papilionaceous with green calyx comprising five each united sepals, and petals (one standard, two wings and two keels). The stamens are in diadelphous (9 + 1) condition. Nine filaments are fused to form a staminal tube while the tenth is free throughout its length. The gynoecium is monocarpellary, with ovules (up to 13) alternately attached to the two placentas. Style normally bends at right angle to the ovary. Stigma is sticky. Pea is strictly self-pollinated in nature. Stigma is receptive to pollen from several days prior to anthesis until 1 day or more after the flower wilts. Pollen is viable from the time anthers dehisce until several days thereafter. For emasculation the flower bud chosen should have developed to the stage just before anther dehiscence, indicated by extension of petals beyond sepals.

Flowers can be emasculated at any time. The first step in emasculation is to tear away with the forceps the tip of the sepal from in front of the keel. The forefinger is positioned behind the flower and thumb in front and a light pressure is applied. This spreads the standard and wings to expose the keel. The exposed keel is slit-open by tips of forceps. Pressure can be applied by the thumb and finger on keel for increased exposure of the pistil and stamens. The 10 stamens are pulled out. Pollen can be obtained throughout the day, preferably from the freshly opened flower. For pollen collection, it is more convenient to pick the male flowers, remove the standard and wings, pull back the keel so that the style protrudes and use the pollen covered stylar brush as an applicator to transfer the pollen to the stigma of the emasculated bud. Older flowers and other flower buds not used in crossing are removed from the peduncle to increase the pod set after crossing.

LEAFY VEGETABLES

AMARANTHUS

Botanical name: *Amaranthus* spp.
Family: Amaranthaceae

Amaranth, popularly known as *chaulai,* is very nutritive and highly suitable crop for kitchen gardening and commercial cultivation. Rapid growth, quick rejuvenation after each harvesting and high yield of edible portion/unit area in limited time as well as high nutritive value are its important features. It is one of the cheapest leafy vegetables in tropical and subtropical parts of the country. It could be a very valuable source for combating under-nutrition and malnutrition in India. This green leafy vegetable constitutes an excellent component of the diet in tropical countries. Many species exist in wild forms and a number of domesticated forms are available in Tamil Nadu. The chromosome number is 2n = 32 in *A. caudatus* and *A. edulis.*

Flowers are monoecious and the inflorescence is a branched compound spike, erect or pendulous, the spike is made up of a number of cymes. Each flower has 3–5 small bracteoles, male flowers with 3–5 stamens; female flowers with 2–3 styles and stigma. Flowers are protogynous; stigma becomes receptive several days before opening of staminate flowers. Dehiscence of anthers and release of pollen grains are maximum between 11 AM or 1 pm. Breeding objectives are to identify suitable vegetable amaranth types with high yield potential coupled with nutritional qualities, to develop an 'erect' and 'clipping' (cutting) types, suited for early harvesting for 'pulling' and continued harvesting by periodical cutting respectively to suit different regions and to develop a dual amaranth types, *i.e.* grain-cum-leafy amaranth.

SPINACH

Botanical name: *Spinacea oleracea*
Family: Amaranthaceae

Spinach is a cool season vegetable. It is an annual crop. The rosette of leaves produced during vegetative phase is used as vegetable. It is cooked with potato, onion, brinjal or alone. Leaves are rich in vitamins A (5,580 IU) and C (28 mg), folic acid (123 mg) and minerals — iron (17.4 mg), calcium (190 mg) and phosphorus (42 mg)/100 g. Spinach is more popular in central and northern India. As a cool season crop it can tolerate frost. Short day conditions coupled with moderately warm temperature are suited for its growth. Under long day and warmer conditions, bolting occurs. Under high humid conditions, leaves become more succulent and tender. Well-drained, loam soil

AMARANTHUS

Variety	Year of release	Breeding method	Pedigree/ parentage	Important traits	Institutes/ universities
CO 1	1968	Selection	It is a selection from a type collected from Tirunelveli (*Amaranthus dubius* Martex Thell.)	It is suitable for growing for tender greens (Mulaikeerai) and immature stems (Thandukkeerai) which are thick and fleshy. The crop duration is 25 days for mulaikeerai and 50–60 days for thandukeerai. It yields 7–8 tonnes of greens/ha at 25 days after sowing. The leaves are broad, thick and dark green in colour. The seeds are very small and black	TNAU, Coimbatore
CO 2	1979	Selection	It is a selection from a germplasm type of Thanjavur, *A. tricolor* L. (syn. *A gangeticus*)	It yields 10.75 tonnes/ha of greens. Seeds are black, bold and exhibit early germination with vigorous growth. The plants are erect with moderate branching	TNAU, Coimbatore
CO 3	1988	Selection	It is a selection from the local type	It is a clipping type (*A. tristis*). It is called Arakkeerai or Killukeerai in Tamil. It lends itself for 10 clippings, commencing from 20 days after sowing and provides a continuous supply of luscious tender green for three months	TNAU, Coimbatore
CO 4	1989	Selection	It is a green cum grain type from	The seeds are rich in protein (15.95%) and essential amino acids like lysine	TNAU, Coimbatore

AMARANTHUS : *Contd.*

Variety	Year of release	Breeding method	Pedigree/ parentage	Important traits	Institutes/ universities
			A. hypochondriacus L.	(7.5 mg/100 g), phenylalanine (5 mg/100 g), leucine (1.2 mg/100 g) and isoleucine (1.8 mg/100 g)	
CO 5	1998	Selection	It is a single plant selection from germplasm (A166-1)	The plants are medium in height with high biomass and nutritive value. The stem and petiole are pinkish red in colour. The leaves are large obovate. It is also ideal for patio or container cultivation. The crop can be harvested at 30 days for mulaikeerai (10 tonnes/ha) and at 50 days for thandukeerai (30 tonnes/h)	TNAU, Coimbatore
Arka Suguna	1996	Selection	Pure line selection from an exotic collection from Taiwan	Light green, succulent stem and broad leaves, first harvest in 25–30 days after sowing and 5–6 cuts in 90 days, moderate ressiatnt to white rust, yield 25–30 tonnes/ha	IIHR, Bangalore
Arka Arunima	1998	Selection	Pure line selection (IIHR 49)	*A. tricolor*, fast growing muticut variety with purple coloured leaves and stem, resistant to white rust under epiphytotic conditions	IIHR, Bangalore
Arun	1992	Selection	Palapoor local	—	KAU, Thrissur

AMARANTHUS : *Contd.*

Variety	Year of release	Breeding method	Pedigree/ parentage	Important traits	Institutes/ universities
Renusree	2006	Selection	—	Green amaranth evolved through selection. High-yielding variety (15.5 t/ha) having green leaves and purple stem with low anti-nutritional factors	KAU, Thrissur
Krishna-sree	2006	Selection	—	Red amaranth evolved through selection. High yielding (14.8 tonnes/ha) with high nutritive value and low anti-nutritional factors	KAU, Thrissur
Chhoti Chaulai	—	Selection	Selection from Indigenous material	*A. blitum*, plants dwarf, succulent small leaves	IARI, New Delhi
Badi Chaulai	—	Selection	Selection from Indigenous material	*A. tricolor*, plants tall, stems are thick and tender, leaves green and large	IARI, New Delhi
Pusa Kiran	1990	Selection	Selection from Indigenous material	Suitable for *kharif*, ready for first picking in 21–25 days after sowing, yield 35 tonnes/ha	IARI, New Delhi
Pusa Kirti	—	Selection	Selection from Indigenous material	Suitable for summer cultivation, ready for first picking in 30–35 days after sowing, yield 50–55 tonnes/ha	IARI, New Delhi

AMARANTHUS : *Contd.*

Variety	Year of release	Breeding method	Pedigree/ parentage	Important traits	Institutes/ universities
Pusa Lal Chauli	—	Selection	Selection from Indigenous material	Suitable for both summer and rainy season, The red dye extracted from this plant can be used as natural food additive and also be used in textile or wollen industry for dyeing.	IARI, New Delhi

SPINACH

Variety	Year of release	Breeding method	Pedigree/ parentage	Important traits	Institutes/ universities
Arka Anupama	2001	Pedigree method	IIHR 10 × IIHR 8	Ressitant to cercospora leaf spot under field conditions, leaves rich in chlorophyll, carotene, vitamin C and iron	IIHR, Bangalore
Ooty 1	1995	Selection	It is a selection from local type	Leaves are green and rich in vitamin A, first picking can be obtained in 45 days after sowing and continued at 15 days intervals for a period of 2 years. It yields potential is 15 tonnes/ha of leaves. The carotene content is high	HRS, TNAU, Ooty

with good fertility is best-suited for spinach, though it can be grown in a wide types of soils. Acidic soils are harmful, the optimum pH being 6–6.5.

In spinach, 4 sex types exist. They are externally male, vegetative male, female and monoecius. Male plants bloom early and hence are poor yielders. Female plants give fairly high yield because they flower late and have longer vegetative phase. There are prickly-seeded varieties and smooth-seeded varieties. Similarly, savoy (wrinkled), semi-savoy and smooth leaved varieties exist.

MORINGA

Botanical name: *Moringa oleifera*
Family: Moringaceae

Drumstick is an important perennial multipurpose vegetable grown widely in India. Popularly known as 'Ganigana', 'Mullakkai', 'Murrugi', 'Sahjan' and 'Muringa', its tender leaves and immature or half-mature fruits are eaten. It is a delicacy in the south Indian households. It is most popular for its distinct, appealing flavouring fruits. The flower buds are also used for culinary purposes. Besides, it has medicinal value also. It is most commonly grown in tropical or semi-tropical conditions. Bush types grow in all types of soils. Sandy red soil or black soil is well-suited. A pH of 6.0–7.5 is ideal. The varieties of drumstick are classified as annual and perennial ones. The annuals are short-duration crops. These are propagated by seeds, while perennials by stem cuttings.

If there is more flower clustering on tree, its yield is low. An individual inflorescence carries 19–126 flowers, average being 53. But it results only in one fruit. Normally 3–5 fruits/inflorescence are obtained. Fruit setting ranges from 1.0 to 2.8%. The fruit setting varies from tree to tree. The fruit length varies from 25 to 100 cm. Each fruit weighs around 230 g containing 10–20 seeds each on an average. Annual drumstick starts bearing from 6th month onwards, providing 200 fruits/plant/annum. Its average yield is 52 tonnes/ha.

ROOT, TUBER AND BULBOUS CROPS

POTATO

Botanical name: *Solanum tuberosum* L.
Family: Solanaceae

Potato is an herbaceous perennial cultivated as an annual, and is susceptible to frost and freezing. In production, it is a cool season

MORINGA

Variety	Year of release	Breeding method	Pedigree/ parentage	Important traits	Institutes/ universities
PKM 1	1989	Selection	It is a pureline selection from seed moringa type (Eppodumvendran of Tirunelveli) for six generations	The pods are 70 cm long, 6.3 cm in girth and weigh about 150 g. The average yield is 218 pods/tree/year, weighing about 35 kg, 52.8 tonnes/ha. The pods attain edible maturity 65 days after flowering	HC&RI, TNAU, Periyakulam
PKM 2	2000	Hybrid derivative	Hybrid derivative developed by cross between MP31 (Eppodumvendran local) × MP28 (Arasaradi local)	The pods are 125.2 cm long with 8.3 cm girth, weighing 280 g. Pods are less seeded, more fleshy and delicious. The average number of fruits is 220/tree/ year	HC&RI, TNAU, Periyakulam

POTATO

Variety	Year of release	Breeding method	Pedigree/ parentage	Important traits	Institutes/ universities
Kufri Kisan (PS 454)	1958	Hybridization and clonal selection	Up-to-date (E) × Sd.16 (E)	Maturity late; tubers white, round, eyes deep and prominent eyebrows and flesh white	CPRI, Shimla

POTATO : *Contd.*

Variety	Year of release	Breeding method	Pedigree/ parentage	Important traits	Institutes/ universities
Kufri Kuber (ON 2236)	1958	Hybridization and clonal selection	(*S. curtilobum* × *S. tuberosum*) × *S. andigena*	Maturity medium; tubers white, oval, tapering towards crown end, eyes medium deep, flesh white; resistant to PLRV and immune to PVY	CPRI, Shimla
Kufri Kumar (S 1758)	1958	Hybridization and clonal selection	Lumbri (E) × Katahdin (E)	Maturity late; tubers white, oval, tapering towards heel end, eyes fleet, flesh white; immune to wart and resistant to charcoal rot	CPRI, Shimla
Kufri Kundan (Hybrid 9)	1958	Hybridization and clonal selection	Ekishirazu (E) × Katahdin (E)	Maturity medium; tubers white, round-oval, flattened, eyes medium deep, flesh creamy; resistant to charcoal rot	CPRI, Shimla
Kufri Red	1958	Clonal selection	Clonal selection of Darjeeling Red Round	Maturity medium; tubers red, round, eyes medium deep and flesh yellow	CPRI, Shimla
Kufri Safed	1958	Clonal selection	Clonal selection of Phulwa	Maturity late; tubers white, round, eyes medium deep and red-purple and flesh light yellow	CPRI, Shimla
Kufri Neela (A 1528)	1963	Hybridization and clonal selection	Katahdin (E) × Shamrock (E)	Maturity late; tubers white, ovoid, eyes medium deep, flesh white; resistant to cyst nematodes	CPRI, Shimla

POTATO : *Contd.*

Variety	Year of release	Breeding method	Pedigree/ parentage	Important traits	Institutes/ universities
Kufri Sindhuri (C 140)	1967	Hybridization and clonal selection	Kufri Red (I) × Kufri Kundan (I)	Maturity late; tubers round, red, eyes medium deep, flesh creamy; moderately resistant to early blight and tolerant to PLRV	CPRI, Shimla
Kufri Alankar (A 3649)	1968	Hybridization and clonal selection	Kennebec (E) × ON-2090 (I)	Maturity medium; tubers white, ovoid, eyes fleet, flesh white; moderately resistant to late blight	CPRI, Shimla
Kufri Chamatkar (ON 1202)	1968	Hybridization and clonal selection	Ekishirazu (E) × Phulwa (I)	Maturity late; tubers white, round, eyes medium deep, flesh yellow; resistant to early blight, charcoal rot and immune to wart	CPRI, Shimla
Kufri Chandra-mukhi (A 2708)	1968	Hybridization and clonal selection	Seedling 4485 (E) × Kufri Kuber (I)	Maturity early; tubers white, oval, slightly flattened, eyes fleet and flesh white	CPRI, Shimla
Kufri Jeevan (SLB/E 427)	1968	Hybridization and clonal selection	M 109-3 (E) × Sd-698-D (E)	Maturity late; tubers white, oval, eyes fleet, flesh white; moderately resistant to late blight	CPRI, Shimla
Kufri Jyoti (SLB/ Z-389 b)	1968	Hybridization and clonal selection	3069 d (4) (E) × 2814 a (1) (E)	Maturity medium; tubers white, oval, eyes fleet, flesh white; moderately resistant to late blight and immune to wart	CPRI, Shimla

POTATO : *Contd.*

Variety	Year of release	Breeding method	Pedigree/ parentage	Important traits	Institutes/ universities
Kufri Khasigaro (SLB/A-67)	1968	Hybridization and clonal selection	Taborky (E) × Sd-698-D (E)	Maturity late; tubers white, round-oval, eyes deep, flesh creamy; moderately resistant to late blight and early blight	CPRI, Shimla
Kufri Naveen (SLB/E-402)	1968	Hybridization and clonal selection	3070 d (4) (E) × Sd. 692-D (E)	Maturity late; tubers white, oval, eyes fleet, flesh white; moderately resistant to late blight and immune to wart	CPRI, Shimla
Kufri Neelamani (A 7871)	1968	Hybridization and clonal selection	Kufri Kundan (I) × 134-D (E)	Maturity late; tubers white, oval, flattened, eyes fleet, flesh white; moderately resistant to late blight	CPRI, Shimla
Kufri Sheetman (C 3745)	1968	Hybridization and clonal selection	Craigs Defiance (E) × Phulwa (I)	Maturity medium; tubers white, oval, eyes fleet, flesh creamy; moderately resistant to charcoal rot and tolerant to frost	CPRI, Shimla
Kufri Muthu (SLB/Z-785)	1971	Hybridization and clonal selection	3046 (1) (E) × M 109-3 (E)	Maturity medium; tubers white, round-oval, eyes medium deep, flesh white; moderately resistant to late blight and immune to wart	CPRI, Shimla
Kufri Lauvkar (A 7416)	1972	Hybridization and clonal selection	Serkov (E) × Adina (E)	Maturity early; tubers white, round, eyes fleet, flesh white; tolerant to heat	CPRI, Shimla

POTATO : *Contd.*

Variety	Year of release	Breeding method	Pedigree/ parentage	Important traits	Institutes/ universities
Kufri Dewa (C 3804)	1973	Hybridization and clonal selection	Craigs Defiance (E) × Phulwa (I)	Maturity late; tubers white, round, eyes pink and medium deep, flesh creamy; tolerant to frost	CPRI, Shimla
Kufri Badshah (JF 4870)	1979	Hybridization and clonal selection	Kufri Jyoti (I) × Kufri Alankar (I)	Maturity medium; tubers white, oval, eyes fleet, flesh white; moderately resistant to late blight, early blight and PVX	CPRI, Shimla
Kufri Bahar (E 3797)	1980	Hybridization and clonal selection	Kufri Red (I) × Gineke (E)	Maturity medium; tubers white, round-oval, eyes fleet, flesh white; immune to wart	CPRI, Shimla
Kufri Lalima (BS/C-1753)	1982	Hybridization and clonal selection	Kufri Red (I) × AG-14 (Wis. × 37) (E)	Maturity medium; tubers red, round, eyes medium deep, flesh white; moderately resistant to early blight and resistant to PVX	CPRI, Shimla
Kufri Sherpa (F 5242)	1983	Hybridization and clonal selection	Ultimus (E) × Adina (E)	Maturity medium; tubers white, round flattened, eyes medium deep, flesh creamy; moderately resistant to early blight and late blight, and immune to wart	CPRI, Shimla
Kufri Swarna (PCN/76-110)	1985	Hybridization and clonal selection	Kufri Jyoti (I) × $(VTn)^2$ 62.33.3 (E)	Maturity medium; tubers white, round-oval, eyes fleet, flesh white; moderately resistant to late blight and early blight,	CPRI, Shimla

POTATO : *Contd.*

Variety	Year of release	Breeding method	Pedigree/ parentage	Important traits	Institutes/ universities
				immune to wart, highly resistant to cyst nematodes	
Kufri Megha (SS/C-562)	1989	Hybridization and clonal selection	SLB/K-37 (I) × SLB/Z-73 (I)	Maturity late; tubers white, round-oval, eyes fleet, flesh white; resistant to late blight	CPRI, Shimla
Kufri Jawahar (JH-222)	1996	Hybridization and clonal selection	Kufri Neelamani (I) × Kufri Jyoti (I)	Maturity medium; tubers white, round-oval, eyes fleet, flesh white; moderately resistant to late blight and immune to wart	CPRI, Shimla
Kufri Sutlej (JI-5857)	1996	Hybridization and clonal selection	Kufri Bahar (I) × Kufri Alankar (I)	Maturity medium; tubers white, oval, eyes fleet, flesh white; moderately resistant to late blight and immune to wart	CPRI, Shimla
Kufri Ashoka (PJ-376)	1996	Hybridization and clonal selection	EM/C-1020 (I) × Allerfrüheste Gelbe (E)	Maturity early; tubers white, ovoid, eyes medium-deep, flesh white	CPRI, Shimla
Kufri Pukhraj (JEX/C-166)	1998	Hybridization and clonal selection	Craigs Defiance (E) × JEX/B 687 (I)	Maturity early; tubers white, oval, slightly tapered, eyes fleet, flesh yellow; resistant to early blight, moderately resistant to late blight and immune to wart	CPRI, Shimla

POTATO : *Contd.*

Variety	Year of release	Breeding method	Pedigree/ parentage	Important traits	Institutes/ universities
Kufri Chipsona-1 (MP/90-83)	1998	Hybridization and clonal selection	MEX 750826 (E) × MS/ 78-79 (I)	Maturity medium; tubers white, oval, eyes fleet, flesh white; resistant to late blight; suitable for chips and French fries	CPRI, Shimla
Kufri Chipsona-2 (MP/91-G)	1998	Hybridization and clonal selection	F-6 (E) × QB/B 92-4 (I)	Maturity medium; tubers white, round, eyes fleet, flesh creamy; resistant to late blight, tolerant to frost and suitable for chips	CPRI, Shimla
Kufri Giriraj (SM/85-45)	1998	Hybridization and clonal selection	SLB/J-132 (I) × EX/A 680-16 (I)	Maturity medium; tubers white, oval, eyes fleet, flesh white; resistant to late blight and immune to wart	CPRI, Shimla
Kufri Anand (MS/82-717)	1999	Hybridization and clonal selection	Kufri Ashoka (I) × PH/F 1045 (I)	Maturity medium; tubers white, oval-oblong, eyes fleet, flesh white; moderately resistant to late blight and immune to wart	CPRI, Shimla
Kufri Kanchan (SE/I-1307)	1999	Hybridization and clonal selection	SLB/Z 405 (A) (I) × Pimpernel (E)	Maturity medium; tubers red, oval-oblong, eyes fleet, flesh creamy; moderately resistant to late blight and immune to wart	CPRI, Shimla
Kufri Arun (MS/92-2105)	2005	Hybridization and clonal selection	Kufri Lalima (I) × MS/82-797 (I)	Maturity medium; tubers red, oval, eyes medium deep, flesh creamy; moderately resistant to late blight	CPRI, Shimla

POTATO : *Contd.*

Variety	Year of release	Breeding method	Pedigree/ parentage	Important traits	Institutes/ universities
Kufri Pushkar (JW-160)	2005	Hybridization and clonal selection	QB/A 9-120 (I) × Spatz (E)	Maturity medium; tubers white, round-oval, eyes medium deep, flesh light yellow; resistant to late blight and immune to wart	CPRI, Shimla
Kufri Shailja (SM/87-185)	2005	Hybridization and clonal selection	Kufri Jyoti (I) × EX/A 680-16 (I)	Maturity medium; tubers white, round-oval, eyes fleet, flesh white; moderately resistant to late blight	CPRI, Shimla
Kufri Chipsona-3 (MP/97-583)	2006	Hybridization and clonal selection	Kufri Chipsona-2 (I) × MP/91-86 (I)	Maturity medium; tubers white, round-oval, eyes fleet, flesh creamy; resistant to late blight, suitable for chips and French fries	CPRI, Shimla
Kufri Surya (HT/92-621)	2006	Hybridization and clonal selection	Kufri Lauvkar (I) × LT-1 (E)	Maturity early; tubers white, oblong, eyes fleet, flesh creamy; immune to wart, resistant to hopper burn and heat tolerant	CPRI, Shimla
Kufri Himalini (SM/91-1515)	2006	Hybridization and clonal selection	I-1062 (E) × Bulk Pollen (4 genotypes)	Maturity medium; tubers white, oval-oblong, eyes shallow, flesh creamy; moderately resistant to late blight	CPRI, Shimla
TPS Population 92-PT-27	2007	Hybridization	83-P-47 (I) × TPS/D-150 (I)	Uniform and high-yielding with resistance to late blight. Both parents flower under short days in the plains. Suitable for cultivation in eastern region	CPRI, Shimla

POTATO : *Contd.*

Variety	Year of release	Breeding method	Pedigree/ parentage	Important traits	Institutes/ universities
Kufri Himsona (MP/94-644)	2008	Hybridization and clonal selection	MP/92-35 (I) × MP/91-G (I)	Maturity medium, tubers white, round-oval, eyes fleet, flesh creamy; resistant to late blight and suitable for chips	CPRI, Shimla
Kufri Sadabahar (MS/93-1344)	2008	Hybridization and clonal selection	MS/81-145 (I) × PH/F-1545 (I)	Maturity medium; tubers white, oblong, eyes fleet, flesh white; resistant to late blight	CPRI, Shimla
Kufri Girdhari (SM/93-237)	2008	Hybridization and clonal selection	SS/C-562 (I) × Bulk Pollen (10 genotypes)	Maturity medium; tubers white, oval-oblong, eyes fleet, flesh pale yellow; resistant to late blight	CPRI, Shimla
Kufri Khyati (J/93-86)	2008	Hybridization and clonal selection	MS/82-638 (I) × Kufri Pukhraj (I)	Maturity early; tubers white, oval, eyes fleet, flesh creamy; moderately resistant to late blight	CPRI, Shimla
Up-to-date	1973	Introduction	UK	Early; tubers white, oval and smooth	—

crop, with optimal average temperature of 50–65°F. In most regions, its yield is more when the plant date falls shortly after last frost. The commercially significant portion of its plant is tuber, which is a swollen underground stem. The swelling of tuber is due to translocation and storage of photosynthates (carbohydrates) which occurs as the aerial portion of the plant reaches maturity. Lodging of aerial plant signifies the filling of tubers and readiness for harvesting. Tubers are used in commercial propagation since the true seed is heterozygous and highly variable, used primarily in crop improvement.

The potato inflorescence is a broad, flat topped cyme. The primary inflorescence is followed by a second and third order blooming as the previous dies (staggered blooming dates). This can be another means by which to estimate the maturity of the plant. Individual flowers are complete, with calyx, corolla, stamens and pistil. Flower colour can range from creamy white to yellow, pink, purple, or striped depending upon cultivar. Bumblebees are primary means of cross pollination; self-pollination most often occurs. A significant amount of potato cultivars are either pollen sterile or fail to set fruits because of some other means.

If established at all, fruits are small (to $1^1/_4$" diameter) and green, resembling a small tomato. The fruit is a berry with seeds in a mucilagenous pulp. Seeds are flattened and ovate, with up to 300 seeds per fruit. Sexual reproduction is the primary mechanism for crop improvement, outside of this, the fruit and seed are of little value.

The breeding objectives are as follows:

- High tuber yield.
- Earliness.
- Photoperiod insensitivity.
- Responsiveness to fertilizer.
- Better keeping quality (resistance/tolerance to shrinkage, rottage and accumulation of sugars, especially reducing sugars and reasonable dormancy).
- Better quality tubers.
- Round, medium sized with shallow eyes and free from greening for general consumption.
- High vitamin C and protein content.
- High specific gravity (dry-matter content) suitable for French fries, chips and dehydrated products.

- Low sugar content for chips and French fries to avoid browning.
- Resistant to diseases and nematodes.

CASSAVA

Botanical name: *Manihot esculenta* Crantz
Family: Euphorbiaceae

Cassava is most important staple food crop growing widely under diverse environmental conditions in tropical regions. The tuberous root crop has its origin in South America. It remains as the most reliable source of food for more than 700 million subsistence farmers in Africa, Asia and Latin America. The sub-Saharan Africa, currently accounts 54% calorie contribution of total world production (FAO, 2005). Cassava, in its calorific contribution remains only fourth after rice, sugarcane and corn with a production of 202 million tonnes. Africa is largest cassava producer (108 million tones) from 12 million ha. In India, it is mainly grown in southern states *viz*. Kerala, Tamil Nadu and Andhra Pradesh. In Kerala, it remains as a major food crop, whereas in Tamil Nadu and Andhra Pradesh it is grown for industrial production of starch and sago. India ranks first and seventh in the productivity (28 tonnes/ha) and production (7 million tonnes). Cassava is fifth important starchy food crop in India after rice, wheat, potato and maize.

Popularly known as tapioca, produces more calories/unit area. Owing to drought tolerance/drought avoidance, wide flexibility to adverse soil, nutrient and management conditions, it is a very important tropical crop. The drought tolerance is mainly due to in-built mechanism to shed or drop the leaves under adverse soil moisture conditions to facilitate slowdown of all vital activities of the plant. As cassava has not determined harvesting time after which it spoils, farmers can have a staggered harvesting rather than on a set date. This adds to the advantage in cassava-based cropping system. Being a photo-insensitive crop, cassava can be profitably cultivated throughout the year with irrigation.

Cassava is monoecious with 36 chromosomes, highly heterozygous due to its out-crossing nature. It is vegetatively propagated mainly by means of stem cuttings and to some extent through seeds. It is monoecious: male and female flowers are located on the same plant. Some varieties flower frequently and regularly, while others flower rarely or not at all. Environmental factors, such as photoperiod and temperature, influence flowering in cassava. Male flowers develop near the tip, while female flowers develop closer to the base of the inflorescence. Though majority of the genotypes of cassava are found

CASSAVA

Variety	Year of release	Breeding method	Pedigree/ parentage	Important traits	Institutes/ universities
H 97	1971	Hybrid	Manjavella × Acc. No. 300	Duration 10 months; average yield 25–35 tonnes/ha; starch 27–31%; cooking quality good	CTCRI, Thiruvananthapuram
H 165	1971	Hybrid	Chadayamangalam Vella × Kalikalan	Duration 8–9 months; average yield 33–38 tonnes/ha; starch 23–25%; cooking quality cood; cyanogen 150–165 mg/g	CTCRI, Thiruvananthapuram
H 226	1971	Hybrid	M 4 × Ethakkakaruppan	Duration 10 months; average yield 30–35 tonnes/ha; starch 28–30%; cooking quality good; cyanogen 180–200 mg/g	CTCRI, Thiruvananthapuram
Sree Visakham (H 1687)	1977	Hybrid	Acc. No. 1501 × S 2312	Duration 10 months; average yield 35–38 tonnes/ha; starch 25–27%; cooking quality good; cyanogens 35–40 mg/g; carotene 466 IU 100/g	CTCRI, Thiruvananthapuram
Sree Sahya (H 2304)	1977	Multiple cross hybrid	Acc. Nos. 468, 174, 3027, 1310, 82, 3939, 3588 and M 4	Duration 10–11 months; average yield 35–40 tonnes/ha; starch 29–31%; cooking quality good; cyanogen 75–85 mg/g	CTCRI, Thiruvananthapuram
Sree Prakash	1987	Selection	Indigenous germplasm collection (S 856)	Duration : 7 months; average yield: 30–35 tonnes/ha; starch 29–31%; cooking quality good; cyanogen : 30–50 mg/g	CTCRI, Thiruvananthapuram

CASSAVA : ***Contd.***

Variety	Year of release	Breeding method	Pedigree/ parentage	Important traits	Institutes/ universities
Sree Harsha (2-14)	1996	Triploidy Breeding	OP-4 (2x) × H 2304 (4x)	Duration : 10 months; average yield: 35–40 tonnes/ha; starch 38–41%; cooking quality good; cyanogen : 40–55 mg/g	CTCRI, Thiruvananthapuram
Sree Jaya	1998	Selection	Indigenous germplasm collection (CI 649) from Kottayam, Kerala	Duration : 6–7 months; average yield: 26–30 tonnes/ha; starch 24–27%; cooking quality excellent; cyanogen : 40–50 mg/g	CTCRI, Thiruvananthapuram
Sree Vijaya	1998	Selection	Indigenous germplasm collection (CI 731) from Thiruvananthapuram, Kerala	Duration : 6–7 months; average yield: 25–28 tonnes/ha; starch 27–30%; cooking quality excellent; cyanogen : 40–60 mg/g	CTCRI, Thiruvananthapuram
Sree Rekha (TCH 1)	2000	Top cross hybrid	TMS 63198 (4)-one cycle of selfing)- R_1-S_1-2; R_1-S_1-2 × H 1687	Duration : 8–10 months; average yield: 45–48 tonnes/ha; starch 28–30%; cooking quality excellent; cyanogen : 49–60 mg/g	CTCRI, Thiruvananthapuram
Sree Padmanabha	2006	Introduction	Exotic germplasm from IITA, Nigeria	Duration : 9–10 months; average yield: 38 tonnes/ha; Starch 26%; cooking quality excellent; cyanogen : 38.20 mg/g	CTCRI, Thiruvananthapuram

CASSAVA : *Contd.*

Variety	Year of release	Breeding method	Pedigree/ parentage	Important traits	Institutes/ universities
Sree Prabha (TCH 2)	2000	Top cross hybrid	TMS 63173 ((4)-one cycle of selfing)-R_1-S_1-72; R_1-S_1-72 × H 1687	Duration : 8–10 months; average yield: 40–45 tonnes/ha; starch 26–29%; cooking quality excellent; cyanogen : 50–85 mg/g	CTCRI, Thiruvananthapuram
CO 1	1977	Selection	It is a clonal selection from a local type (ME 7) collected from Triuchirapalli district	Tubers skin whitish brown with cream rind and white flesh. Starch content is 35% with a total outturn of 10.33 tonnes of starch/ha. The clone yields on an average of 29.97 tonnes of tubers/ha. The clone records very low incidence of mosaic virus ranging from 0 to 8%. The HCN content of tubers is very low *i.e.* 10 µg/g in the flesh and 185 µg/g in the rind	TNAU, Coimbatore
CO 2	1984	Selection	It is a clonal selection (ME 167) from an open-pollinated seedling progenies raised from seeds collected from Thiruvarur district	It exhibits field tolerance to cassava mosaic virus disease. It is a branching type with compactly arranged tubers having brown skin, creamy white rind and creamy white flesh. It flowers profusely. It is suitable for consumption as well as starch industry. It yields 35–37 tonnes/ha in crop duration of 8–8.5 months.	TNAU, Coimbatore

CASSAVA : ***Contd.***

Variety	Year of release	Breeding method	Pedigree/ parentage	Important traits	Institutes/ universities
CO 3	1993	Selection	It is a clonal selection (ME 120-1) from seedling progenies of open-pollinated seeds obtained from IITA, Ibadan, Nigeria	The plants exhibit branching habit. Since the plants are heavily branching, periodical removal of sprouts is necessary. It yields on an average of 42.58 tonnes/ha of tubers under irrigated and 27.31 tonnes/ha of tubers under rainfed conditions respectively.	TNAU, Coimbatore
CO(Tp)-4	2002	Selection	Clonal selection from seedling progenies of hybrid SM-1679	Tubers are long cylindrical with brown skin and white flesh. Starch content 40%. Suitable for both rainfed and irrigated conditions. Duration 300 days and yield of 50 tonnes/ha	TNAU, Coimbatore
MVD 1	1983	Selection	—	It yields 34.5 tonnes/ha in a crop duration of 9 months. The tubers contain 35% starch. It is a non-branching type and exhibits field tolerance to mosaic virus disease. It is suitable for staple food and industrial uses.	State Department of Horticulture, Tapioca Experiment Station, Mulluvadi, Salem, Tamil Nadu

to be diploids, spontaneous occurrence of triploids (3n) and tetraploids (4n) are also reported. These triploids and tetraploids differ from diploid plants in plant vigour and leaf shape and size. Triploids usually grow and yield better than tetraploid and diploid plants.

The main breeding objective of cassava is, high-yielding, high starch hybrids with CMD resistance for industrial areas.

SWEET POTATO

Botanical name: *Ipomoea batatas* (L.) Lam
Family: Convolvulaceae

Sweet potato is an important crop in many parts of the world. It is grown throughout tropical and warm temperate regions of the world. Globally sweet potato is the seventh most important food crop after wheat, rice, maize, potato, barley and cassava with a production of 140 million tones. Sweet potato is the second most important root and tuber crop in the world after potato. Asia is the world's largest sweet potato-producing continent, with a 129 million tonnes annual production. China with 121 million tonnes accounts for 86% of world production.

Of the 45 genera and 1000 species of this family only *Ipomoea batatas* is of economic importance as food. This crop originated from Central or South America. Highly heterozygous crop, it has a chromosome number of 2n (6x) = 90, it is considered to be a natural hexaploid. Sweet potato cultivars are self-incompatible, when self-pollinated it may not set seeds. Sweet potato produced in Asia is mainly used animal feed, the remaining primarily used for human consumption and limited extend as a raw material for industrial purposes as a starch and for alcohol production. The starch is used for manufacturing of adhesives, textile and paper sizing, and confectionery and baking industries.

The flowers of sweet potato are borne solitarily or on cymose inflorescence that grow vertically upward from leaf axils. Each flower has five united sepals, and petals joined together to form a funnel-shaped corolla tube. The stamens five in number, are attached to the base of corolla tube. The style is relatively short and the stigma is white and two-lobed. The ovary consists of two carpels each of which contains one locule. Each locule consists of two ovules and is superior ovary. The pollination is by insects, particularly bees. The main objective of sweet potato breeding in India is for the developing high-yielding and high starch varieties for consumption and developing orange-fleshed sweet potato rich in β-carotene content lines for vitamin A deficiency eradication.

SWEET POTATO

Variety	Year of release	Breeding method	Pedigree/ parentage	Important traits	Institutes/ universities
H 41	1971	Hybrid	Norin × Indigenous cultivar	Duration 120 days; average yield 20–25 tonnes/ha	CTCRI, Thiruvananthapuram
H 42	1971	Hybrid	Vella damph × Triumph	Duration 120 days; average yield 22–25 tonnes/ha	CTCRI, Thiruvananthapuram
H 268 (Varsha)	1983	Double cross hybrid	(Acc. No. 39 × Acc. No. 3) × (Acc. No. 1871 × Acc. No. 1103)	Duration 120 days; average yield 20–22 tonnes/ha	CTCRI, Thiruvananthapuram
Sree Nandini	1987	Seedling selection	Open pollinated seed progeny (76-OP-217) of S.32	Duration 100–105 days; average yield 20–25 tonnes/ha	CTCRI, Thiruvananthapuram
Sree Vardhini	1987	Seedling selection	Open pollinated seed progeny (76-OP-219) of S.13	Duration 100–105 days; average yield 20–25 tonnes/ha	CTCRI, Thiruvananthapuram
Sree Rethna (X-108-2)	1996	Hybrid	S 187 × Sree Vardhini	Duration 90–105 days; average yield 20–26 tonnes/ha	CTCRI, Thiruvananthapuram

SWEET POTATO : *Contd.*

Variety	Year of release	Breeding method	Pedigree/ parentage	Important traits	Institutes/ universities
Sree Bhadra	1996	Seedling selection	Seeds (S 1010) introduced from Nigeria	Duration 90 days; average yield 22 tonnes/ha	CTCRI, Thiruvananthapuram
Sree Arun	2002	Seedling selection	Open pollinated seed progeny RS-III-3) of S.755	Duration 90 days; average yield 20–28 tonnes/ha	CTCRI, Thiruvananthapuram
Sree Varun (56-2)	2002	Seedling selection	Seeds (CIP490056) from CIP, Peru	Duration 100 days; average yield 20–28 tonnes/ha	CTCRI, Thiruvananthapuram
Sree Kanaka	2004	Hybrid	S.187 × H.633	Duration 75–85 days; average yield 13 tonnes/ha; carotene 8.8–10.0 mg/100 g	CTCRI, Thiruvananthapuram
Sankar	1993	Hybrid	H.219 × S.73	Duration 120 days; average yield 14 tonnes/ha	CTCRI, Thiruvananthapuram
Gouri	1993	Hybrid	H.219 × H.42	Duration 110–120 days; average yield 19 tonnes/ha	CTCRI, Thiruvananthapuram

SWEET POTATO : *Contd.*

Variety	Year of release	Breeding method	Pedigree/ parentage	Important traits	Institutes/ universities
Kalinga	2004	Seedling selection	Open-pollinated seed progeny (90/704) of S.285	Duration 105–110 days; average yield 17.2 tonnes/ha	CTCRI, Thiruvananthapuram
Goutam	2005	Seedling selection	Open-pollinated seed progeny (Pol.21-1) of Dhenkanal local, Orissa	Duration 105–110 days; average yield 18.9 tonnes/ha	CTCRI, Thiruvananthapuram
Kishan	2005	Seedling selection	Open-pollinated seed progeny (Pol.13-4) of accession no:1016	Duration 110–120 days; average yield 17 tonnes/ha	CTCRI, Thiruvananthapuram
Sourin	2005	Seedling selection	Open-pollinated seed progeny (Pol.4-9) of accession no:1162	Duration 105–110 days; average yield 16.2 tonnes/ha	CTCRI, Thiruvananthapuram
Kiran	1994	Seedlings selection from OP seeds	—	Early bulking, Duration 110–130 days; average yield 22 tonnes/ha	ANGRAU, Hyderabad

SWEET POTATO : *Contd.*

Variety	Year of release	Breeding method	Pedigree/ parentage	Important traits	Institutes/ universities
Rajendra Sakarkand-92	2001	Seedling selection	—	Spreading, green leaves, 6–7 cylindrical tuber. Red skin and white flesh, suitable for flood prone areas of North Bihar. Duration 75 days and yield of 25 tonnes/ha	RAU, Dholi
Bidhan Jagannath (90/101)	2004	Clonal selection	—	High in carotenoid content. Duration 110–120 days and yield of 21–24 tonnes/ha	BCKV, Kalyani
Birsa Sakharkand (90/704)	2004	Clonal selection	—	High in carotenoid content. Duration 105–110 days and yield of 28 tonnes/ha	BAU, Ranchi
Indira Madhur (IB-90-15-9)	2006	Selection from CIP collections	Selection from CIP-SWA selection	Medium duration. Orange fleshed variety. Duration 120 days and yield of 22 tonnes/ha	IGAU, Raipur
Narendra Shakar-kand-9/ Narendra Malti (NDSP-9)	2004	Clonal selection	—	Duration 120 days and yield of 26 tonnes/ha	NDUAT, Faizabad

SWEET POTATO : *Contd.*

Variety	Year of release	Breeding method	Pedigree/ parentage	Important traits	Institutes/ universities
Narendra Shakar-kand-10 (NDSP-10)	2004	Clonal selection	—	Mid season variety, red skinned, white fleshed tubers. Can be grown as irrigated as well as rainfed. Duration 120 days and yield of 22 tonnes/ha	NDUAT, Faizabad
RNSP-1 (S-30/21)	1999	Selection from open pollinated progeny	—	Duration 110–120 days and yield 22 tonnes/ha	ANGRAU, Rajendra nagar
RNSP-3	2006	Clonal selection	—	Suitable for *kharif* and *rabi* season. Duration 120 days and yield of 15 tonnes/ha	ANGRAU, Rajendra nagar
IGSP-4	2006	Clonal selection	—	Short duration. Duration 90–100 days and yield 25 tonnes/ha	IGAU, Raipur
IGSP-17	2006	Clonal selection	—	Medium duration. Duration 110–120 days and yield of 28 tonnes/ha	IGAU, Raipur
Konkan Ashwini	2000	Selection	Selection from collection of Varor village Tal. Palghar, Dist. Thane	Purple coloured tubers, good in taste, suitable for Rani season. Duration 105 days and yield of 19 tonnes/ha	Dr. BSKKV, Dapoli

SWEET POTATO : *Contd.*

Variety	Year of release	Breeding method	Pedigree/ parentage	Important traits	Institutes/ universities
CO 1	1976	Selection	It is a clonal selection (IB 3) from a type collected from Tiruchirapalli district.	The clone yields on an average 28.33 t/ha of tubers with a range of 22.13–38.54 tonnes/ha. The plants are moderate and less spreading with medium lobbed dark green leaves.	TNAU, Coimbatore
CO 2	1980	Clonal selection	It is a (IB 81) of seedling progeny obtained from open-pollinated seeds of a type (IB 37) in germplasm bank	It yields on an average of 32 tonnes/ha The crop duration is only 110 to 115 days	TNAU, Coimbatore
CO 3	1982	Selection	It is a seedling clone (IB 2837) obtained from the seeds of random mating population of IB 758	Tubers are markedly rich in carotenoids (13.28 mg/100 g) and starch (30.72%). The tubers are medium sized with red skin and deep orange flesh. The root weevil (*Cylas formicarius*) incidence up to 3%.	TNAU, Coimbatore
CO.CIP-1	1999	Selection	It is a clonal progeny of IB 90-10-20 developed through half sib evaluation of open	It yields 31.76 tonnes/ha (tubers). It is adapted to entire Tamil Nadu. The tubers are tolerant to weevil incidence with better market appeal. The tubers have attractive pink skin with yellow	TNAU, Coimbatore

SWEET POTATO : *Contd.*

Variety	Year of release	Breeding method	Pedigree/ parentage	Important traits	Institutes/ universities
			pollinated seedlings, selection from CIP-SWA selection	flesh. Suitable for making cutlet, chips RTS beverage and mash	
Pusa Safed	—	—	—	Tubers white, medium size of excellent quality; maturity 105–120 days. Yield 300 q/ha.	IARI, New Delhi
Pusa Lal	—	—	—	Tubers medium size, thick in the middle, skin red, flesh white, good keeping quality.	IARI, New Delhi
Pusa Sunheri	—	—	—	Flesh light orange, rich in carotene.	IARI, New Delhi
VL-6	1974	Introduction	Introduction from USA (B-219)	Suitable to Uttar Pradesh Hills and plains	VPKAS, Almora
VL Sakar-kand 19	1974	Introduction	Introduced from USA (M 241)	Fruits are characteristic red, round, smooth; skin thick; suitable for processing; plant height 100 cm; maturity 65 DAT; resistant to fruit rot; susceptible to late blight ; yield 200–250 q/ha	VPKAS, Almora

YAMS

Botanical name: *Dioscorea* spp.
Family: Dioscoreaceae

Dioscorea is a genus of over 600 species of flowering plants. It is a native to tropical and warm temperate regions of the world. The vast majority of its species are tropical, with only a few species extending into temperate climates. They are tuberous herbaceous perennials, growing to 2–12 m or more tall. The leaves are spirally arranged, mostly broad heart-shaped. The flowers are individually inconspicuous, greenish-yellow, with six petals; they are mostly dioecious, with separate male and female plants, though a few species are monoecious, with male and female flowers on the same plant. The fruit is a capsule in most species, a soft berry in a few species. Several species, known as yams, are important agricultural crops in tropical regions, grown for their large tubers. Many of these are toxic when fresh, but can be detoxified and eaten. They are particularly important in parts of Africa, Asia, and Oceania.

Cytologically, yams have a basic chromosome number n = 10, but various degrees of polyploidy exist within the same species. The chromosome number ranges wide in different species of *Dioscorea viz. D. bulbifera* (2n = 40–100), *D. esculenta* (2n = 40) and *D. cayenensis* (2n = 140). Within the genus *Dioscorea*, most important species for cultivation in India are: *D. alata* (greater yam), *D. esculenta* (lesser yam) and *D. rotundata* (African yam).

Yams are normally dioecious, with male and female flowers produced on different pants. There are usually more male flowers on each male plant than female flowers on each female plant. Many yam cultivars do not flower at all. The male flowers are borne in panicles produced in the leaf axils. Each male flower is inconspicuous and small with six stamens. The female flower is larger than the male and is borne in spikes in the leaf axils. The ovary has three locules, each of which contains two ovules. There are three stigmas. Even though yam flowers are inconspicuous, their pollination is by insects.

The main breeding objective of yams is to develop high-yielding and good cooking quality tubers for consumption.

ELEPHANT FOOT YAM

Botanical name: *Amorphophallus paeoniifolius* (Dennst.) Nicolson
Family: Araceae (sub Family: Lasioideae)

Elephant-foot yam is widely cultivated in tropical Asia and Africa. It is basically a crop of South-East Asian origin. It occurs in wild form

GREATER YAM

Variety	Year of release	Breeding method	Pedigree/ parentage	Important traits	Institutes/ universities
Sree Keerthi	1987	Clonal selection	Selection (DA-60) from germplasm	Duration 9–10 months; average yield 25–30 tonnes/ha	CTCRI, Thiruvananthapuram
Sree Roopa	1987	Clonal selection	Selection (DA-80) from germplasm	Duration 9–10 months; average yield 25–30 tonnes/ha	CTCRI, Thiruvananthapuram
Sree Shilpa	1998	Hybrid	DA 140 × Sree Keerthi	Duration 8 months; average yield 28 tonnes/ha	CTCRI, Thiruvananthapuram
Sree Karthika	2004	Clonal selection	Selection (DA-199) from germplasm	Duration 9 months; average yield 30 tonnes/ha	CTCRI, Thiruvananthapuram
Orissa Elite	2005	Clonal selection	Selection from Orissa	Duration 6–7 months; average yield 25 tonnes/ha	CTCRI, Thiruvananthapuram
Konkan Ghorkhand	2001	Clonal selection	—	It is high-yielding and erect growing vine produces aerial tubers having purplish green colour stem and simple leaves. Duration 180 days and yield 16 tonnes/ha	Dr BSKKV, Dapoli

GREATER YAM : *Contd.*

Variety	Year of release	Breeding method	Pedigree/ parentage	Important traits	Institutes/ universities
CO 1	1999	Selection	It is a clonal selection from the germplasm introduction	It yields 44.8 tonnes of tubers/ha in a crop duration of 8–8.5 months. The tubers are big with white flesh. Tubers are rich in carbohydrate (28%) and protein (2.5%).	TNAU, Coimbatore
Indu	1998	Clonal selection	TCR-5 (IC-44209)	—	KAU, Thrissur

WHITE YAM

Variety	Year of release	Breeding method	Pedigree/ parentage	Important traits	Institutes/ universities
Sree Priya	1987	Selection	Clonal selection from the African variety, Umidika	Duration 9–10 months; average yield 35 tonnes/ha	CTCRI, Thiruvananthapuram
Sree Subhra	1987	Selection	Clonal selection from the African variety, Iwo	Duration 9–10 months; average yield 35 tonnes/ha	CTCRI, Thiruvananthapuram
Sree Dhanya	1993	Selection	Clonal selection (I-184) from germplasm	Duration 9 months; average yield 20 tonnes/ha	CTCRI, Thiruvananthapuram

LESSER YAM

Variety	Year of release	Breeding method	Pedigree/ parentage	Important traits	Institutes/ universities
Sree Kala	1993	Selection	Selection (DE-55) from African exotic collection — Kombi	Early-maturing (7 months), excellent cooking quality, tubers smooth, oval, average yield 20 tonnes/ha	CTCRI, Thiruvananthapuram
Sree Latha	1983	Clonal selection	Selection from (DE-11) indigenous germplasm	Duration 8 months; average yield 25 tonnes/ha	CTCRI, Thiruvananthapuram
Konkan Kanchan	2001	Clonal selection	—	High-yielding, vine having light green coloured simple alternate leaves, spines at the base of the petiole. Stem is thin, cylindrical and green in colour. Tubers are elongated almost cylindrical with round ends. High-yielding, equal size tubers, white and sweet in taste. Duration 190–210 days with an yield of 18 tonnes/ha	Dr BSKKV, Dapoli

ELEPHANT FOOT YAM

Variety	Year of release	Breeding method	Pedigree/ parentage	Important traits	Institutes/ universities
Sree Padma	1998	Clonal selection	Indigenous (Am15) germplasm collection from Wayanad	Duration 8–9 months; average yield 42 tonnes/ha	CTCRI, Thiruvananthapuram
Sree Athira	2006	Hybrid	Sree Padma × Am 45	Duration 9–10 months; average yield 41 tonnes/ha	CTCRI, Thiruvananthapuram
Gajendra (Kovvur)	1991	Clonal selection	—	Compact plant with medium canopy spread, smooth corms with no cormels, good culinary quality, low acridity, duration 180–210 days and yield of 50 tonnes/ha	ANGRAU, Hyderabad, Andhra Pradesh
Bidhan Kusum	1999	Clonal selection	—	Non-irritant variety, smooth corm with excellent cooking quality, duration 240 days and yield of 50 tonnes/ha	BCKV, Kalyani
Narendra Asha	2006	Clonal selection	—	Early variety, corms non-acrid with good cooking quality, duration 180–210 days and yield of 65 tonnes/ha	NDUAT, Faizabad

in the Philippines, Malaysia, Indonesia, and south-east Asian countries. The genus *Amorphophallus* with 90–170 species occurs in tropical Asia and Africa. Altogether 14 species are reported to occur in India. The *A. paeoniifilius* (Dennst.) Nicolson (syn. *A. campanulatus*), locally known as elephant-foot yam, suran or *jimmikand,* is most popular in our country. It is cultivated in Andhra Pradesh, Gujarat, Maharashtra and Kerala whereas in northern and eastern states its local cultivars are grown in wild forms. The corms are a delicacy and rich in nutrients and it offers considerable scope for adoption as a cash crop due to its high production potential and popularity as a vegetable.

The corms are underground stems which are widely used as vegetable. It is vegetatively propagated by corms. From the central bud of the corm, a solitary dissected tripartite leaf is formed on a long petiole. The petiole (pseudostem) has a distinct pattern of rounded and irregular blotches. The large leaf constitutes the luxuriant outspreading crown like foliage. It is a diploid species having chromosome number of 28 in the somatic cells. It flowers rarely and floral primodium arises from the central bud. The solitary inflorescence developed from the bud consists of a short peduncle of 5–7 cm long and 15–20 cm broad. The spadix holds female flowers at the distal end and terminates in a dark brown spongy sterile appendage. The plants are monoecious. The spadix has tiny flowers : female flowers, no more than a pistil, at the bottom, then male flowers, actually a group of stamens, and then a blank sterile area. This last part, called 'the appendix', consists of sterile flowers, called staminodes, and can be especially large. There is no corolla. There are two main breeding objectives in Amorphophallus, germplasm evaluation and selection and intervarietal hybridization and selection.

The main breeding objective of elephant-foot yam is to develop high-yielding and good cooking quality tubers for consumption.

TARO AND TANNIA

Botanical name: *Colocasia* spp.; *Xanthosoma* spp.
Family: Araceae

Colocasia (taro) is a genus of 6–8 species of flowering plants. It is native to tropical Polynesia and southeastern Asia. Commonly known as "Elephant-ear", or taro, this name is also used for some large-leaved genera, notably *Xanthosoma* (tannia), *Alocasia* (gaint taro), *Cyrtosperma chamissonis* (swamp taro) and Caladium. They are herbaceous perennial plants with a large rhizome on or just below the ground surface. The leaves are large to very large, 20–150 cm long, with a sagittate shape. The elephant's-ear plant gets its name from the leaves, which are shaped like a large ear or shield. They are important food crops only in certain communities in India, south-east Asia, and the Paccific Islands.

The *Colocasia esculenta* and other members of the genus are cultivated for their edible tubers, a traditional starch staple in many tropical areas. The edible types are grown in South Pacific and eaten like potatoes and known as taro, eddo, and dasheen. This famous root vegetable is known as "*Arbi/Arvi*" in the Indian subcontinent where its leaves are also cherished. A favorite Hawaiian dish is made by boiling the starchy underground stem of its plant. In Dakshina Kannada and Udupi districts in Karnataka (India), they are used to make — a popular delicacy; in Kerala (India) they are used to make chembila curry, a tasty delicacy. The stems and roots are also used in preparation of delicacies (ishtu moru curry etc.). In Andhra Pradesh, several delicacies are made either with root or leaves of Chaama. In Gujarat, they are used to make a popular dish called Patra.

The taro, originated in south central Asia, probably in India or Malaysia. During prehistoric times, its cultivation spread to Pacific Islands. Later it was taken to the Mediterranean area and to West Africa, West Indies and tropical parts of America. The chromosome number of taro ranged from 22–42 with 2n = 28. Taro is herbaceous plant consisting of a central corm from which cormels, roots, and shoot arise. The tannia is originated in tropical America where it was first brought under cultivation. From there it spread to south-east Asia, the pacific islands and Africa. Tannia belongs to the genus *Xanthosoma* and most widely cultivated species is *Xanthosoma sagittifolium*.

Flowering in taro is only occasional, and many cultivars have never been known to flower. A plant may bear two or more inflorescences. The peduncle is stout and relatively short. The inflorescence consists of spadix and spathe. The spadix is essentially the spike inflorescence consisting of a central axis on which numerous, small, sessile flowers are attached. The spathe is a large yellowish bract which sheathes the spadix from the base and partly surrounds it. The spathe is about 20 cm long and is rolled inward at its tip. The spadix is 6–14 cm long. The female flowers occur at the base of it, while male flowers occur near the tip. In the region between male and female flowers are a group of sterile flowers. The extreme tip of the spadix has no flowers at all. This sterile appendage of the spadix is much longer in the eddoe types of taro than in the dasheen types. In eddoe, the appendage is longer than the male section of the spadix, while in desheen, it is shorter. Each female

TARO (Colocasia)

Variety	Year of release	Breeding method	Pedigree/ parentage	Important traits	Institutes/ universities
Sree Rashmi	1987	Clonal selection	Indigenous (CI-149) germplasm	Duration 7 months; average yield 18 tonnes/ha	CTCRI, Thiruvananthapuram
Sree Pallavi	1987	Clonal selection	Indigenous collection from Meghalaya (C.221)	Duration 7 months; average yield 16 tonnes/ha	CTCRI, Thiruvananthapuram
Sree Kiran	2004	Hybrid	C.303 × C.383	Duration 6–7 months; average yield 17.5 tonnes/ha	CTCRI, Thiruvananthapuram
Muktakeshi	2002	Clonal selection	Selection from Cuttack, Orissa	Duration 5–6 months; average yield 20 tonnes/ha	CTCRI, Thiruvananthapuram
Pani Saru 2	2005	Clonal selection	Local landrace from Begunia, Orissa	Duration 6–7 months; average yield 13 tonnes/ha	CTCRI, Thiruvananthapuram
Pani Saru 2	2005	Clonal selection	Local landrace from Begunia, Orissa	Duration 6–7 months; average yield 13 tonnes/ha	CTCRI, Thiruvananthapuram

TARO : *Contd.*

Variety	Year of release	Breeding method	Pedigree/ parentage	Important traits	Institutes/ universities
Bidhan Chaitanya	1997	Clonal selection	BCC-1	Plant height 75–80 cm, with entire inclined waxy leaf with blund tip peltate shape and yellowish-greeny colour, branched round to oval cormels. Tubers non-irritant with excellent cooking quality and very mealy and non-aromatic. Duration 180–240 days, yield of 15 t/ha	BCKV, Kalyani, West Bengal
Bidhan Joydeb	1997	Clonal selection	BCC-2	Medium spreading with a plant height of 60–70 cm with entire inclined cup shaped lamina having peltate shape and green in colour. Branched eliptical cormels. Tubers are non-irritant with excellent cooking quality very mealy and non-aromatic. Duration 150–210 days and yield of 12 tonnes/ha	BCKV, Kalyani, West Bengal
Kadma Arvi	2006	Clonal selection	—	Erect, with emerging leaf colour light green, moderate flowering type. Resistant to leaf blight, viral infestation and pests, 18 cormels/plant. Tubers almost non-irritant with excellent cooking quality. Duration 150 days and yield of 17 tonnes/ha	BAU, Ranchi
Ahina Katchu	2004	Clonal selection	—	High yielder better quality tuber. Duration 175 days, yield of 17 tonnes/ha	AAU, Jorhat

TARO : *Contd.*

Variety	Year of release	Breeding method	Pedigree/ parentage	Important traits	Institutes/ universities
PKS-1	2006	Clonal selection	—	Good for chips making. Duration 150 days and yield of 20–31 tonnes/ha	NDUAT, Faizabad
RNCA-1	2006	Clonal selection	—	Medium duration of 180 days and yield of 14–18 tonnes/ha	ANGRAU, Rajendra-nagar
Indira Arvi-1 (BK Col-2)	2003	Clonal selection	—	Erect plant type with emerging leaves with green pigmented margin. Higher number of main marketable cormels (9–10). Cormels are non-acrid with very good cooking quality. Tolerant to water logging condition. Corm shape round and cormel shape club with blunt end. Suitable for cultivation in kharif season. Duration 210–240 days and yield of 22 tonnes/ha	IGAU, Jagdalpur
CO 1	1991	Introduction	It is an introduction from the germplasm	It yields 24.3 tonnes/ha Tubers have high starch content of 22.5% and higher protein content of 2.4%. Tubers have less acidity and good cooking quality	TNAU, Coimbatore
Vallabh Nikki	2007	Selection	Selection from germplasm	High-yielding, resistant to leaf blight, suitable for plains and hills	SVBPUA&T, Meerut

COLOCASIA (Bunda)

Variety	Year of release	Breeding method	Pedigree/ parentage	Important traits	Institutes/ universities
Narendra Bunda-1 (NDB-2)	2001	Clonal selection	—	Plant height 75–125 cm with erect growth habit with clock wise leaf arrangement and erect lamina orientation. Vine colour green with purple streaks on upper and lower portion. Cylindrical corm and eliptical cormels. Resistant to lodging. Corm and cormels are suitable for table purpose as it has very less acridity soft and easy to cook. Duration 180–195 days and yield of 35–40 tonnes/ha	NDUAT, Faizabad
NDB-21	2006	Clonal selection	—	Early maturing. Corms cook easily. Duration 155 days and yield of 45–50 tonnes/ha	NDUAT, Faizabad
NDB-03	2006	Clonal selection	—	Corms cook fast. Duration 121 days and yield of 35–40 tonnes/ha	NDUAT, Faizabad

XANTHOSOMA (Tania)

Variety	Year of release	Breeding method	Pedigree/ parentage	Important traits	Institutes/ universities
Konkan Haritparni	2000	Clonal selection	—	Duration 180 days and yield of 25 tonnes/ha	Dr BSKKV, Dapoli

flower consists essentially of a small ovary bearing a sessile stigma. The ovary has one locule, which contains numerous ovules. Each male flower consists of 2–6 sessile anthers which are fused together. The flowers are fragrant and pollination is by insects, especially flies.

COLEOUS

Botanical name: *Solenostmon rotundifolius* Poir., J.K. Morton.
Family: Lamiaceae

Coleous is an important minor tuber crop grown extensively as a vegetable in most of the homestead gardens in Kelala and Tamil Nadu. It is a small herbaceous bushy annual with succulent stems and aromatic leaves. The plant bears a cluster of heteromorphic tubers with aromatic flavour, which makes it likeable as delicacy among the vegetables. It is generally raised as a monsoon crop and duration is 4–5 months.

Even though profuse flowering is observed during September–November, its crop is completely sterile due to lack of fertile pollen-grains. Highly irregular meiosis and occurrence of desynapsis might have resulted in the complete sterility of the crop. There is a significant difference in tuber size within accessions and not between the accessions.

Its breeding object is to select the photoperiod insensitive, uniform size tuber and high-yielding accession from the germplasm. To develop high yielding and good cooking quality tubers for consumption is main objective for breeding of Coleus.

YAM BEAN

Botanical name: *Pachyrrhizus erosus* (L.) Urban
Family: Fabaceae

Yam bean is leguminous root crop having good nutritional value. The plant is a coarse, hairy, herbaceous twiner with alternate trifoliate leaves. The tubers are simple, turnip like or elongated in shape and flesh of the tuber is white, crispy, juicy, refreshing and sweetish in taste and can be eaten as a raw or cooked. The mature tuber yields starch which is similar to arrow root starch. The pods are poisonous due to the presence of toxic substance 'rotenon'. The powdered seeds are used as an insecticide.

This is a rather, coarse, climbing, herbaceous vine growing from rather large, edible, turnip-shaped, fleshy roots. The leaflets, at least

CHINESE POTATO (COLEUS)

Variety	Year of release	Breeding method	Pedigree/ parentage	Important traits	Institutes/ universities
Sree Dhara	1993	Clonal selection	Selection from indigenous (CP-58) germplasm	Duration 5 months; average yield: 25 tonnes/ha	CTCRI, Thiruvananthapuram
Nidhi	2001	Clonal selection	Clonal selection from NBPGR accession CP 79	—	KAU, Thirssur
Suphala	2006	Mutation	A tissue culture mutant derived from local cultivar	A high yielding (15.93 tonnes/ha) year round cultivable variety with a duration of 120–140 days	KAU, Thrissur
CO 1	1991	Selection	It is a clonal selection from local type introduced from Tenkasi	It yields 31.93 tonnes/ha in crop duration of 180–190 days. The tubers have 21.5% starch. The cooked tubers are tasty and have lesser soil odor	TNAU, Coimbatore

YAM BEAN

Variety	Year of release	Breeding method	Pedigree/ parentage	Important traits	Institutes/ universities
Rajendra Mishrikand-I (RM-I)	1999	Seedling selection from germplasm	—	Good quality, resistant to major pests and diseases, suitable for intercropping with *kharif* maize and arhar. Duration 110–140 days and yield of 35 tonnes/ha	RAU, Dholi

the terminal one, are broader than long, reaching a length of 15 cm and a width of 20 cm; they have a deltoid base and are irregularly and shallowly lobed in the upper half; the lateral leaves are inequilateral. The racemes are up to 45 cm in length, with the lower nodes producing short branches, and the other nodes several flowers each. These flowers are pale blue or blue and white, and 2–2.5 cm long, with the standard about 1.5 cm wide. The pods are about 10 cm long, 10–12 mm wide, flat, and hairy, and contain from 8–10 seeds.

Yam bean is propagated mainly through seeds and it is a self-pollinated crop. Sometimes tubers are used for planting when a particular genotype is desired to be maintained. It grows well on light sandy soil. The plants are pruned once or twice after two months of planting in order to restrict vegetative growth and encourage better tuber development. Crop matures in 6–8 months. Non-flowering plant produces best quality tubers. Hence when grown for tubers, the buds or inflorescence are removed to avoid flowering.

The main breeding objective of yam bean is to develop high-yielding and good cooking/salad quality tuber for consumption.

ONION

Botanical name: *Allium cepa*
Family: Alliaceae

Onion is a popular vegetable grown for its pungent bulbs and flavorful leaves. It is widely grown throughout the world. The bulb is composed of concentric, fleshy, enlarged leaf bases or scales. The outer leaf bases lose moisture and become scaly and the inner leaves generally thicken as bulbs develop. The green leaves above the bulb are hollow and arise sequentially from the meristem at the innermost point at the base of the bulb. The stem is very small and insignificant during vegetative growth. The onion root system is fibrous, spreading just beneath the soil surface to a distance of 30–46 cm. There are few laterals, and total root growth is sparse and not especially aggressive. Therefore, in monoculture, onions tolerate crowding, particularly in loose, friable soils such as peat and muck.

Cultivars differ substantially with respect to threshold daylength required for bulbing. Other factors such as temperature may interact with daylength to modify the bulbing response. In all cultivars, bulbing is accelerated with increasing temperature. Temperature extremes not only affect the rate of bulbing, but also affect the bulb shape. Thick and elongated necks are common in plants exposed to 6°C or lower.

ONION

Variety	Year of release	Breeding method	Pedigree/ parentage	Important traits	Institutes/ universities
N-53	1960	Selection	Local material	Popular var. suitable for *kharif* season all over country. Bulbs globular in shape, scarlet red, medium to large in size, mild in pungency. TSS 11–12%. Poor storer. Maturity 100–110 DAT. Average yield 150–200 q/ha	Department of Agriculture, Maharashtra
N-2-4-1	1985	Selection	Local material	Identified for zones IV, VI and VII. Average bulb diameter is 4–6 cm, globular in shape, pungent in taste. Good storer. Plant matures in 140–145 DAT. TSS 12–13%. Average yield 250–300 q/ ha. Recommended mainly for *rabi* season but can be grown in late *kharif* and *rabi* seasons in Maharashtra	Department of Agriculture, Maharashtra
N-257-9-1	1985	Selection	Local material	Identified for zone VI. Bulbs globose in shape and white in colour. High yielding with good storage. Suitable for *rabi* season in onion growing regions of Maharashtra	Department of Agriculture, Maharashtra
Baswant 780	1986	Selection	Local material	Bulbs attractive red in colour, globular in shape, mild in pungency, less bolting with doubles. TSS 11–12%. Average storer.	MPKV, Rahuri

ONION : *Contd.*

Variety	Year of release	Breeding method	Pedigree/ parentage	Important traits	Institutes/ universities
				Maturity 100–110 DAT. Average yield 200–250 q/ha. Suitable for *kharif* season in Maharashtra	
Phule Safed	1994	Selection	Local material from Kagal	White coloured variety. Bulbs globular in shape with TSS 13%. Suitable for dehydration. Average yield 250–300 q/ha. Storage life is 2–3 months. Recommended for late *kharif* and *rabi* seasons	MPKV, Rahuri
Phule Swarna	2001	Selection	Local material	Bulbs are yellow in colour suitable for export to Europe, Australia and America. Bulb size medium to big. Less pungent, TSS 11.5%, with good storage (4–6 months). Suitable for late *kharif* and *rabi* season. Average yield 240 q/ha	MPKV, Rahuri
Phule Samarth	2004	Selection	Local material	Red bulbs. Suitable for *kharif* and late-*kharif* seasons	MPKV, Rahuri
Pusa Red	1975	Selection	Local material	Short to intermediate day length variety, bulb medium sized, average weight 70–90 g, bronze red in colour, flat to globular in shape, less pungent, good storer. Plants mature 140–145 DAT. TSS 12–13%. Average yield 250 q/ha. Suitable for late *kharif* and *rabi* seasons in Maharashtra	IARI, New Delhi

ONION : *Contd.*

Variety	Year of release	Breeding method	Pedigree/ parentage	Important traits	Institutes/ universities
Pusa White Flat	1975	Selection	Local material	Identified for zone IV, VI and VII. Bulbs attractive white, flattish round. TSS 11–12% with drying ratio 9:1. Maturity 130–135 DAT. Good storer. Suitable for dehydration and green onion. Average yield 325–350 q/ha. Suitable for *rabi*	IARI, New Delhi
Pusa White Round	1975	Selection	Line 106	Identified for zone IV, VI and VII. Bulbs white in colour and roundish flat in shape. TSS 12–13% with drying ratio 8:1. Good storer. Maturity 130–135 DAT. Suitable for dehydration and green onion. Average yield 300–325 q/ha. Suitable for *rabi*	IARI, New Delhi
Pusa Ratnar	1975	Selection	Local material	Developed for zones IV and VII. Bulbs more exposed above ground at maturity, bulbs dark red coloured, obvate to flat, globular, less pungent. Shows neck fall at maturity. Average storer. Plant maturity 125 DAT. TSS 11–12%. Average yield 300 q/ha. Recommended for *rabi* season	IARI, New Delhi
Pusa Madhavi	1987	Selection	Local material of Muzaffarnagar	Identified for Indo-Gangetic plains. Bulbs medium to large in size, light	IARI, New Delhi

ONION : *Contd.*

Variety	Year of release	Breeding method	Pedigree/ parentage	Important traits	Institutes/ universities
				red in colour and flatish round in shape. Good storer. Plants mature in 130–145 DAT. Average yield potential is 300 q/ha. Recommended for *rabi* season	
Early Grano	—	Introduction	USA	Bulbs globular in shape, yellow in colour, very mild in pungency with TSS 6–7%, diameter 7–8 cm. Crop matures 95–110 DAT. Poor storer. Average yield 500–600 q/ha. Suitable for cultivation in plains during late-*kharif* and *rabi* seasons for salad purpose	IARI, New Delhi
Brown Spanish (Long day)	—	Introduction	—	Bulbs brown in colour with mild pungency. TSS 13–14%. Crop matures in 160-180 days and ready for harvesting during August-September. Good storer in hills. Long day type variety for hills. Average yield 280–300 q/ha	IARI, Katrain
Arka Niketan	1987	Mass selection	Local collection, 'IIHR 153'	Identified for zones IV, VII and VIII. Bulbs globular with thin neck, attractive light red colour, uniform, firm, 4–6 cm dia., free from splits, bolters and internal doubles. TSS 12–14%, pungent. Good storer. Matures in 145 DAT. Bulb yield 37–	IIHR, Bangalore

ONION : *Contd.*

Variety	Year of release	Breeding method	Pedigree/ parentage	Important traits	Institutes/ universities
				40 tonnes/ha. Recommended for late *kharif* and *rabi* season. Exported to Gulf and Asian countries	
Arka Kalyan	1987	Selection	Local material of Kalwan taluka of Maharashtra	Identified for IV, VI, VII and VIII zones. Bulbs oval globose shaped, 4–6 cm size with dark red coloured outer scales. TSS 11–12%. Average storer. Matures in 100–110 DAT. Average yield 335 q/ha. Suitable for *kharif* season	IIHR, Bangalore
Arka Bindu	1994	Mass selection	Local rose onion collection 'IIHR 402' from Chickballapur area of Karnataka	Small rose onion, developed for export. Suitable for all three seasons in Karnataka. Bulbs deep pink in colour, small in size with 2.5–3.5 cm diameter, flattish globular in shape, uniform in shape size and maturity, free from bolters and doubles. Highly pungent with 14–16% TSS. Firm bulbs with good storage. Maturity in 100 DAS. Average yield 250 q/ha. Exported to Singapore, Malaysia, Sri Lanka and Bangladesh	IIHR, Bangalore
Arka Pragati	1994	Selection	Local collection IIHR–149 from Nashik	Bulb attractive pink, globular in shape, uniform size with thin neck, highly pungent. Early in maturity, 95–100 DAT. Average yield 200 q/ha. Suitable for *rabi*	IIHR, Bangalore

ONION : *Contd.*

Variety	Year of release	Breeding method	Pedigree/ parentage	Important traits	Institutes/ universities
Arka Pitambar	1996	Pedigree selection	Crossing between two short day local genotypes 'UD 102' (white) and 'IHR 396' (red)	Mildly pungent, medium-sized, globose, firm single centered, thin neck and uniformly yellow coloured bulbs with high TSS, firm bulbs, free from splits, bolters, internal doubles, good keeping quality and high stable bulb yield. Yield 35 tonnes/ha. Identified for southern dry zone of Karnataka. Suitable for export to European Union, USA and Japan	IIHR, Bangalore
Arka Lalima	1996	F_1 hybrid	MS 48 × Sel 14-1-1	Attractive deep red bulbs, uniform, globose and firm. Stores well for 4–5 months. Can be grown in *kharif* and *rabi*. TSS 10–12%. Maturity 140 DAT. Yield 45–50 tonnes/ha. Identified for southern dry zone of Karnataka	IIHR, Bangalore
Arka Kirthiman	1996	F_1 hybrid	MS 65 × Sel 13-1-1	Attractive red bulbs, uniform, globose and firm. Stores well for 4–5 months. Can be grown in *kharif* and *rabi*. TSS 8–10%. Maturity 130 DAT. Yield potential 40–45 tonnes/ha. Identified for southern dry zone of Karnataka	IIHR, Bangalore
Hisar 2	—	Selection	Local material	Bulbs bronze red in colour, globular in shape with tight skin, sweet to pungent	HAU, Hisar

ONION : *Contd.*

Variety	Year of release	Breeding method	Pedigree/ parentage	Important traits	Institutes/ universities
				in taste. TSS 11.5–13.5%. Maturity 120 DAT. Average yield 250 q/ha. Suitable for cultivation during *rabi* in Haryana and Punjab	
HOS 1	2006	Selection	Local material	Red bulbs	HAU, Hisar
Agrifound Dark Red	1987	Selection	Local collection of *kharif* onion grown at Nashik	Identified for plains of Sutlej-Ganga. Bulbs dark red in colour, globular in shape, 4–6 cm in diameter with tight skin, moderately pungent. TSS 12–13%. Maturity in 95–110 DAT. Yield 300–400 q/ha. Average storer. Recommended for *kharif* season all over the country	NHRDF, Nashik
Agrifound Light Red	1993	Mass selection	Local collection of Dindhori area of Nashik	Bulbs globular in shape with tight skin, light red colour and 4–6 cm in diameter. TSS 13%. Good storer. Maturity in 115–120 DAT. Average yield 300–325 q/ha. Recommended for late-*kharif* and *rabi* season all over the country. Can be grown in late-*kharif* season also in Nashik districts of Maharashtra	NHRDF, Nashik
Agrifound White	—	Selection	Local germplasm of white onion	Bulbs globular in shape with tight skin, silvery attractive white colour having	NHRDF, Nashik

ONION : *Contd.*

Variety	Year of release	Breeding method	Pedigree/ parentage	Important traits	Institutes/ universities
			grown during *rabi* season in Nimad area of Madhya Pradesh	4–6 cm diameter. TSS 14–15%, good storer. Maturity 115–120 DAP. Average yield is 200–250 q/ha. Good for dehydration. Suitable for *rabi* season	
Agrifound Rose	—	Selection	Local material from Chickbalapur	Pickling type variety grown in Kolar and Bangalore districts of Karnataka and Cuddapah districts of Andhra Pradesh exclusively for export. Bulbs flattish round in shape, deep scarlet red in colour, 2.3–3.3 cm in diameter. Maturity 95–110 DAS. TSS 15–16%. Average yield 190–200 q/ha. Suitable for *kharif* season in Cuddapah district and all the three seasons in Karnataka	NHRDF, Nashik
Agrifound Red	—	Selection	Local material. Developed at Dindigul	Multiplier onion var. Colour of bulb light red, average size of cluster 7.15 cm and weight 65–67 g. Number of bulblets/cluster 5.79 (avg). Size of bulblet 3.64 cm and weight of single bulblet 8.85 g. Maturity 65–67 DAP. Average yield 180–200 q/ha. Suitable for *kharif* and *rabi*	NHRDF, Nashik
NHRDF Red	2006	Selection	Local material from Patiala area	Dark red bulbs, 5-6 cm dia. TSS 13–14%. Higher yield with better storage.	NHRDF, Nashik

ONION : *Contd.*

Variety	Year of release	Breeding method	Pedigree/ parentage	Important traits	Institutes/ universities
(Line-28)			Punjab	Suitable for *rabi* season. Recommended for Northern, Central and Western India. Maturity 115–120 DAT. Average yield 250–300 q/ha	
VL Piaz 67	1973	Hybrid	BY 2207A/ Shajahanpur local	Plant height 45–50 cm, foliage dark green. Bulbs round with tight, skin, skin, thickness papery, skin of reddish colour, inner scales light reddish in colour. Bulb weight 80–150 g in hills and 50–80 g in plains	VPKAS, Almora
VL Piaz 3	1990	Pedigree	Advance generation cross MS BYG 2207 A (In 13) × In 43 a line derive from a local collection from Uttar Pradesh hills	Medium bulb shape, round, light red, average weight, 80–120 g, average diameter 5.2 cm, white leaf sheath, light pink, maturity mid-season, average yield 200–250 q/ha, dry matter 13.6%, TSS 12%	VPKAS, Almora
Udaipur 101	—	Selection	Local material	Bulbs dark red in colour, flattish globular and sweet with less pungency. TSS 12–14%. Maturity 105–115 DAT. Average yield 200–300 q/ha. Recommended for Rajasthan and adjoining states during *rabi* season	RAU, Rajasthan

ONION : *Contd.*

Variety	Year of release	Breeding method	Pedigree/ parentage	Important traits	Institutes/ universities
Udaipur 102	—	Selection	Local material	Bulbs white in colour, round to flat in shape with low percentage of small bulbs. Diameter of the bulb is 4.5–6.5 cm. Maturity 120 DAT. TSS 12%. Average yield 300–350 q/ha. Suitable for *rabi*	RAU, Rajasthan
Udaipur 103	—	Selection	Local material	Bulbs red in colour, oblate globular bulb, sweet but slightly more pungent. TSS 10.5–13%. Maturity 105–120 DAT. Average yield 250–300 q/ha. Recommended for Rajasthan and adjoining states during *rabi* season	RAU, Rajasthan
Kalyanpur Red Round	—	Selection	Local material	Bulbs bronze red in colour, globular in shape, moderately sweet and moderately pungent. TSS 13–14%. Maturity 105–115 DAT. Good storer. Average yield 250–300 q/ha. Suitable for *rabi*	CSAUAT, Kanpur
Punjab Selection	1975	Selection	Local material	Identified for zones IV, VII and VIII. Plant height 55–65 cm, 6–9 leaves per clump, bulbs red in colour, globular in shape, 5–6 cm diameter, average weight 50–70 g. Bulbs quite firm, good storers. TSS about 14%. Average yield 200 q/ha. Recommended for *rabi* season	PAU, Ludhiana

ONION : ***Contd.***

Variety	Year of release	Breeding method	Pedigree/ parentage	Important traits	Institutes/ universities
Punjab-48	1975	Selection	Local material	Identified for sub humid plains of Sutlej-Ganga. Bulbs flattish round and white in colour. Suitable for dehydration having good texture and flavour. Good storer. Average yield 300 q/ha	PAU, Ludhiana
Punjab Red Round	1993	Selection	Local material	Plants medium tall with green leaves. Bulbs shining red in colour, medium to large in size and round in shape with thin neck. Average yield 300 q/ha. Early maturity. Recommended for *rabi* season	PAU, Ludhiana
Punjab Naroya	1996	Selection	Local material	Red bulbs, medium sized with narrow neck, bulb and seed crop moderately resistant to purple blotch disease, average yield 150 q/ha, suitable for *rabi* season	PAU, Ludhiana
Punjab White	1998	Selection	Local material	White bulbs, suitable for *rabi* cultivation	PAU, Ludhiana
Bhima Raj	2007	Selection	B-780	Bulbs dark red, oval shaped, single centered and thin necked. TSS 10.0–11.0%. Suitable for *kharif* and late-*kharif* in Maharashtra, Karnataka and Gujarat and for rabi in zone VI. Maturity	NRCOG, Rajguru-nagar

ONION : *Contd.*

Variety	Year of release	Breeding method	Pedigree/ parentage	Important traits	Institutes/ universities
				120–125 DAT with no bolters. Average yield 25–30 tonnes/ha, high marketable yield. Yield potential is 40–45 tonnes/ha	
Bhima Super	2007	Selection	B-780	Suitable for *kharif* and late *kharif* in Maharashtra, Karnataka and Gujarat. Bulbs round, red with tapering neck. TSS 10–11%. Maximum number of single centered bulbs. Yield 26–28 tonnes/ha in *kharif* and 40–45 tonnes/ha in late *kharif*. Maturity 110–115 DAT	NRCOG, Rajguru-nagar
Bhima Red	2007	Selection	B-780 (bulb to row selection)	Suitable for *kharif* and late-*kharif* in Maharashtra, Karnataka and Gujarat. Bulbs have attractive red colour, round in shape. TSS 10.0–11.0% with high marketable yield. Average yield 48.0–52.0 tonnes/ha during late *kharif*. Maturity 115–120 DAT during late *kharif*	NRCOG, Rajguru-nagar

R = Red. W = White and Y = Yellow

Aggregatum or Multiplier Onion

Variety	Year of release	Breeding method	Pedigree/ parentage	Important traits	Institutes/ universities
CO 1	1965	Selection	Local material	Multiplier onion variety, medium plant height with green leaves. Medium-sized	TNAU, Coimbatore

ONION : *Contd.*

Variety	Year of release	Breeding method	Pedigree/ parentage	Important traits	Institutes/ universities
				bulblets of red colour. 7–8 bulblets/plant. Average weight of bulblets 55–60 g/clump. Average yield 100 q/ha. Adapted through Tamil Nadu. Maturity 90 DAP. Fairly pungent with medium TSS. Suitable for *kharif* and *rabi*	
CO 2	1975	Selection	Local material	Multiplier onion var. Plant height medium with light green cylindrical foliage. Moderately bigger size bulblets of crimson colour. 7–9 bulblets/clump. Shorter duration than CO-1. Matures in 65 DAP. Average yield 120 q/ha. Pungent with high TSS. Adaptability through out Tamil Nadu. Good in storage, moderately resistant to thrips and alternaria blight. Suitable for *kharif* and *rabi*	TNAU, Coimbatore
CO 3	1979	Selection	Local material	Multiplier onion var. Plants taller than CO 1 and CO 2. Foliage light green in colour with erect cylindrical leaves. Bulblets pink in colour, bigger than CO 2 with good consumer appeal. 8–10 bulblets/ plant, weighing 75 g/clump. Maturity 65 DAP. Moderately resistant to thrips. Average yield 160 q/ha. TSS 13%. Good	TNAU, Coimbatore

ONION : *Contd.*

Variety	Year of release	Breeding method	Pedigree/ parentage	Important traits	Institutes/ universities
				storer, no sprouting up to 120 days. Suitable for *kharif* and *rabi*	
CO 4	1982	Selection	Local material	Plants taller with cylindrical green leaves. 8–10 bulblets/plant with light brown colour bulbs weighing 90 g/clump. Maturity 60–65 DAP. Average yield 180 q/ha. Moderately resistant to thrips. Suitable for *kharif* and *rabi*	TNAU, Coimbatore
CO.On 5	2001	Pedigree method	Mass pedigree method of selection	This variety has the ability of free flowering and seed set throughout Tamil Nadu. It possess high bulb yield 18.9 tonnes/ha in a crop duration of 90 days	TNAU, Coimbatore
MDU 1	1979	Selection	Local material. Developed at Madurai Campus	Plant with medium height and leaves, cylindrical, light to dark green in colour. Bulbs uniform, bigger in size with bright red in colour. Adapted in Southern districts of Tamil Nadu. Weighing 75 g/clump, 10–11 bulblets/plant. Average yield 150 q/ha. Maturity 60–75 DAP. Good storer. Tolerant to lodging due to thick erect leaves. Suitable for *rabi*	AC&RI, TNAU, Madurai

At one time, all onion cultivars were open-pollinated, and many of these cultivars are still offered by seed companies. The discovery of male sterility in onion led to a rapid change to F_1 hybrids, possibly due to simplicity and low-cost of seed production. Hybrids have higher yield, larger and more uniform bulb sizes compared to open-pollinated cultivars. Multiplier onions are cultivars of *A. cepa* of group *aggregatum* with white flesh and yellow or brown scales. These are distinguishable from the shallot by the latters red scales and more delicate flavor.

GARLIC

Botanical name: *Allium sativum* L.
Family: Alliaceae

Garlic is one of the most popular spices in the world. The subterranean reserve structure derived from a leaf is the used plant part. It has a strong and characteristic odour, which is markedly different in fresh and fried state. The origin of garlic is in Central Asia, but cultivated all over the world, in Spain, France, Egypt, Bulgaria, Hungary, USA, Mexico and Brazil. It features in the mythology, religion and culture of many nations. Arab legend has it that garlic grew from one of the devil's footprints as he left the Garden of Eden. It is reported that in ancient Egypt, the workers who had to build the great pyramids were fed their daily share of garlic, and the Bible mentions garlic as a food the Hebrews enjoyed during their sojourn in Egypt. In Europe, garlic has been a common spice since the days of the Roman Empire, and it was extensively used from India to East Asia even before the Europeans arrived there. In Chinese mythology, garlic has been considered capable of warding off the Evil Eye, the symbol of misfortune and ill fate. After the Age of Exploration, its use spread rapidly to Africa and both Americas.

It is used as a flavouring, vegetable and medicinal herb that has accumulated superstitions over the centuries. Some cuisines like salads and sauces are fond of raw garlic. In parts of Austria, salads are prepared with vinegar, oil and squeezed garlic, and raw garlic appears in quite a multitude of Mediterranean sauces. Freshly grated garlic is served in liberal amounts to spring rolls and soups in Northern Vietnam. Raw garlic may also be pickled in vinegar or olive oil. Since some of garlic's aroma is extracted by the liquid, pickled garlic is usually very mild. Herbal vinegar is commonly made with one or two garlic cloves per litre vinegar. Usage of fried or cooked garlic is, however, much more common. On heating, the pungency and strong odour get lost and the aroma becomes more subtle and less dominant, harmonizing perfectly with ginger, pepper,

GARLIC

Variety	Year of release	Breeding method	Pedigree/ parentage	Important traits	Institutes/ universities
Godavari	1987	Clonal selection	Local collection from Jamnagar	Medium size and pinkish white colour having 22–25 cloves/bulb. Maturity 140–145 DAP, yield 100–105 q/ha. Suitable for *rabi*	MPKV, Rahuri
Sweta	1987	Clonal selection	Collection from Gujarat	Bulbs bigger in size, silvery white colour having 25–26 cloves/bulb. Maturity 130–135 days and yield 100–105 q/ha. Suitable for *rabi*	MPKV, Rahuri
Phule Baswant	2007	Clonal selection	Local collection	Purple coloured bulbs. Suitable for *rabi*	MPKV, Rahuri
Pusa Sel 10	—	Clonal selection	Local collection	Suitable for *rabi*	IARI, New Delhi
HG 1	—	Clonal selection	Local collection	White bulbs. Suitable for *rabi*	HAU, Hisar
HG 2	—	Clonal selection	Local collection	White bulbs. Suitable for *rabi*	HAU, Hisar
G 1	1991	Mass selection	Local collection obtained from Delhi (Azadpur) market	Bulbs compact, silvery white skin with creamy flesh. Sickle shaped cloves. The variety tolerant to insect pests and diseases like purpleblotch, stemphylium	NHRDF, Nasik

GARLIC : *Contd.*

Variety	Year of release	Breeding method	Pedigree/ parentage	Important traits	Institutes/ universities
				blight and onion thrips. TSS 38%, dry matter 39.5%, good storer. Yield 150–175 q/ha. Recommended for cultivation in *rabi*, all over the country	
G 41	1989	Mass selection	Local collection obtained from Bihar Sharif area in Bihar	Bulbs, compact, silvery white with creamy flesh. Big elongated cloves 20–25 numbers. Susceptible to purple blotch and stemphylium blight common in Northern parts. TSS 41%, dry matter 42.78%, good storer with average yield of 130 q/ha. Recommended for *rabi* cultivation in areas where there is not much problem of purpleblotch or stemphylium blight	NHRDF, Nasik
Yamuna Safed 2 (G 50)	1996	Mass selection	Local collection obtained from Karnal area in Haryana	Bulbs compact, attractive white creamy flesh, TSS 38–40%, dry matter 40–41%. Average yield 150–200 q/ha. Recommended for *rabi* in Northern parts	NHRDF, Nasik
G 282	1990	Mass selection	Local collection obtained from Dindigul, Tamil Nadu	Variety has done very well in Northern parts and also in Central parts of India. Leaves wider than other varieties. Bulbs creamy white and bigger sized, 15–16 number of cloves/bulb. TSS 38.42%, dry	NHRDF, Nasik

GARLIC : *Contd.*

Variety	Year of release	Breeding method	Pedigree/ parentage	Important traits	Institutes/ universities
				matter: 39–43%, medium storer. Average yield 175–200 q/ha. Suitable for *rabi*. Variety is suitable for export	
Agrifound Parvati (G 313)	1992	Selection	Exotic collection obtained from Hongkong market	Long day type suitable for cultivation in mid and high hills of Northern states. Bulbs bigger sized (5–6.5 cm), creamy white colour with pinkish tinge. 10–16 cloves, tolerant to common diseases. Average yield 175–225 q/ha, medium storer, suitable for export. Suitable for *rabi*	NHRDF, Nasik
G 323	1990	Mass selection	Local collection obtained from Badlapur, Jaunpur district of Uttar Pradesh	Plants vigorous with wider, green leaves. Bulbs silvery white and bigger sized, no. of cloves around 30 to 35, TSS 40–42% and dry matter 41–42%. Average yield 175–200 q/ha. Recommended for growing in north India during *rabi*	NHRDF, Nasik
VL 7	2004	Clonal selection	Selection from EC 158250	White bulbs. Suitable for *rabi*	VPKAS, Almora
ARU 52		Clonal selection	Local collection	White bulbs. Suitable for *rabi*	ARU, Almora

GARLIC : *Contd.*

Variety	Year of release	Breeding method	Pedigree/ parentage	Important traits	Institutes/ universities
Punjab Garlic 1	1993	Clonal selection	Local collection	Bulbs uniform, large and attractive white, resistant to purple blotch disease, average yield 38 q/ha, suitable for *rabi*	PAU, Ludhiana
Garlic 56-4 (T-56-4)	—	Selection	Local collection	Bulbs smaller in size and white in colour having 25–35 cloves/bulb. Yield 80–100 q/ha	PAU, Ludhiana
GG 1	—	Clonal selection	Local collection	—	GAU, Gujarat
GG 2	—	Clonal selection	Local collection	—	GAU, Gujarat
GG 10	—	Clonal selection	Local collection	—	GAU, Gujarat
DARL 52	—	Clonal selection	Local collection	—	DARl, Pithoragarh
Ooty 1	—	Clonal selection	Local collection	Bulbs bigger in size with dull white colour. Each bulb possesses 20–25 cloves. Yield potential 17 tonnes/ha in 120–130 days	TNAU, Coimbatore

chillies and many other spices. Therefore, it is an essential ingredient for nearly every cuisine of the world. Garlic products include garlic butter, purée, dried flakes and garlic salt.

Garlic is a perennial of the lily family. It grows to a height of about 60 cm. It has short, flat upright leaves of 15–30 cm. The tall single flower stem bears spherical head of pale pink or greenish-white blooms, often mixed with tiny bulbils. The subterranean white-skinned bulb or corn is subdivided into numerous 'cloves'. It contains a wealth of sulphur compounds; most important for the taste is allicin (diallyl disulphide oxide), which is produced enzymatically from alliin (*S*-2-propenyl-L-cysteine sulfoxide) if cells are damaged; its biological function is to repel herbivorous animals. Allicin is desactivated to diallyl disulphide; therefore, minced garlic changes its aroma if not used immediately. In the essential oil from steam destillation, diallyl disulphide (60%) is found besides diallyl trisulphide (20%), diallyl sulfide, ajoene and minor amounts of other di- and polysulphides.

ORNAMENTAL CROPS

TUBEROSE

Botanical name: *Polianthes tuberosa*
Family: Amaryllidaceae

Tuberose is a perennial plant. Its extracts is used as a middle note in perfumery. The common name derives from the Latin *tuberosa*, meaning swollen or tuberous in reference to its root system. It consists of about 12 species. *Polianthes* means "grey flower". It is a night-blooming plant thought to be native to Mexico along with every other species of *Polianthes*. It is a prominent plant in Indian culture and mythology. Its flowers are used in wedding ceremonies, garlands, decoration and various traditional rituals. Its Hindi name is "Rajnigandha". The name Rajnigandha means "night-fragrant". In parts of south India, it is known as "Sugandaraja", which translates to "king of fragrance/smell". In Tamil Nadu, it is called as Sambangi or nilasambangi and traditionally used in all type of garlanding, especially in south Indian marriages.

It is a half hardy,herbaceous perennial with fibrous roots, 60–120 cm high, having a bulb like tuberous rootstock and fleshy leaves linear, grass like foliage. The foliage is narrow at the base and wider at the top and is arranged in a rosette at the base. The tuberose grows in elongated spikes up to 45 cm (18 inch) long that produce clusters of fragrant waxy white flowers that bloom from the bottom towards the top of the spike.

TUBEROSE

Variety	Year of release	Breeding method	Pedigree/ parentage	Important traits	Institutes/ universities
Shringar	2006	Hybridization and selection	Mexican Single × Pearl Double	Single	IIHR, Bangalore
Suasini	2006	Hybridization and selection	Mexican Single × Pearl Double	Double	IIHR, Bangalore
Prajwal	2000	Hybridization and selection	Shringar × Mexican Single	Single	IIHR, Bangalore
Vaibhav	2001	Hybridization and selection	Mexican Single × IIHR-2	Double	IIHR, Bangalore

The chief centres of its commercial production, are Devanahalli, Tumkur, Kolar, Belgaum and Mysore (Karnataka); Pune, Nashik, Ahmednagar and Thane districts (Maharashtra); Guwahati and Johrat (Assam); Udaipur, Ajmer and Jaipur (Rajasthan); Navsari and Valsad (Gujarat); Coimbatore and Madurai districts (Tamil Nadu); East Godavari and Guntur (Andhra Pradesh) and Midnapur district (West Bengal).

CHRYSANTHEMUM

Botanical name: *Chrysanthemum* spp.
Family: Asteraceae

Chrysanthemum is major floral crop in the world. It is cultivated as cut flower on commercial scale as well as for its aesthetic value in gardens. It is a perennial herb valued for its cut flowers and potted plants. It is a popular decorative species, cultivated widely by amateur growers. Flower varies in size and colour. Its cultivars are perfect for floral preparations, garland making and hair decoration. The large flowers are appreciated for their exhibition value. Chrysanthemum also has medicinal importance. It is considered a sedative and anti-hypertensive. The plant is useful in cold, headache, bronchitis, whooping cough, hysteria, fungal infection and stomachic. It is also known as *guldaudi* and *sevanti*.

Modern chrysanthemums are much more showy than their wild relatives. The flowers occur in various forms, and can be daisy-like, decorative, pompons or buttons. This genus contains many hybrids and thousands of cultivars developed for horticultural purposes. In addition to traditional yellow, other colorus are available, such as white, purple, and red. The most important hybrid is *Chrysanthemum* × *morifolium* (syn. *C. grandiflorum*), derived primarily from *C. indicum* but also involving other species.

Chrysanthemums are broken into two basic groups, Garden Hardy and Exhibition. Garden Hardy Mums are perennials capable of being wintered over in ground in most northern latitudes. Exhibition varieties are not usually as sturdy. Garden hardies are defined by their ability to produce an abundance of small blooms with little if any mechanical assistance (*i.e.* staking) and withstanding wind and rain. Exhibition varieties on the other hand require staking, over-wintering in a relatively dry cool environment, sometimes with the addition of night lights. Exhibition varieties can be used to create many amazing plant forms; large disbudded blooms, spray forms, as well as many artistically trained forms.

CHRYSANTHEMUM

Variety	Year of release	Breeding method	Pedigree/ parentage	Important traits	Institutes/ universities
Chandra-kant	1990	Hybridization	Angela × G.P.67	White, garland, floral decoration, pot plant, bedding plant	IIHR Bangalore
Chandrika	1994	Hybridization	Angela × G.P.1	White, garland, floral decoration, making bouquet	IIHR Bangalore
Indira	1980	Hybridization	Lord Doonex-14 × Flirt × Valentine-1	Golden yellow, cut flower, pot plant Bangalore	IIHR
Kirti	1994	Hybridization	Angela × G.P.1	White, floral decoration, pot plant, bedding plant	IIHR Bangalore
Nilima	1992	Hybridization	Flirt × Valentine	Red-purple (70.8) fades to red-purple (62.C), cut flower, floral decoration, making bouquet	IIHR Bangalore
Pankaj	1990	Hybridization	Lord Doonex-14 × Flirt × Valentine-1	Light pink fades to pale pink, cut flower, floral decoration, making bouquet	IIHR Bangalore
Rakhee	1980	Hybridization	Open pollinated seedling of Lord Doonex	Primrose yellow with red stripes on ray florets, pot plant	IIHR Bangalore
Ravikiran	1992	Hybridization	Flirt × Valentine	Greyed-red (178.B) fades to greyed-orange (174.C), cut flower, pot plant	IIHR Bangalore

CHRYSANTHEMUM : *Contd.*

Variety	Year of release	Breeding method	Pedigree/ parentage	Important traits	Institutes/ universities
Red Gold	1980	Hybridization	Flirt × Valentine	Greyish orange fades to golden yellow, Cut flower, making bouquet	IIHR Bangalore
Yellow Gold	1992	Hybridization	Induced mutant of 'Flirt'	Yellow (33.A) having greyed-orange (173.B) stripes fades to yellow (2.B), Garland, cut flower, making bouquet	IIHR Bangalore
Yellow star	1990	Hybridization	Flirt × Valentine-22	Yellow, garland, floral decoration, making bouquet	IIHR Bangalore
Usha Kiran	2001	Induced mutation	Kirti	Yellow, garland, floral decoration	IIHR Bangalore
Pusa Anmol	2005	Mutation	An induced radiation mutant of hybrid variety Ajay (evolved at NBRI, Lucknow)	Highly floriferous, bushy variety, spray group, relatively thermo-and photo-insensitive	IARI, New Delhi
Pusa Centenary	2005	Mutation	An induced radiation mutant of a very popular commercial exotic cut flower variety Thai Chen Queen that produces light	Vigorous growing, flowers possess brilliant yellow incurving ray and disc florets which are fleshy and delicate, suitable for cut flower and exhibition purpose	IARI, New Delhi

CHRYSANTHEMUM : *Contd.*

Variety	Year of release	Breeding method	Pedigree/ parentage	Important traits	Institutes/ universities
			orange flowers, which are suitable for cut flower purpose.		
Shanti	2000	Selection	Open-pollinated seedling selection	Small flowered, white decorative type, cut flower and garland variety	NBRI, Lucknow
Y2K	2000	Selection	Open-pollinated seedling selection	Small flowered, white Anemone type mini chrysanthemum, plants dwarf, bushy, profuse blooming	NBRI, Lucknow
Sadbha-vana	2000	Selection	Open-pollinated seedling selection	Small flowered, red yellow, open disc double Korean type mini-chrysanthemum	NBRI, Lucknow
Kargil 99	2000	Selection	Open-pollinated seedling selection	Small flowered, spoon-shaped, purplish mauve florets, leaves variegated	NBRI, Lucknow

Chrysanthemum blooms are divided into 13 different bloom forms by the US National Chrysanthemum Society, Inc., which is in keeping with the international classification system. The bloom forms are defined by the way in which the ray and disk florets are arranged. The bloom forms are Irregular Incurve, Reflex, Regular Incurve, Decorative, Intermediate Incurve, Pompon, Single/Semi-Double, Anemone, Spoon, Quill, Spider, Brush and Thistle and Exoti.

Chrysanthemum blooms are composed of many individual flowers (florets), each one capable of producing a seed. The disk florets are in the center of the bloom head, and the ray florets are on the perimeter. The ray florets are considered imperfect flowers, as they only possess the female productive organs, while the disk florets are considered perfect flowers as they possess both male and female reproductive organs.

MARIGOLD

Botanical name: *Tagetes* spp.
Family: Asteraceae

The *Tagetes* is a genus of 52 species of annual and perennial herbaceous plants. They are native to the area stretching from the southwestern USA into Mexico and south throughout South America. African marigolds (usually referring to cultivars and hybrids of *T. erecta*, although this species is not native to Africa), or French marigolds (usually referring to hybrids and cultivars of *T. patula*, many of which were developed in France although the species is not native to that country). At least one species is a naturalized weed in Africa, Hawaii, and Australia.

Different species vary in size from 0.05 to 2.2 m tall. They have pinnate green leaves, and white, golden, orange, yellow, to an almost red floral heads typically (0.1) to 4–6 cm diameter, generally with both ray florets and disc florets. The foliage has a musky/pungent scent, though some later varieties have been bred to be scentless. It is said to deter some common insect pests (although it is recorded as a food plant for some Lepidoptera larvae including Dot Moth), as well as nematodes. *Tagetes* are hence often used in companion planting. The *T. minuta* (Khakibush), originally from South America, has been used as a source of essential oil, known as tagette, for the perfume industry as well as a flavourant in the food and tobacco industries in South Africa, where the species is also a useful pioneer plant in the reclamation of disturbed land. Some of the perennial species are deer, rabbit, rodent and javalina resistant.

MARIGOLD

Variety	Year of release	Breeding method	Pedigree/ parentage	Important traits	Institutes/ universities
Pusa Narangi Gainda	—	—	—	It takes 125–136 days for flowering from sowing. Plants are 73.30 cm tall, vigorous and uniform; foliage dark green; flowers orange coloured, carnation type, double, compact, 7.8 cm in size and disc florets present. It gives an yield of 349 q/hectare fresh flowers. It is most suitable for loose flower production for garland making and religious offerings and for carotenoid production	IARI, New Delhi
Pusa Basanti Gainda	—	—	—	It is late and takes 135–145 days for flowering, but flowering duration is long (40–45 days). Plants are 58.80 cm tall, vigorous and uniform, foliage dark green, flowers sulphur yellow coloured, carnation type, double, compact 6.9 cm in size and 58/plant in number, disc florets present but invisible. The rows of whorls in the capitulum range from 7 to 10. It is ideal for growing both in pots and beds in the gardens	IARI, New Delhi
MDU 1	1986	Selection	—	The plants are medium-tall with moderate branching habit. The plant produces on an average 97 flowers, weighing 561.40 g/plant, with an estimated yield of 41.54 tonnes/ha. The flowers are large with a stalk length of 8.39 cm. The light orange colour petals are compactly arranged and each flower has 210 petals	AC&RI (TNAU), Madurai

The common name, "marigold", is derived from "Mary's Gold", and the plant is associated with the Virgin Mary in Christian stories. The marigold was regarded as the flower of the dead in pre-Hispanic Mexico, parallel to the lily in Europe, and is still widely used in the Day of the Dead celebrations.

The petals of *Tagetes* are rich in the orange-yellow carotenoid lutein and as such extracts of *T. erecta* are used as a food colour. The marigold is also widely cultivated in India and Thailand, particularly the species *T. erecta*, *T. patula*, and *T. tenuifolia*. Vast quantities of marigolds are used in garlands and decoration for weddings, festivals, and religious events.

GLADIOLUS

Botanical name: *Gladiolus* spp.
Family: Iridaceae

Gladiolus is also called the sword lily, the most widely-used English common name for these plants is simply gladiolus. These attractive, perennial herbs are semi-hardy in temperate climates. They grow from rounded, symmetrical corms, that are enveloped in several layers of brownish, fibrous tunics. Gladiolus is very much liked for its majestic spikes containing attractive, elegant and delicate florets. These florets open in sequence over a longer duration and hence has a good keeping quality of cut spikes. There is a wide range of colours — self or bicolour with or without central mark varying from white to darkest crimson. The spikes of gladiolus are mainly used for garden and interior decoration, and for making bouquets. West Bengal, Maharashtra, Uttar Pradesh, Punjab, Haryana and Andhra Pradesh are major gladiolus-growing states.

Their stems are generally unbranched, producing 1–9 narrow, sword-shaped, longitudinal grooved leaves, enclosed in a sheath. The lowest leaf is shortened to a cataphyll. The leaf blades can be plane or cruciform in cross section.The fragrant flower spikes are large and one-sided, with secund, bisexual flowers, each subtended by 2 leathery, green bracts. The sepals and petals are almost identical in appearance, and are termed tepals. They are united at their base into a tube-shaped structure. The dorsal tepal is the largest, arching over the three stamens. The outer three tepals are narrower. The perianth is funnel-shaped, with the stamens attached to its base. The style has three filiform, spoon-shaped branches, each expanding towards the apex. The ovary is 3-locular with oblong or globose capsules, containing many, winged brown, longitudinally dehiscent seeds. In their center must be noticeable the specific pellet like

GLADIOLUS

Variety	Year of release	Breeding method	Pedigree/ parentage	Important traits	Institutes/ universities
Aarti	1980	Hybridization	Shirley × Meloidy	Poppy red (40.D) with reddish purple (60.A) and canary yellow (9.C) blotch and mandarin red, cut flower, flower arrangements, bouquet, breeding plant	IIHR, Bangalore
Apsara	1980	Hybridization	Black Jack × Friendship	Ruby red (61.A) with barium yellow (10.D) flecks in throat, cut flower, bouquet, flower arrangements	IIHR, Bangalore
Darshan	1983	Hybridization	Watermelon Pink × Shirley	Red Purple (63.C) with Red Purple (64.D) margin, Biotech White (155.A), Cut flower, flower arrangements, bouquet, breeding plant	IIHR, Bangalore
Dhiraj	1994	Hybridization	Beauty spot × Psittacinus hybrid	Red purple (60.D) with (60.B) margin, biotech yellow (3.C) and red-purple (58.A), cut flower, flower arrangements, bouquet, bedding plant	IIHR, Bangalore
Kumkum	1992	Hybridization	Watermelon Pink × Lady John	Red (47.C) with red (46.C) margin having red (45.A) spots, biotech yellow (3.D), cut flower, flower arrangements, bouquet, bedding plant	IIHR, Bangalore
Meera	1990	Hybridization	G.P.1 × Friendship	Snow white, cut flower, bouquet, flower arrangements	IIHR, Bangalore

GLADIOLUS : *Contd.*

Variety	Year of release	Breeding method	Pedigree/ parentage	Important traits	Institutes/ universities
Nazrana	1979	Hybridization	Black Jack × Friendship	Carsinal red (53.B) with barium yellow (10.D) flash in throat, cut flower, bouquet, flower arrangements	IIHR, Bangalore
Poonam	1986	Hybridization	Geliber Herald × R.N.121	Dresden yellow (5.0) with mimosa yellow (8.C) blotch, cut flower, bouquet, flower arrangements	IIHR, Bangalore
Sapna	1986	Hybridization	Green Wood Pecker × Friend-ship	Barium yellow (10.D) with primrose yellow (4.A) blotch and Mandarin Red (41.D) tinge on margins, cut flower, bouquet, flower arrangements	IIHR, Bangalore
Sagar	1990	Hybridization	Meloidy × Wild Rose	Red (49.A) having Red (48.B) magin with Red (43.B) and yellow (5.D) blotch, cut flower, bouquet, flower arrangements	IIHR, Bangalore
Shakti	1990	Hybridization	Watermelon Pink × Wild Rose	Red (51.A) with red (53.C) margin having white (155.A) lines and blotch, Cut flower, flower arrangements, bouquet, bedding plant	IIHR, Bangalore
Shobha	1980	Mutation breeding	Induced mutant of Wild Rose	Shell Pink (37.C) with Empire yellow (11.D) throat, cut flower, bouquet, flower arrangements	IIHR, Bangalore

GLADIOLUS : *Contd.*

Variety	Year of release	Breeding method	Pedigree/ parentage	Important traits	Institutes/ universities
Sindur	1993	Hybridization	Watermelon Pink × Lady John	Red (44.C) with red (45.B) spots. Blotch Red (46.A) with yellow (2.D) lines and splash, flower arrangements, bouquet	IIHR, Bangalore
Tilak	1993	Hybridization	Watermelon Pink × Lady John	Red (49.A) with red (53.B) specks. Lower lip Red (45.A) with yellow (2.D) lines, cut flower, flower arrangements, bouquet	IIHR, Bangalore
Agni Rekha	1980	Selection	An open-pollinated seedling of Sylvia	Midseason, spikes 85 cm long, florets 18 per spike, fire red with saffron yellow markings and scarlet stripes, 9.3 cm across, good multiplier	IARI, New Delhi
Anjali	1997	Hybridization	A hybrid of Sancerre × Rose Spire	Spikes 90–100 cm long. Florets. 17–19 per spike, pinkish white	IARI, New Delhi
Archana	1997	Hybridization	Hybrid of Creamy Green × American Beauty	Spikes 67 cm long. Florets 16, scarlet pink with yellow dusting on falls, mid-season	IARI, New Delhi
Bindiya	1997	Selection	Open-pollinated seedling of Ratna Butterfly	Spikes 64 cm long, florets 18, yellowish orange	IARI, New Delhi

GLADIOLUS : *Contd.*

Variety	Year of release	Breeding method	Pedigree/ parentage	Important traits	Institutes/ universities
Chandani	1997	Hybridization	A hybrid of Green Woodpecker × White Butterfly	Spikes 80–90 cm long, florets 14–15/ spike, creamy white, early in maturity	IARI, New Delhi
Chirag	1997	Hybridization	Hybrid of Cygnet × Little Fawn	Spikes 71 cm long, florets 18, orange with yellow tip, early in maturity	IARI, New Delhi
Dhan-vantari	1997	Hybridization	A hybrid of Junior. Prom × Lucky Sta	Spikes 71 cm long, florets 17, light yellow	IARI, New Delhi
Gunjan	2000	—	—	Florets soft orange with deeper splashing having fan-shaped colouring on 2 side falls. Spikes 87 cm long with 16 florets per spike. Floret diameter 11.0 cm. Early mid-season	IARI, New Delhi
Jyotsna	2002	Selection	A selection from open pollinated seedlings of the variety Melodie	Florets orange (41 D) in colour with darker streaks on petals. Lip colour 2 side falls of 41 A orange which turn 10 D yellow. It takes 98 days to flowering and produces a plant height of 121 cm. Spike length more than 103 cm with 18 florets per spike each with a diameter of more than 11 cm. It is a very good multiplier each mother corm producing 2.4 daughter corms	IARI, New Delhi

GLADIOLUS : *Contd.*

Variety	Year of release	Breeding method	Pedigree/ parentage	Important traits	Institutes/ universities
Kamini	2000	—	—	Floret colour orange-red with fan-shaped purple red lip on light yellow base on 2 side falls. Spikes 88 cm long with 14 florets/spike, floret diameter 11.0 cm, early mid-season	IARI, New Delhi
Lohit	2000	—	—	Floret colour red with white midribs on 2 side falls. Spikes 125 cm long with 16 florets per spike. Floret diameter 11.8 cm. Early mid-season	IARI, New Delhi
Mohini	2000	—	—	Floret colour red-purple with fan-shaped deep purple colour on yellow on 2 side falls. Spike length 92 cm with 15 florets per spike. Floret diameter 9.5 cm. Early mid-season	IARI, New Delhi
Mayur	1980	Selection	An open-pollinated seedling of cv. Sylvia	Mid- and late season. Spikes 75 cm long. Florets Lilac Purple with dark purple throat, 20/spike. Very good multiplier	IARI, New Delhi
Neelam	1995	Hybridization	Hybrid of Sylvia × Patricia	Spikes 100 cm long Florets 16–19/spike, deep mauve in colour. Mid-season in flowering	IARI, New Delhi

GLADIOLUS : *Contd.*

Variety	Year of release	Breeding method	Pedigree/ parentage	Important traits	Institutes/ universities
Pusa Suhagin	1987	Selection	An open-pollinated seedling of cv. Sylvia	Midseason, requires 65 days to flower. Spikes 91 cm in length. Florets ruby red with barium yellow streaks on the tip and beetroot purple throat, 18 per spike. Cormel production good	IARI, New Delhi
Rangmahal	2000	—	—	Florets ruffled, red-purple and compactly arranged. Spikes 84 cm long with 16 florets per spike. Floret diameter 10.5 cm. Mid-season	IARI, New Delhi
Sanjevani	—	Hybridization	A hybrid between Purple Burma and Lucky Sta.	Florets 16/spike, white in colour. A mid-season variety	IARI, New Delhi
Shabnam	2000	Selection	A selection from open pollinated seedlings of the variety Snow Princess	Florets white in colour with fan-shaped dusted mauve lips on 2 side falls. It takes about 100 days to flowering. Spikes more than 100 cm with 17 florets/spike	IARI, New Delhi
Shagun	2000	—	—	The plant of this variety is erect. The flower colour is dazzling red with creamy midrib on two side falls in throat. Petals are wavy. The leaves are dark green and crop is very attractive	IARI, New Delhi

GLADIOLUS : ***Contd.***

Variety	Year of release	Breeding method	Pedigree/ parentage	Important traits	Institutes/ universities
				and healthy. The number of florets per spike is 17. It takes 100–150 days for flowering	
Shringarika	2000	—	—	The flowers of this variety are purple with white midrib on wing petals and the two side falls. Petal tips are yellow mixed with purple having slightly frilled petals. Spikes are thick, but requires staking. It is a mid-season variety and takes 100–120 days for flowering. The number of florets/spike is around 14	IARI, New Delhi
Sukanya	2000	—	—	Floret colour white with scarlet ring in the lip. Spikes 98 cm long with 18 florets. Floret diameter 9.7 cm. Early mid-season	IARI, New Delhi
Swapnil	2000	—	—	Floret colour red with white lip. Spikes 109 cm long with 17 florets of 11.4 cm diameter. Early blooming	IARI, New Delhi
Swarnima	2000	—	—	Florets coppery yellow. Spikes 102 cm long with 16 florets/spike. Floret diameter 11.3 cm. Mid-season	IARI, New Delhi

GLADIOLUS : ***Contd.***

Variety	Year of release	Breeding method	Pedigree/ parentage	Important traits	Institutes/ universities
Sarang	1997	Selection	An open-pollinated seedling of White Oak	Spikes 80 cm long. Florets 21 with purple red colour. Mid-season in flowering	IARI, New Delhi
Shweta	1997	Hybridization	Hybrid of Wind Song and Pink Frost	Spikes 65 cm long. Florets 17/spike, white. Mid-season in flowering	IARI, New Delhi
Suchitra	1980	Hybridization	Raised from a cross of Sylira × Jo Wagenaar	Spikes 86 cm in length. Florets camellia rose with stripes of vermilion and blotch and throat of dianthus purple, 20/spike. Good multiplier	IARI, New Delhi
Sunayna	1997	Hybridization	Hybrid of George Mazure and Melody	Spikes 90–100 cm long. Florets 16–18/ spike, pink colour. Mid-season in maturity	IARI, New Delhi
Suryakiran	1995	Hybridization	A hybrid of George Mazure and Melody	Spikes 90–100 cm long. Florets 16–17/ spike, pink in colour	IARI, New Delhi
Urmi	2002	Hybridization	A hybrid between Berlew and Heady Wine	Florets red-purple (71 C) with pale dusted fan-shaped throat on 2 side falls. It takes about 100 days for flowering and produces a plant height of 105 cm. Spikes are more than 88 cm with 16 florets/spike each with a diameter of 10	IARI, New Delhi

GLADIOLUS : *Contd.*

Variety	Year of release	Breeding method	Pedigree/ parentage	Important traits	Institutes/ universities
				cm. Good multiplier each mother corm produces 2 replacement corms	
Urmil	2000	—	—	Florets violet with creamy throat. Spikes 91 cm long with 13 florets/spike. Floret diameter 11.5 cm. Early blooming	IARI, New Delhi
Urvashi	2002	Selection	A selected from open pollinated seedlings of the variety red Softglow	Florets soft red purple (74C), lip fan shaped (blunt tipped) deep purple on 2 side falls on creamy base. Mid-season bloomer requiring 107 days. Flowering span 31 days and spike durability 12 days. Spike length 80 cm with 17 florets of 11 cm. About 7 florets remain open at a time. Good multiplier producing 2 replacement corms from each mother corm	IARI, New Delhi
Vandana	1997	—	—	Orange coloured variety with 95 cm long spikes having 16 florets of 13.0 cm diameter. Early mid-season	IARI, New Delhi
KKL 1	1993	Selection	Selection from cv. American Beauty	The flower colour is an attractive red purple with white flushed throat. The plant grows to a height of 116 cm. The mean spike length is 89.4 cm with an	HRS (TNAU), Kodaikanal

GLADIOLUS : *Contd.*

Variety	Year of release	Breeding method	Pedigree/ parentage	Important traits	Institutes/ universities
				average of 16.2 florets/spike. The average floret size is 13.5 cm. The florets are open shaped. Each spike weighs on an average of 92.3 g with a vase life of 12.1 days. The selection yields on an average of 21.1 spikes and 19.5-corms/sq.m. The spikes are ready for first picking in 90 days after planting. The corms are harvested from 150 days after the leaves started yellowing and drying. It yields 2,11,100 flower spikes and 1,95,000 corms/ha	

structure which is the real seed without the fine coat. In some seeds this structure is wrinkled and with black color. These seeds are unable to germinate.

These flowers are variously coloured, pink to reddish or light purple with white, contrasting markings, or white to cream or orange to red. Gladioli have been extensively hybridized and a wide range of ornamental flower colours are available from many varieties. The main hybrid groups have been obtained by crossing between four or five species, followed by selection: Grandiflorus, Primulines and Nanus. They make very good cut flowers. However, due to their height, the cultivated forms frequently tend to fall over in the wind if left on the plant.

ROSE

Botanical name: *Rosa* spp.
Family: Rosaceae

Rose is a perennial flowering shrub or vine of the genus *Rosa*. Over 100 species, in its family, some form a group of erect shrubs, and climbing or trailing plants, with stems that are often armed with sharp thorns. Most are native to Asia, with smaller number of species native to Europe, North America, and North-West Africa. Natives, cultivars and hybrids are all widely grown for their beauty and fragrance. The leaves of most species are 5–15 cm long, pinnate, with (3–) 5–9 (–13) leaflets and basal stipules; the leaflets usually have a serrated margin, and often a few small prickles on the underside of the stem. The vast majority of roses are deciduous but a few (particularly in Southeast Asia) are evergreen or nearly so. The flowers of most species roses have five petals, with the exception of *Rosa sericea*, which usually has only four. Each petal is divided into two distinct lobes and is usually white or pink, though in a few species yellow or red. Beneath the petals are five sepals (or in case of some *Rosa sericea*, four). These may be long enough to be visible when viewed from above and appear as green points alternating with the rounded petals. The ovary is inferior, developing below the petals and sepals.

The aggregate fruit of the rose is a berry-like structure called a rose hip. Rose species that produce open-faced flowers are attractive to pollinating bees and other insects, thus more apt to produce hips. Many of the domestic cultivars are so tightly petalled that they do not provide access for pollination. The hips of most species are red, but a few (*e.g. Rosa pimpinellifolia*) have dark purple to black hips. Each hip comprises an outer fleshy layer, the hypanthium, which

ROSE

Variety	Year of release	Breeding method	Pedigree/ parentage	Important traits	Institutes/ universities
Dr. G.S. Randhawa	1993	Hybridization	Queen Elizabeth × American Heritage	Peach pink fading to salmon pink, garden display and loose flower	IIHR, Bangalore
Kiran	1993	Hybridization	Big Red × Blue Moon	Tyrian purple fading to rose bengal with white steaks on some petals, exhibition and loose flower	IIHR, Bangalore
Nishkant	1993	Selection	Spontaneous mutant of *Rosa multiflora*	Thornless root stock, used as root stock	IIHR, Bangalore
Hybrid Teas					
Abhisarika	1977	Mutation	A mutant of the popular HT rose Kiss of Fire	Apricot colour changing to deep pink on aging with silvery reverse of petals and deep pink to light purplish streaks	IARI, New Delhi
Akash Sundari	1982	—	—	Lilac pink, deeper reverse, displaying an attractive silvery lilac. High centred buds open into very large fully double flowers with slight fragrance. Exhibition type flowers, last long even when cut	IARI, New Delhi
Aravalli Princess	1983	Selection	Raised from a natural cross of 'Sonia'	Lovely porcelain pink, high centred blooms, which unfurl slowly to reveal a very large number of petals. Symmetrically and beautifully arranged.	IARI, New Delhi

ROSE : *Contd.*

Variety	Year of release	Breeding method	Pedigree/ parentage	Important traits	Institutes/ universities
				The outermost petals sometimes have a flush of deep pink	
Aruna	1968	Selection	Seedling of 'Independence'	Medium sized, double flowers of a most dazzling shade of velvety orange scarlet, produced mostly singly, but in small clusters in warm weather. Growth is vigorous and plenty of flowers are produced	IARI, New Delhi
Ashirwad	1984	—	—	A superb bright current red with golden-based petals. Elegant buds open to large and double flowers, which are high centred with broad recurved petals showing a lovely form, slightly fragrant	IARI, New Delhi
Anurag	1980	Hybridization	A hybrid seedling of 'Sweet Afton' × 'Gulzar'	Shapely blooms of high centre and tyrian rose colour, which attain exhibition size and are borne singly on sturdy shoots freely. Medium and balanced bush with upright growth. Highly fragrant	IARI, New Delhi
Apsara	1983	Hybridization	Raised from a cross between 'Sonia' and 'Sabine'	Exquisite flesh pink with lovely salmon shadings. Immaculate buds are borne on very long, very upright stems, there being one flower to each stem. The buds	IARI, New Delhi

ROSE : *Contd.*

Variety	Year of release	Breeding method	Pedigree/ parentage	Important traits	Institutes/ universities
				open slowly into beautiful medium-sized many petalled blooms, which last for a long time. Fragrant. Good for cut flowers	
Arjun	1980	Hybridization	A hybrid seedling of 'Blithe Spirit' × 'Montezuma'	Porcelain rose colour, large, beautifully shaped blooms are borne on very long shoots (87 cm) of medium thickness emerging from the base. Produces on an average 13 blooms/bush during winter flush which are of export quality, for their long lasting quality and borne on long shoots. Extra tall, upright growing and hardy plant. Tolerant to most diseases and pests. Suitable for growing for export of cut flowers	IARI, New Delhi
Bhim	1970	Hybridization	A hybrid seedling of 'Charles Mallerin' × Delhi Princess'	Flowers very large, bright red, with many petals and some fragrance. Stands up to cold nights. Highly recommended variety	IARI, New Delhi
Chambe di Kali	1983	—	—	Very large high centered buds of very clean pink reminiscent of 'princess Margaret of England' open into large blooms borne on long straight stems. Large shining leaves. Vigorous and floriferous	IARI, New Delhi

ROSE : ***Contd.***

Variety	Year of release	Breeding method	Pedigree/ parentage	Important traits	Institutes/ universities
Charu-gandha	1972	Hybridization	A hybrid seedling of 'Delhi princess' × 'Eiffel Tower'	Crimson-red with velvety sheen on petals, very large, well-formed blooms are borne singly on strong shoots in winter. Highly fragrant	IARI, New Delhi
Chitra	1995	Selection	A striped and spotted bud-sport of the popular and floriferous Hybrid Tea rose variety 'Janina'	It has creamish-white, white and golden yellow coloured prominent stripes and spots on the vermilion orange base of attractive petals. Golden-yellow colour on reverse of these petals makes these flowers extremely charming. These flowers are not affected by winter and open fully showing their beautiful yellow centre. During sprint flush, flowers are produced in large bunches and the young coppery foliage develops different shades of green and brown pigments in the form of stripes and spots	IARI, New Delhi
Chitra-lekha	1972	Hybridization	A hybrid seedling of 'Montezuma' × Baccara'	Ruby red colour of flowers giving an effect of smoky colour in cold winter. Large-sized flowers are borne freely on long and strong shoots, vigorous	IARI, New Delhi
Chitwan	1971	Hybridization	A hybrid seedling of 'Western Sun' × 'Golden Splendor'	Beautiful, greenish-yellow bud open to primrose yellow, large shapely flowers, borne very freely and singly in large	IARI, New Delhi

ROSE : *Contd.*

Variety	Year of release	Breeding method	Pedigree/ parentage	Important traits	Institutes/ universities
				numbers on compact bushes. Vigorous with light green foliage. Good for bedding	
Dark Boy	1965	Selection	Seedling of 'Nigrette'	Velvety, very dark maroon red, almost black in cool weather. The medium-sized blooms are freely produced. A new addition to garden for those who like very dark roses	IARI, New Delhi
Delhi Apricot	1964	—	—	Apricot-yellow to light coppery-orange buds open to decorative blooms which keep their colour. Vigorous and floriferous	IARI, New Delhi
Delhi Debutante	1964	—	—	Clear medium-pink, the buds are beautifully shaped and open to medium-large flowers. Delightful fragrance	IARI, New Delhi
Delhi Pastel	1964	Selection	A seedling of 'Una Wallace'	Large pale pastel pink of exhibition form. Growth, foliage and form of flower reminiscent of the well-known variety 'Una Wallace'. Which is of a much deeper colour	IARI, New Delhi
Delhi Sunshine	1963	Selection	Seedling of 'Mme Charles Sauvage'	Deep unfading cream. Sometimes flushed pink outside. imbricated.	IARI, New Delhi

ROSE : *Contd.*

Variety	Year of release	Breeding method	Pedigree/ parentage	Important traits	Institutes/ universities
				somewhat flattened large blooms. The arrangement of the petals is extremely regular and attractive and the edges are frilly. Good in warmer weather in Delhi	
DiI-Ki-Rani	1935	—	—	Very lovely shade of etherial pink with a tinge of lavender. The buds are long and beautiful and open into large high centred blooms of' exquisite shape. As the bud opens, very large petals arrange themselves to form an exhibition quality bloom. Every flower produced is perfect. The bush is vigorous and floriferous	IARI, New Delhi
Dilruba	1984	—	—	A glorious blackish-velvety crimson. High centred blooms with plenty of petals and substances open from elegant buds. Highly fragrant flowers last well and have good keeping quality when cut. Flowers are large and full thick and semi variegated leaves are added attraction. Sturdy erect plants are very hardy	IARI, New Delhi
Diva Swapna	1981	Hybridization	Raised from 'Sonia' Cross	Dream-like blending of shimmering silvery pink and white are found in	IARI, New Delhi

ROSE : *Contd.*

Variety	Year of release	Breeding method	Pedigree/ parentage	Important traits	Institutes/ universities
				this rose, which combines exhibition size with elegance of form and strong fragrance. The lovely buds open into flowers with reflexing petals which are good for cutting and exhibition	
Dr Benjamin Pal	1993	Hybridization	A cross between 'Sweet Afton' and 'First Prize'	It is long pointed deep pink buds open to exquisitely formed large flowers of lighter shade in the centre. This similar colour of inner side of the petals produces a biocolour effect in the opened flowers. The base of the petals is paler in colour. Most of the flowers with 45 or more petals are produced on very long and sturdy stems. Vigorous and upright growing plants. Slightly fragrant	IARI, New Delhi
Dr Bharat Ram	2006	Hybridization	A cross between Sweet afton and Ganga	High centered, pointed buds of peach pink colour are borne on long stems on vigorous bush. Apricot flowers having pink shade are produced in big bunches during spring flush. Profuse flowering. Slightly fragrant	IARI, New Delhi
Dr Homi Bhabha	1968	Selection	Seedling of 'Virgo'	Exquisite buds of classic shape open into very large, pure-white flowers	IARI, New Delhi

ROSE : *Contd.*

Variety	Year of release	Breeding method	Pedigree/ parentage	Important traits	Institutes/ universities
				having about 60 petals, so that the open flower does not show an empty centre. Stands low night temperatures in winter and warm weather in early summer better than most white varieties. Needs protection from thrips	
Dr R.R. Pal	1983	—	—	A distinctively different variety. Colour ranging from very deep rose red to dark velvety red. Huge shapely classic flowers with plenty of petals commanding immediate attention. Growth and habit are excellent	IARI, New Delhi
Dr B.P. Pal	1980	Selection	Seedling No. 350 C	Exquisitely formed, large blooms of deep pink colour, with high centre. Individual bloom carried majestically on long and straight shoots. Flowers especially attractive for their luminescent colour having compact shape with good lasting quality. Abundant flowering on vigorous and upright growing bush. Slightly fragrant. Blooms not affected by frost	IARI, New Delhi
Dr M.S. Randhawa	1989	Hybridization	Raised from a cross between	The flower is large and double with fifty or more petals, the outside of	IARI, New Delhi

ROSE : *Contd.*

Variety	Year of release	Breeding method	Pedigree/ parentage	Important traits	Institutes/ universities
			'Sabine' and 'Kiss of Fire'	which is creamy white, and the inside splashed or edged deep pink on a whitish ground. Pleasing fragrance. Floriferous. Young shoots of vigorous upright plant are very glossy plum purple with 5–7 leaflets	
Dulhan	1983	Hybridization	An offspring of 'Bonne Nuit' and 'Ena-Harkness'	Its habit is free and robust like the parents. The large shapely flowers are fragrant and the attractive colour varies from rose red to red with the base of the petals cream. As the flower ages white streaks and splashes may appear, being prominent at the edges of the petals. A tall vigorous plant, flowers freely	IARI, New Delhi
Eastern Princess	1984	—	—	Colour is a novel orange coral with gold base, adding to the lustre. Beautiful elegant buds open slowly keeping the high pointed centre. Large petals reflex beautifully as the flower opens. Strongly scented. Vigorous plants with young leaves purplish red. Very free flowering	IARI, New Delhi
Golconda	1968	Selection	Seedling of 'Mme Charles Sauvage'	Colour varies from pale yellow to deep apricot, being usually deep in the centre.	IARI, New Delhi

ROSE : *Contd.*

Variety	Year of release	Breeding method	Pedigree/ parentage	Important traits	Institutes/ universities
				Flowers large and attractive. Pleasing light fragrance. Vigorous growth	
Gulbadan	1976			Buds light yellow and peach pink shaded with dark red, open to large blooms of creamy white edged and flushed with pink a very striking effect. Fragrant	IARI, New Delhi
Ganga	1970	Selection	Seedling of 'Sabine'	Medium-sized flowers of beautiful shape are borne singly on upright stems quite freely. The colour varies from pale yellow to gold. Healthy and vigorous, with handsome foliage	IARI, New Delhi
Golden Afternoon	1983	—	—	Glorious colour, golden apricot and orange on the upper surface of petals and rich coppery orange on the outside. Strong spicy fragrance and handsome glossy leaves. Very floriferous. The flowers open quickly in warm weather	IARI, New Delhi
Gulzar	1971	Hybridization	A hybrid seedling of 'Kiss of Fire' × 'Prelude'	Deep magenta red, exquisitely formed blooms are borne majestically on strong shoots. In spring flush, the flowers are normally home in bunches and tend to become more bluish in colour. Vigorous and compact bush. Fragrant	IARI, New Delhi

ROSE : *Contd.*

Variety	Year of release	Breeding method	Pedigree/ parentage	Important traits	Institutes/ universities
Hans	1970	Hybridization	A hybrid seedling of 'Message' × 'Vigro'	White, shapely blooms are borne singly on long shoots. Upright growing bush with light green foliage. The flowers are not affected by cold injury	IARI, New Delhi
Haseena	1979	Hybridization	A hybrid seedling of 'Youki San' × Balinese'	Perfectly shaped long buds open to large, lovely blooms of soft pink with occasionally a light lilac flush. Flowers are long lasting and fragrant. Medium height	IARI, New Delhi
Homage	1986	Hybridization	Sonia' × 'Princess Margaret of England'	Cerise red buds open into large, shapely, double flowers of deep warm pink. Reverse of petals and also the base of paler in colour. The back of petals is washed with pink lavender and at a later stage may become clear lilac pink. The number of petals is about 30. The flowers that are borne on long stems keep their shape over a long period. The flowers are high centered and of exhibition quality and they retain their attractiveness over a long period. The plant is vigorous and flowers freely	IARI, New Delhi

ROSE : *Contd.*

Variety	Year of release	Breeding method	Pedigree/ parentage	Important traits	Institutes/ universities
Indian Princess	1980	Hybridization	A hybrid sseedling of 'Super Star' × 'Granada'	Large long buds open to well-shaped blooms with many petals. The colour, a vibrant carmine vermilion is striking. The outer petals are light red while the inner ones are carmine pink with silvery edge. Well scented. Strong growth	IARI, New Delhi
Jawahar	1980	Hybridization	A hybrid seedling of 'Sweet Afton' × 'Delhi Princess'	Greenish white buds developing into creamy white, very large, high centered and well formed blooms. The flowers are borne singly as well as in small bunch of two or three on a strong shoot. In cold weather, the blooms attain very good size and depth with an effect of spirally arranged petals, which look very attractive. Upright, tall growing bush with light green foliage. Very good exhibition rose, vigorous with strong fragrances	IARI, New Delhi
Jawani	1985	Selection	Seedling of 'Mme Charles Sauvage'	Buds and flowers of the richest shades of gold and apricot. Flowers are of decorative type and attract attention by the unique lovely colour. Delicious tea type fragrance. Spreading flexible branches. Lovely for half standards	IARI, New Delhi

ROSE : *Contd.*

Variety	Year of release	Breeding method	Pedigree/ parentage	Important traits	Institutes/ universities
Kanakangi	1968	Selection	Seedling of 'Mme Charles Sauvage'	Buds and flowers of the richest shades of gold and apricot. Flowers are of decorative type and attract attention by the unique lovely colour. Delicious tea type fragrance. Spreading flexible branches. Lovely for half standards	IARI, New Delhi
Kulu Belle	1972	Selection	Seedling of 'Sabine'	Very long tapering buds open into large flowers of exquisite shape and *form.* Dark pink colour. A seedling of Sabine with handsome foliage	IARI, New Delhi
Kamla Devi Chatto-padhyay	1989	Hybridization	Raised from a cross between 'Sonia' and 'Kathleen O'Rourke'	The large double flowers are of a brilliant orange salmon colour, open slowly and are long lasting and very fragrant. The plant is vigorous with good foliage and the blooms are borne on long upright stems	IARI, New Delhi
Lal Makhmal	1983	Selection	A seedling of the famous, 'Samourai' ('Scarlet Knight' in the USA)	Very dark velvety red, which could be described as ebony red, with a scarlet sheen. The large symmetrical flowers are borne on long upright stems and attract instant attention because of the rich glowing colour	IARI, New Delhi

ROSE : *Contd.*

Variety	Year of release	Breeding method	Pedigree/ parentage	Important traits	Institutes/ universities
Lalima	1978	Hybridization	A hybrid seedling of 'Picture' × 'Jour'	Deep rose pink buds open to large perfect blooms of light crimson red, the colour being uniform and very pleasing. Slightly fragrant blooms are freely produced on a bush of medium height on upright stems. Vigorous plant with adequate foliage	IARI, New Delhi
Madhumati	1973	Selection	Seedling of 'General Mac Arthur'	Large, deep rose pink flowers of good shape are freely produced on a medium to tall plant. Remarkably rich fragrance of old rose type	IARI, New Delhi
Madhushala	1973	Selection	Seedling of 'Africa Star'	Large, double flowers of a clear shade of medium deep mauve. Bushy growth, but more upright than that of 'Africa Star'	IARI, New Delhi
Meghdoot	1972	Selection	Seedling of 'Virgo'	The long urn shaped buds open to large flowers of shining white, the petals recalling the silky texture of a poppy flower. The plants are very vigorous and tall	IARI, New Delhi
Mrs. K.B. Sharma	1989	Hybridization	Raised from cross 'White Masterpiece'	Very long, white buds open to large white flowers very gracefully in a style	IARI, New Delhi

ROSE : *Contd.*

Variety	Year of release	Breeding method	Pedigree/ parentage	Important traits	Institutes/ universities
			and 'Michele Meilland'	reminiscent of the famous 'Pristine'. Petals large and long. Lightly scented. Leaves medium green colour	
Madhosh	1975	Mutation	A mutant of 'Gulzar'	Deep magenta red with mauve coloured faint and broad streaks on the petals	IARI, New Delhi
Maharani	1986	Selection	Raised from open pollinated seeds of the old variety 'President	—	IARI, New Delhi
Macia	1933	—	—	Large, very long buds of deep pink colour develop into very large blooms borne on long, sturdy shoots, which retain flower on the plant for a long time. Fully open blooms are also attractive. Upright, tall and vigorous growing bush	IARI, New Delhi
Mechak	1979	Selection	A seedling of 'Samourai'	As the name suggests, dark and desirable, beautiful long shapely buds of blackest red, open to high centered, perfectly shaped flowers of deep ebony red with blackest velvety shading. Producing lasting flowers, very freely	IARI, New Delhi

ROSE : *Contd.*

Variety	Year of release	Breeding method	Pedigree/ parentage	Important traits	Institutes/ universities
Mother Teresa	1994	Hybridization	A hybrid seedling raised from a cross between 'Sweet Afton' and 'First Prize'	Long pointed soft pink buds open to large exquisitely formed flowers of pinkish white colour, which is later stage change to almost white with pink edges. The vigorous and upright growing tall plants bear flowers freely on long and sturdy stems. Flowers can withstand cold weather and open fully exposing beautiful centre. Flowers are fragrant	IARI, New Delhi
Mridula	1975	Hybridization	A hybrid seedling of 'Queen Elizabeth' and seedling of 'Sir Henry Segrave'	Very long, pinkish white and shapely buds open to soft white blooms of large size. Upright and very vigorous growing bush with abundant rich green foliage. Fragrant	IARI, New Delhi
Mrinalini	1973	Hybridization	A hybrid seedling of 'Pink Parfait' × 'Christian Dior'	Phlox pink, long pointed buds open to very large flowers, vigorous bushed. Slightly fragrant	IARI, New Delhi
Nazneen	1969	Selection	Seedling of 'Queen Elizabeth'	The large flowers are of a very delicate, translucent shade of pale pink suggesting very great delicacy. Flowers profusely produced on long upright stems either singly or in small clusters. Plant upright with good foliage	IARI, New Delhi

ROSE : ***Contd.***

Variety	Year of release	Breeding method	Pedigree/ parentage	Important traits	Institutes/ universities
Nayika	1975	—	—	Gorgeous, vibrant and eye-catching coral orange with reflexed petals. Lasting flowers with high centre and slight perfume. Free bloomer	IARI, New Delhi
Nandini	1983	—	—	Large high centred flowers with exquisite bi-colouring phlox pink on the upper surface of petals, and light creamy yellow on the outside. Medium-sized plant. Floriferous	IARI, New Delhi
Nehru Centenary	1989	Hybridization	A hybrid from the cross 'Christian Dior' × 'Avon'	Dark red, large pointed buds, open to high centered, large blooms, which are borne freely on long, straight and strong shoots. Slightly fragrant. Dark green and large foliage makes the strong and vigorous growing bushes very attractive	IARI, New Delhi
Nishada	1982	—	—	Very long and shapely bud of impressive size open slowly to very large flowers borne on long stems, the colour is rich pink with orange shading, the latter being specially noticeable in the earlier stages. Very fragrant. The plants are vigorous and spreading with neat foliage	IARI, New Delhi

ROSE : *Contd.*

Variety	Year of release	Breeding method	Pedigree/ parentage	Important traits	Institutes/ universities
Nurjehan	1980	Hybridization	A hybrid seedling of 'Sweet Afton' × 'Crimson Glory'	Rose bengal coloured well-formed blooms of very large size and high centre are borne singly on medium and strong shoots. Highly fragrant. Medium spreading bush	IARI, New Delhi
Pahadi Dhun	1981	Selection	A seedling of 'Lady X'	Medium-sized exquisitely shaped high centred blooms with reflexing petals are borne on long stems. Dark purple buds open to silvery flowers lightly bruised at edges with deeper mauve, reverse of petals is clear pink mauve: pleasant fragrance, very free flowering	IARI, New Delhi
Pale Hands	1965	Selection	Seedling of 'Me Gredy's Ivory'	Very large flowers ranging in colour from ivory white to shades of pale yellow, buff and peach. The buds are exquisitely shaped and can be used for exhibition or decoration	IARI, New Delhi
Pat Rani	1981	—	—	A two-tone pink, long tapering buds opening to perfect flowers with good shape, substance and reflexed petals; regular blooms borne on long stems with great freedom	IARI, New Delhi

ROSE : *Contd.*

Variety	Year of release	Breeding method	Pedigree/ parentage	Important traits	Institutes/ universities
Pink Sport of Montezuma	1980	Selection	A bud sport of the rose variety 'Montezuma'	Spinel pink coloured flowers. Except the change in flower colour of the parent variety from vermilion to spinel pink, other plant and flower characters are same as of the parent variety	IARI, New Delhi
Pusa Christina	1975	Mutation	A mutant of 'Christian Dior'.	Fuchsia pink with yellow base of petals	IARI, New Delhi
Poornima	1971	Selection	Seedling of 'Fernand Aries'	The buds are long and open into large blooms with many petals. The buds are light yellow but the open flower is often deeper in colour	IARI, New Delhi
Preyasi	1991	Hybridization	A hybrid seedling of 'Chandrama' × 'Queen Elizabeth'	Fuchsine pink, shapely, high centered large sized flowers are produced in great abundance. Tall and vigorous growth bush. Highly fragrant	IARI, New Delhi
Priyadarshini	1986	Hybridization	A hybrid seedling of 'Pink Parfait' and 'First Prize'	Exquisite flowers of rhodamine pink colour suffused with deeper pink towards the edges of petals produce a bicolour effect. The flowers are produced in great profusion on medium and long shoots on compact bushes. Flowers are slightly fragrant	IARI, New Delhi

ROSE : *Contd.*

Variety	Year of release	Breeding method	Pedigree/ parentage	Important traits	Institutes/ universities
Pusa Sonia	1968	Selection	Seedling of 'McGredy's Yellow'	Bright golden-yellow flowers of very attractive shape, borne in singles, freely on long and sturdy stems. During hot months, the colour of flowers becomes light yellow, medium bush with attractive and abundant foliage	IARI, New Delhi
Pusa Sonora	1934	Hybridization	A hybrid seedling of 'Queen Elizabeth' × 'First Prize'	Soft pink, high centered, stable blooms are produced on sturdy stems. Vigorous and upright growing bush	IARI, New Delhi
Pusa Christina	1975	Mutation	An induced mutant of cv. Christian Dior	Flowers fuchsia pink with yellow base of petals	IARI, New Delhi
Pusa Mohit	2002	Hybridization	Cross between 'Suchitra' (a pink floriferous cultivar) and an exotic variety 'Christian Dior' (a red coloured cultivar)	A vigorous growing hybrid tea with recurrent blooming habit produces almost thornless shoots. It produces long pointed buds with red petals having lighter shade on the reverse side. It is tolerant to black spot disease and is suitable as a cut flower variety	IARI, New Delhi
Pusa Mansij	2004	Hybridization	It is a hybrid tea rose originated as	The vigorous growing bushes can reach a height of 80–85 cm during winter	IARI, New Delhi

ROSE : *Contd.*

Variety	Year of release	Breeding method	Pedigree/ parentage	Important traits	Institutes/ universities
			a natural bud-sport Raja Surendra Singh of Nalagarh (scarlet Knight × Montezuma)	season and 100–125 cm during spring. The plants have moderately glossy, dark green foliage. After 45 days of pruning the plant produces long flowering shoots bearing individual long, pointed flower buds, which are orange in colour. These buds open to double high centered flowers which are dazzling orange salmon pink in colour with prominent white stripes and spots on both sides of the petals. The centre of flowers is filled with yellow anthers with pinkish filaments providing a beautiful contrast when the flowers open fully. It is ideal as a cut flower and for exhibition purpose besides garden display and is recommended for growing in the Northern plains	
Raat-Ki-Rani	1975	Selection	A seedling of 'Scarlet Knight'	Large pointed buds open to velvety crimson red blooms of good quality. Free flowering and vigorous growing variety	IARI, New Delhi
Raja Surendra Singh of Nalagarh	1977	Hybridization	A hybrid seedling of 'Scarlet Knight' × 'Montezuma'	Large-sized, good, high centred flowers of classic shape of dazzling orange salmon pink colour. Buds are extremely beautiful and fully double large flowers	IARI, New Delhi

ROSE : *Contd.*

Variety	Year of release	Breeding method	Pedigree/ parentage	Important traits	Institutes/ universities
				with imprecated petals have long lasting quality	
Raktima	1991	Hybridization	A hybrid seedling of 'Pink Parfait' × "Sugandha'	Shining red, high centred double flowers are produced in singles on strong and straight stems. Sweet fragrance. Recurrent blooming habit tending to produce quality flowers even in off-season. Very tall and upright growing bushes	IARI, New Delhi
Rampa Pal	1975	—	—	The buds and flowers are large and of classic shape. Outer petals pink with white base, while the inner core gives the appearance in early stages of having been dipped in orange-scarlet. The colour blending of pink, white and yellow is most beautiful to behold. Free flowering. Vigorous	IARI, New Delhi
Rosy Evening	1935	Hybridization	An excellent rose from 'Invitation' and 'Anvil Sparks'	Flowers are very attractive having coppery pink to flame orange pink colour. High centered buds open to well shape, large full flowers freely on long stems. Suitable for exhibition and also for bedding	IARI, New Delhi

ROSE : *Contd.*

Variety	Year of release	Breeding method	Pedigree/ parentage	Important traits	Institutes/ universities
Scented Bowl	1965	Selection	A seedling of 'General McArthur'	Glowing red with many petals. The fragrance is strong and very attractive. A useful addition to the really well. Scented roses	IARI, New Delhi
Shanti Pal	1989	Hybridization	Raised from a cross between 'Sonia' and 'Kathleen O' Rourke'	large, high-centred double flowers have a beautiful shade of salmon pink with sometimes a touch of coral. The buds open slowly into large perfectly shaped scented blooms which are produced freely on long stems. Plants vigorous and well clothed with leaves	IARI, New Delhi
Sharmili	1986	Selection	A seedling of the famous French rose 'Rose Gaujard'	Beautiful, long, high centred pointed buds of milky white flushed with pink are deeper at petal edges. The attractive flowers are long lasting and are produced on strong stems. The flowers are moderately fragrant. They are able, to withstand the cold nights, characteristic of some areas in the plains of north-west India. Upright vigorous bush of medium height	IARI, New Delhi
Sweet Innocence	1980	—	—	Shapely, pure white blooms with delicious fragrance. Gives adequate number of flowers all through the	IARI, New Delhi

ROSE : *Contd.*

Variety	Year of release	Breeding method	Pedigree/ parentage	Important traits	Institutes/ universities
				season. Buds are large, long and shapely opening to very large high centred double flowers	
Sandeepini	1983	—	—	Vibrant glowing pink, which is almost red in cool weather. Buds of classic shape open into large, very fragrant flowers. Vigorous plants with good foliage repeat flowering freely	IARI, New Delhi
Rajhans	1983	—	—	An impressive new white rose. Long, shapely buds open into very large white flowers with a large number of petals, which may go up to 100. These are beautifully arranged. The most outstanding attribute is very sweet, very strong fragrance of the flowers. These are borne freely on a vigorous plant	IARI, New Delhi
Raj Kumari	1975	Hybridization	A hybrid seedling of 'Charles Mallerin' × 'Delhi Princess'	Long and slender buds open to large well-formed deep pink flowers. Compact and vigorous growth bush	IARI, New Delhi
Rakta-gandha	1975	Hybridization	A hybrid seedling of 'Christian Dior' × seedling of 'Carrousel'	Shining red, large pointed buds open to large high centered flowers. The flowers are tolerant to winter injury. Slight fragrance	IARI, New Delhi

ROSE : *Contd.*

Variety	Year of release	Breeding method	Pedigree/ parentage	Important traits	Institutes/ universities
Raktima	1991	Hybridization	A hybrid seedling of 'Pink Parfait' × 'Sugandha'	Shining red, high centered double flowers are produced in singles on strong and straight stems. Sweet fragrance. Recurrent blooming habit tending to produce quality flowers even in off-season. Very tall and upright growing bushes	IARI, New Delhi
Rangshala	1969	Selection	Seedling of 'Margaret Spaull'	Beautiful buds of light Saturn-red colour open to blooms of charming shades of peach, apricot and amber yellow colours. Perfectly formed, medium sized flowers are borne in great profusion on compact bushes. Recurrent blooming. Medium and balanced bush	IARI, New Delhi
Ranjana	1975	—	—	A pleasing rich geranium lake colour. Urn shaped large buds open to shapely large, full, double flowers of exhibition quality. Fragrant	IARI, New Delhi
Ratnaar	1985	—	—	An exhilarating cool pink, large, high centered buds give arise to very large, well shaped blooms whose petals reflex	IARI, New Delhi

ROSE : ***Contd.***

Variety	Year of release	Breeding method	Pedigree/ parentage	Important traits	Institutes/ universities
				beautifully. Bushes are vigorous and very free flowering. Suitable for bedding and exhibition	
Shreyasi	1991	Hybridization	A hybrid seedling of 'Pink Parfait' × 'Christian Dior'	Bicoloured flowers are of tyrian purple with mimosa yellow base and spinel red on reverse. Medium sized flowers are very attractive and good for flower arrangement. Pleasing fragrance, upright growing bush. Flowers stand up well even in severe cold	IARI, New Delhi
Sir C.V. Raman	1989	Hybridization	Raised from a cross between 'Scarlet Knight' and 'First Prize'	Characterized by a vigorous plant with good foliage and upright stems is most outstanding feature of this rose. Brilliant orange scarlet colour of the large double flowers. In very cold weather, the outer side of the petals may take on a dark shade of red. The fragrant, long-lasting flowers are freely produced into the hot weather	IARI, New Delhi
Soma	1980	Hybridization	A hybrid seedling of 'Chandrama' × 'Surekha'	Shades of solferino purple, large, well formed are borne freely. A vigorous growing bush	IARI, New Delhi

ROSE : *Contd.*

Variety	Year of release	Breeding method	Pedigree/ parentage	Important traits	Institutes/ universities
Sugandhini	1969	—	—	Seedling of 'Margaret Spaull'. Light rhodamine pink, exquisitely formed well-filled blooms. Profuse flowering. Medium compact and vigorous growing bush. Good for standards and bedding. Fragrant	IARI, New Delhi
Sujata	1971	Selection	Seedling of 'Sabine'	Deep, Crimson red, wellformed, high centered large blooms are borne freely on medium shoots. Vigorous bush with dark green and leathery foliage. Fragrant	IARI, New Delhi
Surabhi	1975	—	—	A hybrid seedling of 'Oklahoma' × 'Delhi Princess'. Phlox pink, long tapering buds open slowly. Large-sized, attractive shape and long lasting quality of flowers	IARI, New Delhi
Surekha	1969	Selection	A seedling of 'Queen Elizabeth'	Moderately large, exquisitely formed buds are produced continuously on long stems and open to long lasting, coral red, many petalled blooms of perfect form in abundance on very healthy, tall growing and vigorous bush	IARI, New Delhi
Surkhab	1976	—	—	A beautiful red and white bi-coloured rose. Flowers are large and of good	IARI, New Delhi

ROSE : Continued ....

Variety	Year of release	Breeding method	Pedigree/ parentage	Important traits	Institutes/ universities
				shape and borne on strong and vigorous plants. The flowers are slightly fragrant and very freely produced	
Uttam	1969	Selection	Seedling of 'Elite'	Attractive, pointed buds and delicately pastel pink coloured, well-formed, medium-sized blooms are borne freely on long stems. Floriferous, vigorous with abundant foliage	IARI, New Delhi
Uma Rao	1989	—	—	Long, pointed buds open to large and very double flowers of apricot yellow colour with shades of pink and deep pink in the edges of petals. Medium growing bush. Floriferous	—
Vasant	1980	Hybridization	A hybrid seedling of 'Sweet Afton' × 'Delhi Princess'	Chinese yellow with tinge of neyron rose on edges of petals. The flowering is continuous throughout the year. The flowers are borne singly in winter and in small bunches in spring flush, very freely on straight stems in large numbers. The plant grows to an average height of 100cm. Healthy and vigorous bush. Ideal for bedding	IARI, New Delhi

ROSE : *Contd.*

Variety	Year of release	Breeding method	Pedigree/ parentage	Important traits	Institutes/ universities
White Nun	1968	Selection	A seedling of 'Virgo'	Pure white equisitely-shaped buds reminiscent of its parent 'Virgo'. open tog listening flowers with large petals. Petals moderate in number. Flowers produced in small clusters in warm weather. Upright growth	IARI, New Delhi
			Floribundas		
Akash Nartaki	1983	Selection	Raised from 'Sterling Silver'	The attractive clean silvery lilac colour is reminiscent of that famous variety. Semi-double flowers are freely produced in small clusters. The large petals are distinctly wavy and slightly frilled adding to the charm of the fragrant flowers. The colour is retained to the last and petals drop off cleanly. The plant is low growing	IARI, New Delhi
Azzez	1965	—	—	Exquisitely shaped buds of a shade of salmon or coral pink with lighter reverse at certain season, reminding one of a smaller edition of "Miss Ireland" The open flowers are attractive and reasonably resistant to both cold and heat	IARI, New Delhi

ROSE : *Contd.*

Variety	Year of release	Breeding method	Pedigree/ parentage	Important traits	Institutes/ universities
Arunima	1975	Selection	A seedling of 'Frolic'	Deep pink flowers are produced in small bunches. Long lasting flowers. Suitable for flower arrangement. Very compact bushes	IARI, New Delhi
Banjaran	1969	—	—	An aptly named, most attractive vigorous floribunda with well spaced clusters of double or semi-double flowers. The colour is a blend of gold and flame red. The red softens to flame pink in later stages. Handsome foliage. As colourful as the vestments of gypsy women	IARI, New Delhi
Belle of Punjab	1965	Hybridization	A hybrid seedling of 'Montezuma' × 'Flamenco'	An entrancing shade of' warm pink. The beautifully shaped buds open to flowers of medium size. The flowers are produced in singles or small clusters in winter and in larger clusters in warm weather. Upright and vigorous with handsome foliage	IARI, New Delhi
Celestial Star	1965	—	—	The medium-sized plants produce clusters of exquisite HT shaped flowers about one-and-a-half to two inches in diameter. The colour is most attractive	IARI, New Delhi

ROSE : *Contd.*

Variety	Year of release	Breeding method	Pedigree/ parentage	Important traits	Institutes/ universities
				approaching that of the famous 'Super Star', but perhaps, just a shade deeper. A very free bloomer	
Chamba Princes	1967	—	—	Single to semi-double blooms of most beautiful shape are clear pink with large petals and golden anthers	IARI, New Delhi
Chingari	1976	Selection	Seedling of 'Charleston'	An extremely colourful blend of yellow and deep velvety red. Yellow is more prominent in early stages, but later the red deepens and spreads and there is only an occasional streak of yellow. Flowers profusely produced and foliage is handsome	IARI, New Delhi
Chandrama	1980	Hybridization	A hybrid seedling of 'White Bouquet' × 'Virgo'	Dawn pink buds opening to large, moon-light white, double flowers borne in clusters of 4 - 6 flowers/cluster on sturdy shoots. The average number of flowers produced on a bush in the winter flush is 44. The plant grows to an average height of 100 cm, with a healthy and balanced growth. Good for garden display as well as exhibition purposes	IARI, New Delhi

ROSE : *Contd.*

Variety	Year of release	Breeding method	Pedigree/ parentage	Important traits	Institutes/ universities
Chitchor	1972	Selection	A seedling of 'Pink Parfait'	The captivating HT-type, double flowers are white very lightly suffused with pink and deep pink on edges; the centre is sometimes illuminated by a pale golden glow. Very vigorous and floriferous	IARI, New Delhi
Deepak	1977	—	—	Medium sized. Dazzling orange and flame red flowers are produced in large clusters. Plant is of medium height, very vigorous and free flowering. Very attractive	IARI, New Delhi
Delhi Daintiness	1963	—	—	Single to semi-double flowers in clusters with large flaring petals, most attractively shaped and waved. Colour light pink, deep on reverse, developing attractive 'freckles' as the petals age. Medium height with very handsome young foliage	IARI, New Delhi
Delhi Maid	1963	—	—	Single flowers, flame-orange with gold towards the base, produced both singly and in clusters. Extremely attractive colour, especially in cooler weather.	IARI, New Delhi

ROSE : *Contd.*

Variety	Year of release	Breeding method	Pedigree/ parentage	Important traits	Institutes/ universities
				Petals slightly wavy. Medium, tall, vigorous plant with dark green leaves	
Delhi Pink Powder puff	1965	—	—	Clusters of the loveliest shade of soft pink. A large number of petals arranged in a most pleasing manner. Upright, vigorous growth with handsome foliage	IARI, New Delhi
Delhi Princess	1963	—	—	Cerise red buds open to sparkling deep warm pink flowers which come in clusters, in amazing abundance. Every flower is well-formed. The foliage is bronzy red in the young shoots. Stands both cold nights and warm days. Lightly tea-scented. An outstanding variety for bedding and half-standards. Very vigorous	IARI, New Delhi
Delhi Rosette	1965	—	—	A beautiful floribunda with small, pretty flowers of brilliant orange scarlet. Plant vigorous, upright, and compact. Foliage abundant, dark green, glossy	IARI, New Delhi
Delhi Sherbet	1963	Selection	A seedling of 'Gruss an Teplitz'	Deep rose pink flowers produced in great profusion, both singly and in small	IARI, New Delhi

ROSE : *Contd.*

Variety	Year of release	Breeding method	Pedigree/ parentage	Important traits	Institutes/ universities
				clusters. Highly fragrant. Very vigorous, and useful for a low fragrant hedge	
Deepika	1975	Selection	A seedling of 'Shepherd's Delight'	Pointed small buds open to very showy flowers of mandarine red with lighter reverse and yellow base. Brown in small bunches	IARI, New Delhi
Deepshikha	1975	Hybridization	A hybrid seedling of 'Sea Pearl' × 'Shola'	Dutch vermilion double and open flowers are borne in small and big clusters. Produces good flowers in summer and rainy season also	IARI, New Delhi
Delhi Brightness	1963	—	—	Semi-double. The buds are brilliant geranium lake and remind one of a deeper coloured 'Jiminy Cricket'. The open flowers are of an attractive orange pink, produced in large well-spaced clusters. Bush medium tall, vigorous with deep bronzy foliage in early stages	IARI, New Delhi
Dr. S.S. Bhatnagar	1994	Hybridization	A hybrid seedling raised from a cross between 'Oklahoma' and 'White Christmas'	A floribunda rose with bright and velvety red colour. Medium sized double flowers of long lasting quality are produced in number of 4 to 6, flowers in great profusion throughout the season. Vigorous	IARI, New Delhi

ROSE : *Contd.*

Variety	Year of release	Breeding method	Pedigree/ parentage	Important traits	Institutes/ universities
				and upright growing bushes. Slightly fragrance and good for bedding	
Fugitive	1965	Selection	Seedling of 'Mrs. Oakley Fisher'	Large clusters of single 5-petalled flowers adorn this floribunda. The buds of a beautiful flame, colour lightens on opening when shades of buff and pale yellow chase each other	IARI, New Delhi
Himangini	1968	Selection	A seedling of 'Saratoga'	Ivory white, medium sized flowers are produced in huge trusses. The profusion is so great that the plant gets completely covered with a mass of white flowers of very long lasting quality. The flowers at opening have a light buff centre changing to uniform ivory white colour, which later fades to greenish white. Recurrent blooming. Vigorous. Recommended for bedding, borders and as a cut flower for its very long lasting quality	IARI, New Delhi
Jantar Mantar	1982	—	—	A red floribunda rose, 25–30 petals make excellent blooms in a cluster of 4–6 long stems. Named after the Jantar Mantar observatory at New Delhi. Highly	IARI, New Delhi

ROSE : *Contd.*

Variety	Year of release	Breeding method	Pedigree/ parentage	Important traits	Institutes/ universities
				floriferous, vigorous bush with bronze green healthy foliage	
Kumkum	1972	Selection	Seedling of 'Orangeade'	The single to semi-double flowers are brone in large clusters of the most dazzling orange scarlet imaginable	IARI, New Delhi
Kavita	1972	Selection	A seedling of 'Margaret Spaull'	The open flowers come in attractive shades of orient pink with light yellow base of the petals. Vigorous growing bush	IARI, New Delhi
Lahar	1991	Hybridization	A hybrid seedling of 'Pink Parfait' × 'Ganga'	Beautiful mimosa yellow with tinge of spinel pink flowers give a multi-colour effect. Almost recurrent blooming. HT type flowers are borne in big bunches in great profusion. Upright, vigorous growing bush. Tolerant to diseases. Flowers stand up very well to low temperature in winter	IARI, New Delhi
Loree	1968	Selection	Seedling of 'Frolic'	Extremely pale pink double flowers borne on a plant of medium height in large clusters. The colour deepens slightly in a very cool weather and becomes almost white in warm weather	IARI, New Delhi

ROSE : *Contd.*

Variety	Year of release	Breeding method	Pedigree/ parentage	Important traits	Institutes/ universities
Manmatha	1989	Hybridization	Raised from a cross between 'Sonia' and 'Princess Margaret of England'	HT type floribunda blooms profusely. The basic colour of the petals is smoky pink *or* silvery mauve, but as the flower opens the edges become flushed with red. Vigorous plants have attractive leaves	IARI, New Delhi
Madhura	1979	Hybridization	A hybrid seedling of 'Kiss of Fire' × 'Goud Vilinder'	This charming rose has H.T. shaped buds of lovely blend of light yellow and pink. Flowers freely produced on single and small clusters on a vigorous plant. Honey fragrance	IARI, New Delhi
Manasi	1991	Selection	A seedling of 'Frolic'	Light pink flowers having shades of cream and lilac colour are produced freely in very big bunches covering almost the whole plant with flowers. Medium growing bushes. Very floriferous	IARI, New Delhi
Mohini	1970	Hybridization	A hybrid seedling of 'Sea Pearl' × 'Shola'	An unusual chocolate brown colour with yellow tinge at the base of petals at low temperature and dazzling orange at high temperature. Dark green and very glossy foliage, medium growing bush, floriferous	IARI, New Delhi

ROSE : *Contd.*

Variety	Year of release	Breeding method	Pedigree/ parentage	Important traits	Institutes/ universities
Nutkhut	1969	Hybridization	A hybrid seedling of 'Rumba' × 'Cocorico'	Large clusters of pompon like flowers of a lovely coral red. Vigorous plant with attractive leaves	IARI, New Delhi
Navneet	1971	Hybridization	A hybrid seedling of 'Prelude' × 'Africa Star'	Creamy white, beautiful, semi-double flowers are produced in small and big bunches on long and strong shoots. The profusion of flowering is great. Vigorous and spreading bush with large and green foliage. Good for bedding	IARI, New Delhi
Nav-Sadabahar	1980	Selection	A bud sport of 'Sadabahar'	Except for the white streaks on the original deep pink colour of the flowers of the parent variety 'Sadabahar', other plant and flower characteristics are the same as those of the parent variety	IARI, New Delhi
Neelambari	1975	Hybridization	A hybrid seedling of 'Blue Moon' × 'Africa Star'	Beautifully formed pointed buds develop to a full bloom of classic rose form and magenta rose colour. The flowers are borne freely in singles as well as in small and big bunches. The variety is attractive for its bluish colour	IARI, New Delhi
Orange Cup	1965	Selection	A seedling of 'Cocorico'	Very large trusses of most brilliant orange-scarlet. Grow tall and upright;	IARI, New Delhi

ROSE : *Contd.*

Variety	Year of release	Breeding method	Pedigree/ parentage	Important traits	Institutes/ universities
				very vigorous and free flowering. Makes a wonderfully, brilliant hedge	
Punch	1966	—	—	'Ruby-red, semi-double flowers with a brilliant sheen are very freely produced. The foliage is especially handsome being a shining red-bronze in the early stage. Vigorous, upright growth. Good as a hedge, 'and also for individual planting as a shrub	IARI, New Delhi
Parwana	1974	—	—	The buds and flowers are a beautiful combination of light to medium yellow with red on edges. The flowers are produced freely in singles and small clusters	IARI, New Delhi
Paharan	1971	Selection	A seedling of 'Anna Wheatcroft'	Large clusters of exquisite buds open into small, pretty, fully double flowers of pleasing shades varying from pearly white to pink, reminiscent of the complexion of the hill maid, after whom it is named. Plants of medium height	IARI, New Delhi
Prema	1970	Hybridization	A hybrid seedling of 'Sea Parl' × 'Shola'	Soft pink with deep pink edges, very shapely flowers are borne profusely in big bunches. The flowers have very long	IARI, New Delhi

ROSE : *Contd.*

Variety	Year of release	Breeding method	Pedigree/ parentage	Important traits	Institutes/ universities
				lasting quality thus making the entire bush a bouquet of flowers. Medium compact growing bush. Tolerant to most diseases and pests	
Pusa Abhishake	2002	Selection	Striped and spotted bud sport of an indigenous, floriferous and hardy variety 'Jantar Mantar'	A vigorous floribunda with shining foliage produces big bunches having 25–30 flowers. Longitudinal stripes of creamy white, white and light yellow colours appear on red background. Tolerant to powdery mildew and black spot diseases. Suitable for garden display and exhibition purposes	IARI, New Delhi
Pusa Komal	2006	Hybridization	A hybrid between an exotic variety Pink Parfait and an indigenous variety Suchitra	It is a highly floriferous bushy variety, whose short to medium stature bushes are completely thornless. The variety produces as many as 60 flowers, which are mostly borne in clusters. The pink coloured buds are pointed in shape and unfurl in to double flowers. It produces very attractive pink coloured flowers, petals (numbering 50–60) delicate and are light pink in colour. Moderately tolerant to insect pest like thrips and powdery mildew and black spot diseases.	IARI, New Delhi

ROSE : *Contd.*

Variety	Year of release	Breeding method	Pedigree/ parentage	Important traits	Institutes/ universities
				The blooms are mildly fragrant and are suitable for growing in pots and beds. It produces 40–45 flowers during winter 60–65 flowers during spring season. The blooms remain fresh for 8–10 days in field conditions	
Pusa Barahmasi	2000	Selection	A seedling of Sadabahar	Attractive deep pink flowers, flowering almost throughout the year in bunches. Good flowering is obtained even in off season, *i.e.* during summer and rainy season. Very hardy plants. Good multiplier. Good for garden, exhibition and making standards. Tolerant to die back, powdery mildew and black spot	IARI, New Delhi
Pusa Manhar	2002	Hybridization	Developed by crossing two indigenous varieties 'Jantar Mantar' (red) and 'Lahar' (yellow with pink edges)	A medium growing floribunda. The plant produces hybrid tea type flowers, which are almost double with 100–115 petals. The petals are creamy white having magenta shades on outer petals and edges of petals. It is tolerant to powdery mildew and black sopt diseases and is good for garden display, exihibition purposes, etc.	IARI, New Delhi

ROSE : *Contd.*

Variety	Year of release	Breeding method	Pedigree/ parentage	Important traits	Institutes/ universities
Pusa Maskan	2002	Hybridization	Hybrid between 'Pink Parfait' × 'Alinka'	An attractive colourful vigorous growing floribunda produces large clusters of flowers, which are creamy white having pink shade and red coloured edges	IARI, New Delhi
Pusa Pitambar	2000	Hybridization	A cross between Jantar Mantar and Banjaran	Very attractive golden yellow coloured HY type flowers having very sweet fragrance. During spring flush produce very big bunches having 15–20 flowers on a stem. Very good for garden and exhibition purposes tolerant to powdery mildew and black spot	IARI, New Delhi
PuPusa Virangana	2000	Selection	A seedling of Jantar Mantar	A very large, deep red flowers are produced on very vigorous, tall and spreading bushes. Very good for garden and making standards. 6–12 flowers in a bunch	IARI, New Delhi
PuPusa Shatabdi	2006	Hybridization	Hybrid between two exotic varieties Jadis and Century Two	A vigorous growing variety producing strong, straight stems with bronze to reddish brown pigmentation in the initial stages. The flowers are born on long shoots, which are straight and strong. The pink coloured buds are pointed in shape and are very big in size often	IARI, New Delhi

ROSE : *Contd.*

Variety	Year of release	Breeding method	Pedigree/ parentage	Important traits	Institutes/ universities
				reaching 6–8 cm in length and unfurl in to double flowers. Very attractive light pink coloured flowers last almost 10 days in vase. The petals (numbering 35–40) are fleshy and are pink in colour. It is moderately tolerant to powdery mildew and leaf spot diseases. The blooms are mildly fragrant and are suitable for cut flower and exhibition purposes. It produces 20–30 flowers during winter and 35–40 flowers during spring season. The blooms remain fresh for 8–10 days in field conditions and have a vase life of 10–12 days when harvested at bud stage	
Pusa Urmil	2002	Selection	Bud sport of Jantar Mantar	Produces unusual brown pink coloured flowers in big clusters of 40–45 flowers during spring flush. It produces vigorous bushes and is tolerant to powdery mildew disease	IARI, New Delhi
Pusa Ranjana	2002	Hybridization	A floribunda type hybrid seedling of Pink Parafit × Iceberg	Produces dwarf and compact bushes having novel type of growth habit, long lasting dark pink flowers born in big clusters, profuse flowers covering entire bush, plants are vigorous with bright	IARI, New Delhi

ROSE : *Contd.*

Variety	Year of release	Breeding method	Pedigree/ parentage	Important traits	Institutes/ universities
				green foliage, good for pot and bedding purpose	
Ragini	1972	Selection	A seedling of 'Queen Elizabeth'	Delightful clear rose pink. The flowers are produced in well-shaped clusters	IARI, New Delhi
Rajbala	1975	—	—	Light cerise pink, large single blooms are produced in bunches. Beautiful flowers and attractive foliage make a very good combination. Very vigorous bush clothed with large, light green abundant foliage	IARI, New Delhi
Rose Sherbet	1962	Selection	A seedling of the well-known rose 'Gruss an Teplitz'	Very free flowering, producing large numbers of flowers in small clusters. Colour glowing deep pink, beautifully scented and stands up satisfactorily to cold nights in winter. Good for mass planting or for rose hedge	IARI, New Delhi
Rupali	1971	Hybridization	A hybrid seedling of 'Sweet Afton' × 'Delhi Princess'	Deep rose madder, large flowers come in bunches of amazing size. Every flower is beautifully formed. Sometimes the bunches are borne on very long and strong shoots coming from the base. Very vigorous and tall growing bush with clean foliage. Slightly fragrant	IARI, New Delhi

ROSE : *Contd.*

Variety	Year of release	Breeding method	Pedigree/ parentage	Important traits	Institutes/ universities
Sailoz Mookherjea	1974	—	—	Absolutely dazzling, pure cadmium orange HT type buds open into long lasting flowers with firm petals. The clusters are large. Vigorous plant and healthy foliage	IARI, New Delhi
Sandhya Bela	1971	Selection	A seedling of 'Zambra'	Medium sized, fully double flowers in striking new shade of orange. The plant is of medium height and moderate growth	IARI, New Delhi
Stanza	1967	—	—	Single to semi-double small blooms in numerous clusters, very free flowering. The pink colour becomes distinctly lighter towards the centre of the flowers	IARI, New Delhi
Suhashini	1972	Selection	A seedling of 'Queen Elizabeth'	It is a very majestic floribunda giving abundant bloom of rich transluscent coral-pink. Flowers are of perfect shape and form with pointed petal edges, very full and large, borne in clusters of 4 or 5 carried on stiff strong sterns. Plants of medium height and vigour	IARI, New Delhi
Suryakiran	1979	—	—	A HT type floribunda. Large well-shaped flowers of brilliant orange aging to salmon orange, come in large clusters on upright stems. Very vigorous	IARI, New Delhi

ROSE : *Contd.*

Variety	Year of release	Breeding method	Pedigree/ parentage	Important traits	Institutes/ universities
Suryodaya	1968	Selection	Seedling of 'Orangeade'	Dazzling orange, double, well-shaped flowers. Borne in singles and clusters of 4 to 6 flowers; free flowering. Tall and strong growing bushes	IARI, New Delhi
Sadabahar	1969	Selection	A seedling of 'Frolic'	Deep pink buds open to flowers of medium size. The flowers are produced in clusters in such a great number that the plant remains covered with mass of flowers almost throughout the season. Ideal for bedding and border, makes very beautiful half-standards. Very vigorous. Flowers stand well to severe winters. Tolerant to most of the diseases.	IARI, New Delhi
Saroja	1984	Selection	An enchanting, bicolour floribunda form of 'Rose Gaujard'	A deep cherry pink with a silvery base and reverse. Beautiful buds and blooms are elegantly borne singly or in trusses. Very lasting and repeat quickly	IARI, New Delhi
Shabnam	1975	Selection	A seedling of 'Baby Sylvia'	White, compact, small flowers are produced in enormous clusters. The colour of the flowers becomes icy-white in the spring flush. Upright and vigorous growing bush	IARI, New Delhi

ROSE : *Contd.*

Variety	Year of release	Breeding method	Pedigree/ parentage	Important traits	Institutes/ universities
Shola	1969	Selection	A seedling of 'Anna Wheatcroft'	Dutch vermilion buds open to shades of sparking orient red, medium-sized, long lasting and borne in clusters in with large and green foliage. Good for bedding. Amazing abundance. Low and spreading bush	IARI, New Delhi
Shringar	1972	Hybridization	A hybrid seedling of 'Eiffel Tower' × 'Suryodaya'	The flowers are deep camellia rose with lighter base and reverse of petals. Semi-double flowers are beautifully formed and borne freely both singly and in clusters. The flowers hold up remarkably well on the plant for their good lasting quality	IARI, New Delhi
Sindoor	1980	Hybridization	A hybrid seedling of 'Sea Pearl' × 'Suryodaya'	The long and pointed buds open to geranium lake, well formed, double and large blooms on long shoots in single and small clusters during winter flush and big clusters of 15 flowers in great abundance. Tall and vigorous growing bush with clean and attractive foliage	IARI, New Delhi
Suchitra	1972	Hybridization	A hybrid seedling of 'Lady Frost' × 'Swati'	Rhodamine pink with mimosa yellow base and silvery white reverse, beautifully formed double flowers are produced in singles and large clusters.	IARI, New Delhi

ROSE : *Contd.*

Variety	Year of release	Breeding method	Pedigree/ parentage	Important traits	Institutes/ universities
				Good for bedding. The Floriferousness and continuous flowering habit make it a prized garden bush	
Suryakiran	1979	—	—	A H.T. type floribunda. Large well-shaped flowers of brilliant orange aging to salmon orange, come in large clusters on upright stems. Very vigorous	IARI, New Delhi
Temple Flame	1965	Selection	A seedling of 'Orangeade'	Brilliant orange colour, which becomes flushed with red in cool weather. The growth and the colour of the flowers is reminiscent of 'Orangeade'. Extremely vigorous and floriferous. The flowers are produced singly and in small clusters	IARI, New Delhi
Tarang	1989	Selection	'Queen Elizabeth'	HT type, medium-sized flowers are produced in cluster. The light yellow colour is set off by a salmon pink flush. The plants are slightly above medium height and flowers are quite attractive and are produced in clusters. The seed parent is 'Queen Elizabeth' from which the new variety inherits its vigour and freedom of flower. The attractive leaves are bronzy and very glossy	IARI, New Delhi

ROSE : ***Contd.***

Variety	Year of release	Breeding method	Pedigree/ parentage	Important traits	Institutes/ universities
Usha	1975	Selection	A seedling of 'Orangeade'	The small flowers of deep camellia rose colour with lighter reverse are produced in very big clusters. Long lasting quality of flowers	IARI, New Delhi
				Polyanthas	
Swati	1968	Selection	Seedling of 'Winifred coulter'	Deep carmine pink, dainty, well-shaped buds with silvery reverse open to semi-double, small white flowers having deep pink edge of the petals produced in clusters. Low, compact bush with mass of flowers. Excellent for edging, pots and bedding	IARI, New Delhi
				Miniatures	
Delhi Starlet	1963	Selection	Seedling of 'Goudvlinder'	Produces large number of small flowers resembling 'Goudvlinder'. The buds are of deep yellow and the quickly opening flowers are of a light yellow or cream colour. Dwarf and compact with handsome new foliage	IARI, New Delhi

ROSE : *Contd.*

Variety	Year of release	Breeding method	Pedigree/ parentage	Important traits	Institutes/ universities
			Climbing and Rambling Roses		
Climbing Sadabahar	1991	Selection	A climbing bud sport of famous floribunda variety 'Sadabahar'	Deep pink flowers are produced in clusters in such a great number that the whole plant gets covered with mass of flowers in April and May	IARI, New Delhi
Delhi White Pearl	1963	Selection	A seedling of 'Prosperity'	The flowers are pearly white, full with a faint musky fragrance. The flowers come in truly enormous clusters several times in a year. It surpasses 'Prosperity' in vigour and freedom of flowering. It can also be grown as a shrub	IARI, New Delhi
YCD-1 Rose	1985	Selection	It is a clonal selection from open pollinated seedling progenies	Yellow flowers, 100–120 flowers/plant/ year vase life — 7 days	HRS, TNAU, Yercaud
YCD-2 Rose	1991	Selection	Clonal progeny of seedling collection (1043), from the bulk population	Buds are yellow with scarlet margin, changes to red at full bloom, 200 flower/plant/year	HRS, TNAU, Yercaud
YCD-3 Rose	1994	Selection	Selection from open pollinated progeny	Crimson flowers with 9.5 cm diameter, 150 flowers/plant/year	HRS, TNAU, Yercaud

CARNATION

Variety	Year of release	Breeding method	Pedigree/ parentage	Important traits	Institutes/ universities
IIHRP 1	2006	Mutation breeding (*in vitro* mutagenesis)	Accession CG 109	Dark red coloured flowers with blunt flower margin	IIHR, Bangalore

contains 5–160 seeds (technically dry single-seeded fruits called achenes) embedded in a matrix of fine, but stiff, hairs. Rose hips of some species, especially the *Rosa canina* and *Rosa rugosa*, are very rich in vitamin C. The hips are eaten by fruit-eating birds such as thrushes and waxwings, which then disperse the seeds in their droppings. Some birds, particularly finches, also eat the seeds.

Rose prickles are typically sickle-shaped hooks, which aid the rose in hanging onto other vegetation when growing over it. Some species such as *Rosa rugosa* and *R. pimpinellifolia* have densely packed straight spines, probably an adaptation to reduce browsing by animals, but also possibly an adaptation to trap wind-blown sand and so reduce erosion and protect their roots (both of these species grow naturally on coastal sand dunes). Despite the presence of prickles, roses are frequently browsed by deer. A few species of roses only have vestigial prickles that have no points.

The important classes of roses are Hybrid Tea, Polyantha, Floribunda, Grandiflora, Miniature and Climbing/Rambling.

CARNATION

Botanical name: *Dianthus caryophyllus*
Family: Caryophyllaceae

Carnation flower is valued for its excellent keeping quality, wide array of colours and forms, and ability to rehydrate after continuous transportation. Carnation is indigenous to the Mediterranean areas. Due to high cost of production inside greenhouses in Europe and USA, its cultivation is shifting to more naturally-growing regions where they are being produced at lesser cost. In India, carnation culture is in a budding stage. Very few commercial varieties are seen in the market. The Sim race of carnation was first introduced in India in 1980. In India, it is cultivated in 50 ha, including annual types. Moderate climatic control measures that are economical can deliver quality carnations at the internationally competitive prices year round. Carnations are now being grown commercially in Solan, Shimla, Mandi, Kullu, Chandigarh, Ludhiana, Delhi, Gurgaon, Pune, Bangalore and Kalimpong. A few private growers are also exporting carnations. The midhills of Himachal Pradesh, Jammu and Kashmir, Uttar Pradesh, West Bengal, Nilgiri hills, Bangalore and Nasik are potential areas for its commercial cultivation.

The *Dianthus caryophyllus* probably native to the Mediterranean region but its exact range is unknown due to extensive cultivation for the last 2,000 years. It is wild ancestor of the garden carnation.

Although originally applied to the species *Dianthus caryophyllus*, the name carnation is also often applied to some of the other species of *Dianthus*, and more particularly to garden hybrids between *D. caryophyllus* and other species in the genus.

It is a herbaceous perennial plant growing to 80 cm tall. The leaves are glaucous greyish green to blue-green, slender, up to 15 cm long. The flowers are produced singly or up to five together in a cyme; they are 3–5 cm diameter, and sweetly scented; the original natural flower colour is bright pinkish-purple, but cultivars of other colours, including red, white, yellow and green, have been developed.

Standard and spray types of carnations are grown commercially. Standard types are disbudded to a terminal single flower, producing a large flower on a sturdy long stem. In spray carnations, the center apical bud is removed, allowing the upper buds to develop, these may be on a relatively short stem providing a tight cluster or longer stems originating lower down the stem providing a more open spray. The standard carnations are more in demand in Indian markets, while in the world trade sprays (miniature) supersede standards. The latest development are microcarnations for polybowls. Wholesalers also dye white carnations to green, blue, yellow and other shades, a process known as tinting.

JASMINE

Botanical name: *Jasminum* spp.
Family: Oleaceae

Jasmine is a evergreen semi-vining shrub. It grows to the height of 8–10 feet. The flower is about one inch. It has oval green rich leaves, which have five to nine leaflets. Each leaflet is 2–3 inches long. Jasmine flowers are white in colour. Its stems are slender, trailing, green and glaborous. Its oil is used for making perfumes and incense. Its flowers are used to flavour Jasmine tea and other herbal or black tea. Its oil is also used in creams, shampoos and soaps. In India, its flowers are stringed together to make garlands. Women in India wear this flower in their hair. Some communities even use this flower to cover the face of the bridegroom. The *Jasminum sambac* is a species of jasmine native to southern Asia, in India, Philippines, Myanmar and Sri Lanka. Common names include *Arabian Jasmine*, *Mogra* (Hindi and Marathi), *Mallikâ* (*Sanskrit*), Kampupot, Melati (*Malay and Indonesian Language*), Sampaguita (*Filipino*), Mallepuvvu (*Telugu*), Mallikaipu (*Tamil*), Mallige (*Kannada*) and Kaliyan (*Urdu*).

JASMINUM

Variety	Year of release	Breeding method	Pedigree/ parentage	Important traits	Institutes/ universities
			Jasminum auriculatum		
Parimullai	1972	Clonal selection	It is a clonal selection from a Germplasm	The plants exhibit resistance to gall mite. The yield is 7800 kg of flower buds/ha. The buds are white with moderate corolla tube length (1.25 cm). The concrete recovery is 0.29%	TNAU, Coimbatore
CO 1	1980	Clonal selection	Secondary clonal selection from a local type	The flower buds are white and bold with long corolla tube (1.50 cm) than Parimullai. The yield is 8800 kg of flower buds per hectare in a year. The jasmine concrete recovery is 0.34%	TNAU, Coimbatore
CO 2	1988	Clonal selection	—	The length of corolla tube is 1.70 cm while it is 1.5 cm in CO 1. It yields on an average of 11,198 kg of fresh flower buds as against 8,825 kg in CO 1. It exhibits complete field tolerance to the phyllody disease and gall mite infestation	TNAU, Coimbatore
			Jasminum grandiflorum		
Arka Surabi	1993	Clonal selection	Local *Jasminum grandiflorum*	Yield – 10 tonnes/ha, Drought tolerant High yield of essential oils 0.35% concrete recovery	IIHR Bangalore

JASMINUM : ***Contd.***

Variety	Year of release	Breeding method	Pedigree/ parentage	Important traits	Institutes/ universities
CO 1	1980	Clonal selection	It is a secondary clonal selection from germplasm collection	The average flower yield is 10,144 kg/ha in a year. The flower buds are pink tinged with long corolla tube. It is suitable for oil extraction with a concrete recovery of 0.29%. The concrete yield is 29.42 kg/ha	TNAU, Coimbatore
CO 2	1991	Mutation	It is an induced mutant (I.M.3) developed by treating the vegetative cuttings of CO 1 Pitchi with gamma rays @ 1.5 kR	This mutant is characterised by bold pink buds. The flower bud is 4.14 cm in length as against 4.00 cm in CO 1. The 100 buds weight is 10 g. This variety is amenable for earlier and quicker tying of buds in garland making. It yields on an average 11.68 tonnes of flower buds/ha	TNAU, Coimbatore
			Jasminum sambac		
Arka Aradhana	2000	Selection	Single mogra	Yield – 8 tonnes/ha, Double whorled, Used as fresh flowers and concrete extraction	IIHR Bangalore
			Jasminum multiflorum		
Arka Arpan	1999	Selection	*Jasminum multiflorum* var. *pubescence*	Yield – 6 tonnes/ha Pinkish buds, mildly fragrant flowers	IIHR Bangalore

It is an evergreen vine or shrub reaching up to 1–3 m tall. The leaves are opposite or in whorls of three, simple (not pinnate, like most other jasmines), ovate, 4–12.5 cm long and 2–7.5 cm broad. The flowers are produced in clusters of 3–12 together, strongly scented, with a white corolla 2–3 cm diameter with 5–9 lobes. The flowers open at night, and close in the morning. The fruit is a purple-black berry 1 cm in diameter.

The *Jasminum grandiflorum* is a species of jasmine native to South Asia. In India, its leaves are widely used as an Ayurvedic herbal medicine and its flowers are used to adorn the coiffure of women. It is closely related to, and sometimes treated as merely a form of *Jasminum officinale*. It is a scrambling deciduous shrub growing to 2–4 m tall. The leaves are opposite, 5–12 cm long, pinnate with 5–11 leaflets. The flowers are produced in open cymes, the individual flowers are white having corolla with a basal tube 13–25 mm long and five lobes 13–22 mm long. The flower's fragrance is unique and sweet.

CHINA ASTER

Botanical name: *Callistephus chinensis*
Family: Asteraceae

The *Callistephus* is a genus of flowering plants. The genus includes only one species, *C. chinensis* (China Aster). Native to China, it is an annual plant, growing to 20–80 cm tall with branched stems. The leaves are alternate, 4–8 cm long, ovate, and coarsely toothed. The flower heads are variable, with either all ray florets or an outer ring of ray florets surrounding central disc florets; ray florets are white to purple, disc florets, if present, usually yellow. Their colours vary from white to creamy-yellow, pink, blue, red and purple. Its plants grow do well in beds, borders or pots and are a favourite as cut flowers because of their longevity.

It is a popular ornamental plant in gardens, and numerous cultivars are available; the cultivars are grouped by size, with very dwarf (up to 20 cm), dwarf (20–40 cm), intermediate (40–60 cm) and tall (60–80 cm). In Japan, this species is very important in the floriculture industry for cut flowers.

ORCHIDS

Family: Orchidaceae

A certain mystique has surrounded orchids for centuries. Orchids are undoubtedly some of the most exquisite and abundant members of

CHINA ASTER

Variety	Year of release	Breeding method	Pedigree/ parentage	Important traits	Institutes/ universities
Kamini	1992	Hybridization	An advanced pedigree selection of the cross AST-6 × AST-36	Red Purple (73.A), loose flower, cut flower, making bouquet, pot plant, floral decoration	IIHR, Bangalore
Poornima	1992	Hybridization	An advanced pedigree selection of the cross AST-29 × AST-3	Snow white loose flower, cut flower, making bouquet, pot plant, floral decoration	IIHR, Bangalore
Shashank	1994	Selection	An advanced pedigree selection of the cross Local Pink × AST-2	Creamy white, loose flower, cut flower, making bouquet, pot plant, floral decoration	IIHR, Bangalore
Violet cushion	1994	Selection	An advanced pedigree selection of the cross Local Pink × AST-2	Violet (83.A) loose flower, cut flower, making bouquet, pot plant, floral decoration	IIHR, Bangalore

ORCHIDS

Variety	Year of release	Breeding method	Pedigree/ parentage	Important traits	Institutes/ universities
Deep Blush	2006	Hybridization	Hybrid between D. Nagoya Pink × D. (Candy Stripe × Tomie Drake)	Long arching inflorescence with 9–10 flowers. Flowers are large, magenta coloured, double shaded and striped with full appearance	KAU, Thirissur
Lemon Glow	2006	Hybridization	Hybrid between D. Chiangmai Pink D. (Candy Stripe × Tomie Drake)	Long, arching inflorescence with 10–12 flowers. Flowers are medium-large, flat, squarish in appearance, thick, glossy, light pink towards the outer side and greenish white towards the centre	KAU, Thirissur
Master Delight	2006	Hybridization	Hybrid between D. Sonia - 17 × D. (Candy Stripe × Tomie Drake)	Long, arching inflorescence with 10–12 flowers. Flowers are very thick, glossy and flat with pointed petals and sepals giving a stellar appearance. Flower colour is dark purple with very faint stripes and sepal tips are white	KAU, Thirissur
Pink Cascade	2006	Hybridization	Hybrid between D. Nagoya Pink × D. (Candy Stripe × Tomie Drake)	Long, arching inflorescence with 8–10 flowers. Flowers are very thick, glossy, large, light pink with prominent dark pink stripes throughout	KAU, Thirissur
Velvet Soft	2006	Hybridization	Hybrid between D. Rungnapa × D. (Candy Stripe × Tomie Drake)	Long, arching inflorescence with 10–12 flowers. Flowers are thick, glossy, large, deep purplish magenta and striped with white operculum and full appearance	KAU, Thirissur

the plant kingdom. A moderate estimate would suggest about 25,000 species in plant family Orchidaceae. India is home to about 1,700 species of orchids, of which about 800 are found in the north-eastern region of the country. North-eastern, Himalayas are not only rich in number of species, but more importantly many of them rank at the top of the list of ornamentally important ones. The Sikkim Himalaya, comprising hills of Sikkim and Darjeeling, harbours about 450 species and the centre of origin for important species like *Cymbidium*. However, many of these, once abundant, species are threatened or might have already disappeared in the wild. There is immense scope for improving orchids in India, because a large number of species are native to our country. Many of them have already proved to be important parent plants, contributing in the production of several outstanding hybrids in the world. Due to diversity of environmental condition in India, it is possible to grow all types of orchids in suitable places without the control of environment.

The commercial orchids are both terrestrial and epiphytic, with an abundance in epiphytic. Monopodial (having single-stemmed growth) and sympodial (having the appearance of multi-stemmed growth) are equally preferred for commercial cultivation. Sympodial types *(Cymbidium* and *Dendrobium)* rank high in the export market. Among sympodial ones, a major share or the area in Kerala is occupied by *Dendrobium* hybrids. They can be successfully tried in foothills of north-eastern states. Other sympodial genera suitable for Kerala are *Cymbidium* (at high altitudes), *Oncidium* and *Cattleya*. The *Vanda, Arachnis* and *Phalaenopsis* are monopodial genera that flourish under our conditions. Intergeneric monopodial hybrids — *Aranda*, *Assocenda* and *Mokara* — also perform well.

Flower

Orchidaceae are well-known for the many structural variations in their flowers. Some orchids have single flowers but most have a racemose inflorescence, sometimes with a large number of flowers. The flowering stem can be *basal*, that is produced from the base of the tuber, like in *Cymbidium*, apical, meaning it grows from the apex of the main stem, like in *Cattleya*, or *axillary*, from the leaf axil, as in *Vanda*. As an apomorphy of the clade, orchid flowers are primitively zygomorphic (bilaterally symmetrical), although in some genera like *Mormodes*, *Ludisia*, *Macodes* this kind of symmetry may be difficult to notice.

The orchid flower, like most flowers of monocots has two whorls of sterile elements. The outer whorl has three sepals and the inner whorl has three petals. The sepals are usually very similar to petals (and thus called *tepals*), but may be completely distinct. The upper medial petal, called *labellum* or lip, is always modified and enlarged. The inferior ovary or pedicel usually rotates 180 degrees, so that the labellum, goes on the lower part of the flower, thus becoming suitable to form a platform for pollinators. This characteristic, called resupination occurs primitively in the family and is considered apomorphic (the torsion of the ovary is very evident from the picture). Some orchids have secondarily lost this resupination, *e.g. Zygopetalum* and *Epidendrum secundum*. The normal form of the sepals can be found in *Cattleya*, where they form a triangle. In *Paphiopedilum,* lower two sepals are fused together into a synsepal, while the lip has taken the form of a slipper.

Orchid flowers primitively had three stamens, but this situation is now limited to the genus *Neuwiedia*. The *Apostasia* and the Cypripedioideae have two stamens, the central one being sterile and reduced to a staminode. All of the other orchids, the clade called *Monandria*, retain only the central stamen, the others being reduced to staminodes. The filaments of the stamens are always adnate (fused) to the style to form cylindrical structure called the *gynostemium* or column. The stigma is very asymmetrical as all of its lobes are bent towards the centre of the flower and lay on the bottom of the column. Pollen are released as single grains, like in most other plants, in the Apostasioideae, Cypripedioideae and Vanilloideae. In other subfamilies the anther, carries and two pollinia. A pollinium is a waxy mass of pollen grains held together by the glue-like alkaloid viscin, containing both cellulosic stands and mucopolysaccharides. Each pollinium is connected to a filament which can take the form of a *caudicle*, like in *Dactylorhiza* or *Habenaria* or a *stipe*, like in *Vanda*. Caudicles or stipes hold the pollinia to the *viscidium*, a sticky pad which sticks the pollinia to the body of pollinators.

Pollination

Orchids have developed highly specialized pollination systems and thus the chances of being pollinated are often scarce. This is why orchid flowers usually remain receptive for very long periods and why most orchids deliver pollen in a single mass; each time pollination succeeds thousands of ovules can be fertilized. Pollinators are often visually attracted by the shape and colours of the labellum. The flowers may produce attractive odours. Although absent in most species, nectar may be produced in a spur of the labellum, on the point of the sepals or in the septa of the ovary, the most typical position amongst the Asparagales.

In orchids that produce pollinia, pollination happens as some variant of the following. When the pollinator enters into the flower, it touches a viscidium, which promptly sticks to its body, generally on the head or abdomen. While leaving the flower, it pulls the pollinium out of the anther, as it is connected to the viscidium by the caudicle or stipe. The caudicle then bends and the pollinium is moved forwards and downwards. When the pollinator enters another flower of the same species, the pollinium has taken such position that it will stick to the stigma of the second flower, just below the rostellum, pollinating it. The possessors of orchids may be able to reproduce the process with a pencil or similar device.

BOUGAINVILLEA

Botanical name: *Bougainvillea* spp.
Family: Nyctaginaceae

Bougainvillea is a genus of flowering plants. It is native to South America, from Brazil west to Peru and south to southern Argentina. Different authors accept between four and 18 species in the genus. The name comes from Louis Antoine de Bougainville, an admiral in the French Navy who discovered its plant in Brazil in 1768. Its plants are thorny, woody, vines growing anywhere up to 1–12 m tall, scrambling over other plants with their hooked thorns. The thorns are tipped with a black, waxy substance. They are evergreen where rainfall occurs all year, or deciduous if there is a dry season. The leaves are alternate, simple ovate-acuminate, 4–13 cm long and 2–6 cm broad. The actual flower of the plant is small and generally white, but each cluster of three flowers is surrounded by three or six bracts with bright colors associated with the plant, including pink, magenta, purple, red, orange, white, or yellow. The *Bougainvillea glabra* is sometimes referred to as "paper flower" because its bracts are thin and papery. The fruit is a narrow five-lobed achene.

Numerous cultivars and hybrids have been selected, including nearly thornless shrubs. Some cultivars are sterile, and are propagated from cuttings.

GERBERA

Botanical name: *Gerbera* spp.
Family: Asteraceae

It is a genus of ornamental plants. Named in the honour of the German naturalist Traugott Gerber, a friend of Carolus Linnaeus, it has approximately 30 species as wild, extending to South America,

BOUGAINVILLEA

Variety	Year of release	Breeding method	Pedigree/ parentage	Important traits	Institutes/ universities
Chitravati	1979	Hybridization and selection	Lalbaugh × Red Glory	Mandarin Red (40.C), standards and hedges	IIHR, Bangalore
Dr H.B. Singh	1977	Hybridization and selection	Trinidad × Formosa	Light violet-purple, standards and hedges	IIHR, Bangalore
Jawahar Lal Nehru	1975	Selection	Spontaneous bud mutant from seedling of cv. Lalbagh	Claret Rose (50.A) fading to orange-red, leaves with mimosa yellow variegation along the margin, standards	IIHR, Bangalore
Purple Wonder	1979	Hybridization and selection	Formosa × Trinidad	Light violet purple, Standards, hedges and pot plant	IIHR, Bangalore
Sholay	1977	Seedling selection	Seedling selection of open-pollinated population from Red Glory	Delft rose (46.D), standards, hedges and pot plant	IIHR, Bangalore
Usha	1977	Seedling selection	Seedling selection of open-pollinated population from 'Lady Hope'	Magenta rose (64.C) fading to mandarin red (40.C), standards, hedges and pot plant	IIHR, Bangalore

GERBERA

Variety	Year of release	Breeding method	Pedigree/ parentage	Important traits	Institutes/ universities
YCD 1	1992	Clonal selection	—	The flowers are double in form with cherry red colour. Flowers are large (9.11 cm diameter) with moderately prominent disc. Flowers are borne on long (47–79 cm) and thick stalk. Free from the disorders like bent neck, petal necrosis (during vase life) and temporary wilting in field are absent. Plants flower earlier (within in 45 days after planting) and produce about 60 flowers/plant in a year. Flowers have a retentivity of 8 days on plant with a vase-life of 7 days. The variety is suitable for use as cut flower, raising as borders in garden and for pot cultivation. It is suitable for growing in the hill ranges of Tamil Nadu situated at an altitude of 1,000–2,000 m	HRS (TNAU), Yercaud
YCD 2	1995	Selection	—	It blooms throughout the year with peak flowering during May–June. The flowers are attractive, rosy pink coloured, borne on long stalk without bend. The flowers have a vase-life of 15 days in hills and 10 days in plains. The variety yields about 80 flowers/clump in a year and suitable for cultivation in hilly regions of Tamil Nadu	HRS (TNAU), Yercaud

Africa and tropical Asia. The first scientific description of a Gerbera was made by J.D. Hooker in *Curtis's Botanical Magazine* in 1889 when he described *Gerbera jamesonii*, a South African species also known as Transvaal daisy or Barberton Daisy.

Gerbera species bear a large capitulum with striking, two-lipped ray florets in yellow, orange, white, pink or red colours. The capitulum, which has the appearance of a single flower, is actually composed of hundreds of individual flowers. The morphology of its flowers varies depending on their position in the capitulum. The flowers can be as small as 7 cm (Gerbera mini 'Harley') in diameter or up to 12 cm (Gerbera 'Golden Serena').

Gerbera is very popular and widely used as a decorative garden plant or as cut flowers. The domesticated cultivars are mostly a result of a cross between *Gerbera jamesonii* and another South African species *Gerbera viridifolia*. The cross is known as *Gerbera hybrida*. Thousands of cultivars exist. They vary greatly in shape and size. Colors include white, yellow, orange, red, and pink. The centre of the flower is sometimes black. Often the same flower can have petals of several different colors.

Gerbera is also important commercially. It is the fifth most used cut flower (after rose, carnation, chrysanthemum, and tulip) in the world. It is also used as a model organism in studying flower formation. Gerbera contains naturally-occurring coumarin derivatives.

PLANTATION CROPS

COCONUT

Botanical name: *Cocos nucifera*
Family: Palmae

Coconut is a commercial crop in India. Kerala, Tamil Nadu, Andhra Pradesh and Karnataka are major coconut-producing states in India. Kerala accounts for 54.7% of the total area and 42.3% of the production, followed by Tamil Nadu and Karnataka. The coconut palm is the most useful palm in the world. Every part of the tree is useful to human life for some purpose or the other. Hence, coconut palm is endearingly called 'kalpavriksha' meaning the tree of heaven. The copra obtained by drying the kernel of coconut is the richest source of vegetable oil containing 65–70% oil. The coconut provides a nutritious source of meat, juice, milk, and oil that has fed and nourished populations around the world for

COCONUT

Variety	Year of release	Breeding method	Pedigree/ parentage	Important traits	Institutes/ universities
VHC 1	1982	Hybridization	ECT × Dwarf Green	Early-bearer, mean annual nut yield 98 with a range of 80–145 nuts/palm/year	CRS, TNAU, Veppankulam
VHC 2	1988	Hybridization	ECT × MYD	Yields 107 nuts/paim/year with an oil content of 69%	CRS, TNAU, Veppankulam
VHC 3	2000	Hybridization	ECT × MOD	Good yield compared to other hybrids @ 127 nuts/palm/year. Copra content is 162 g/nut with 70% oil	CRS, TNAU, Veppankulam
Aliyar-nagar Tall	2001	Selection	Arasampatti Tall	Tall variety with circular crown shape, starts flowering at the age of 48 months and yielding at 66 months	CRS, TNAU, Veppankulam
Pratap	1987	Selection	Banawali Green Round	Tall palm with semicircular canopy and green colour round shaped nuts	KKV, Dapoli
Godavari Ganga	1992	Hybridization	ECT × GBGD	High yielder, 140 nuts/palm/year	ANGRAU, Hyderabad
Double century	1995	Selection from Philippines Ordinary	—	It grows to the height of 10–12 m, good yielder. Annual nut yield ranged from 90 to 200.The mean copra content is 198 g/nut with 66% oil	ANGRAU, Hyderabad

COCONUT : *Contd.*

Variety	Year of release	Breeding method	Pedigree/ parentage	Important traits	Institutes/ universities
Gauthami Ganga	2007	Selection from Gangabondam Green Dwarf	—	Dwarf palm with semi circular canopy reaches the height of 2.19 m with oblong shaped green colour fruits, starts yielding at the age of 50 months. Suitable for tender coconut	ANGRAU, Hyderabad
Kamrupa	1999	Selection from Assam Green Tall	—	Tall palm with circular canopy and 9 m height with regular bearing habit. Nut shape is oblong and green colour, copra yield 62.5 g/nut	AAU, Guwahati
Kahikuchi Coconut Hybrid	2007	Hybridization	MYD × WCT	Semi tall hybrid with circular canopy, starts bearing at age of 64 months with oval-shaped nut and yellowish green colour	AAU, Guwahati
Kalyani Coconut 1	2007	Selection from Jamaica Tall	—	Oil content is higher than the present varieties; moderately tolerant to drought; average yield 80 nuts/palm/year	BCKVV, Kalyani
Kera Keralam	2007	Selection from West Coast Tall	—	Tall coconut variety. It is found to be promising and adaptive to wide ranging cultivation zones; comes to flowering in 58 months	TNAU, BCKVV

COCONUT : *Contd.*

Variety	Year of release	Breeding method	Pedigree/ parentage	Important traits	Institutes/ universities
Konkan Bhatye Coconut Hybrid 1	2007	Hybridization	GBGD × ECT	Tall palm with semicircular canopy bearing at 66 months with green colour oval shaped fruit	DBSKKV
Kera Bastar	2007	Selection from Fiji Tall	—	Tall palm with circular canopy with oblong shaped yellowish green fruits which starts yielding at the age of 80 months	DBSKKV, ANGRAU, IGKV
Chandra Sankara	1984	Hybrid with Chowghat Orange Dwarf as female parent and West Coast Tall as male parent	Chowghat Orange Dwarf × West Coast Tall	High nut, oil and copra yield; annual yield of 110 nuts/palm, 22.8 kg copra/ palm, 68% oil in copra	CPCRI, Kasargod
Chandra Laksha	1984	Hybrid with Laccadive Ordinary Tall as female parent and Chowghat	Laccadive Ordinary Tall × Chowghat Orange Dwarf	High nut, oil and copra yield; annual yield of 109 nuts/palm, 21 kg copra/ palm, 69% oil in copra; relatively tolerant to drought	CPCRI, Kasargod

COCONUT : *Contd.*

Variety	Year of release	Breeding method	Pedigree/ parentage	Important traits	Institutes/ universities
		Orange Dwarf as male parent			
Chandra Kalpa	1985	Selection from Laccadive Ordinary Tall	Laccadive Ordinary Tall	High nut, oil and copra yield; average nut yield is 97 nuts/palm with copra output of 18.9 kg/palm/year and oil content of 72%	CPCRI, Kasargod
Chowghat Orange Dwarf	1991	Selection from Chowghat Orange Dwarf	Chowghat Orange Dwarf	Very good quality of tender nut water. The volume of tender nut water is around 350 ml. The tender nut water contains a total sugar content of 7 g/100 ml, reducing sugars — 4.7 g/100 ml, amino acids — 1.8 mg/100 ml, sodium — 20 ppm and potassium — 2003 ppm	CPCRI, Kasargod
Kera Sankara	1991	Hybrid with West Coast Tall as female parent and Chowghat Orange Dwarf as male parent	West Coast Tall × Chowghat Orange Dwarf	High nut, oil and copra yield; Under rain fed condition, mean annual yield is 106 nuts/palm; copra yield is 21 kg/palm/ year; 68% oil in copra	CPCRI, Kasargod

COCONUT : *Contd.*

Variety	Year of release	Breeding method	Pedigree/ parentage	Important traits	Institutes/ universities
Kera Chandra	1995	Selection from Philippines Ordinary Tall	Philippines Ordinary Tall	High nut, oil and copra yield; mean yield of 110 nuts/palm/year; copra out turn of 20.8 kg/palm/year with oil content of 66%	CPCRI, Kasargod
Kalpa Pratibha	2007	Selection from Cochin China Tall	IND 016S CPCRI Coconut Selection-4 (CCS-4)	High yield; relatively tolerant to drought; good quality of tender nut water; annual yield of 91 nuts/palm, 23.3 kg copra/palm (under rain fed condition); 67% oil in copra; contains 448 ml of tender nut water; oil extracted from the copra of this variety has 47.81% of lauric acid	CPCRI, Kasargod
Kalpa Mitra	2007	Selection from Java Tall	IND 022S CPCRI Coconut Selection-5 (CCS-5)	High yield; relatively tolerant to drought; annual yield of 80 nuts/palm, 19.3 kg copra/palm (under rain fed condition); 66.50% oil in copra; oil extracted from copra of this variety contains 47.88% of lauric acid	CPCRI, Kasargod
Kalpa Dhenu	2007	Selection from Andaman Giant Tall	IND 006S CPCRI Coconut Selection-6 (CCS-6)	High yield; relatively tolerant to drought; annual yield of 86 nuts/palm, 20.8 kg copra/palm (under rain fed condition); 65.50% oil in copra; oil extracted from the copra of this variety is rich in lauric acid (50.26%)	CPCRI, Kasargod

COCONUT : *Contd.*

Variety	Year of release	Breeding method	Pedigree/ parentage	Important traits	Institutes/ universities
Kalpa Raksha	2007	Selection from Malaysian introductions	CPCRI Coconut Selection-7 (CCS-7)	High nut and oil yield; annual nut yield of 65 nuts/palm; oil content of 65.5% with yield potential of 2.44 tonnes of copra/ha; good quality (290 ml) of sweet tender nut water; easy to climb since the variety is semi-tall; field resistance to root (wilt) disease; Amenable for high density planting	CPCRI, Kasargod
Laksha-ganga	1989	Hybrid with Laccadive Ordinary Tall as female parent Ganga-bondam Green Dwarf as male parent	Laccadive Ordinary Tall × Ganga-bondam Green Dwarf	The mean yield is 108 nuts/plam/year; annual per palm copra yield is 21kg and oil content in the copra is 69%	KAU, Vellanikara
Keraganga	1989	Hybrid with West Coast Tall as female	West Coast Tall × Gangabondam Green Dwarf	The average annual yield is 100 nuts/ palm; annual copra outturn is 20 kg/ palm and the oil content in copra is 69%	KAU, Vellanikara

COCONUT : *Contd.*

Variety	Year of release	Breeding method	Pedigree/ parentage	Important traits	Institutes/ universities
		parent and Ganga-bondam Green Dwarf as male parent			
Ananda-ganga	1989	Hybrid with Andaman Ordinary Tall as female parent and Ganga-bondam Green Dwarf as male parent	Andaman Ordinary × Gangabondam Green Dwarf	Average annual yield is 95 nuts/palm; annual copra outturn of 21 kg/palm; copra oil content of 68%	KAU, Vellanikara
Kerasree	1992	Hybrid with West Coast Tall as female parent and Malayan Yellow Dwarf as male parent	West Coast Tall × Malayan Yellow Dwarf	Average annual yield is 112 nuts/palm, with copra content of 216 g/nut. The oil content in the copra is 66%	KAU, Vellanikara

COCONUT : *Contd.*

Variety	Year of release	Breeding method	Pedigree/ parentage	Important traits	Institutes/ universities
Kerasow-bhagya	1993	Hybrid with West Coast Tall as female parent and Straits Settlement Apricot as male parent	West Coast Tall × Strait Settlement Apricot	Annual nut yield is 130 nuts/palm; copra content is 195 g/nut with oil content of 65%	KAU, Vellanikara
Kerasagara	2006	Selection from Seychelles (SE Asia)	Seychelles tall	Mean yield 99 nuts/palm/year; annual copra outturn 20 kg/palm; oil content in copra is 67.8%	KAU, Vellanikara

generations. On many islands coconut is a staple in the diet and provides the majority of the food eaten. Nearly one-third of the world's population depends on coconut to some degree for their food and their economy. Among these cultures the coconut has a long and respected history.

It is highly nutritious and rich in fibre, vitamins, and minerals. It is classified as a "functional food" because it provides many health benefits beyond its nutritional content. Coconut oil is of special interest because it possesses healing properties far beyond that of any other dietary oil and is extensively used in traditional medicine among Asian and Pacific populations. Pacific Islanders consider coconut oil to be the cure for all illness. The coconut palm is so highly valued by them as both a source of food and medicine that it is called "The Tree of Life." Only recently has modern medical science unlocked the secrets to coconut's amazing healing powers.

The *Cocos* is a monotypic genus. The evidence is in favour of its origin somewhere in South-East Asia and it is rather speculative to further specify the area of origin. From the centre of origin, it moved eastwards to pacific and further into America and towards the West to India and Madagascar over the sea. Over the years, tremendous variability has been generated in its crop due to cross-pollination. Two distinct varieties are generally recognized. The talls (typica) occupy most of the area. As the name indicates, they are tall palms growing up to 30 m and take 6–7 years to fruit. They yield 80-100 nuts/year and the nuts give 150–200 g good quality copra with 65–70% oil.

Cultivars like west Coast Tall (WCT), East coast Tall (ECT), Laccadive ordinary (LCT), Andaman ordinary (ADOT), Straight Settlement Apricot (SSAT) etc. fall in this group. Dwarfs (nana), on the other hand, do not grow over 10 m and come to bearing in 4 years after planting. The leaves are smaller. Female flowers that are produced in large numbers are retained. There is a high amount of selfing due to overlapping of male and phases in the female inflorescences and, hence, are more homozygous. Although, they yield over 150 nuts, there is a tendency for alternate bearing. The nuts are smaller containing 90–120 g copra per nut and the oil content is only about 65%. Dwarfs are grown mainly for tender nut and ornamental purposes. Two dwarf cultivars from Chowghat orange Dwarf (COD) and Chowghat Green Dwarf (CGD). Gangabondam (GBGD) is another dwarf from Andhra Pradesh. Intermediate types like Ayiramkachi from Tamil Nadu has also been reported.

ARECANUT

Botanical name: *Areca catechu*
Family: Palmae

Arecanut or betel nut or *supari* is chewed both as raw nut and after processing. While ripe arecanut is favoured in Assam, Kerala and northern parts of West Bengal, *chali* is more popular is Western and Northern parts of India. Processed green nut *kalipak* is the choice in Karnataka and Tamil Nadu. It is an important crop with a prominent role in social, cultural and economic activities in India. Owing to medicinal properties, it is used in treating leucoderma, cough, fits, worms, anaemia and obesity. Arecanut is of utmost importance in many religious ceremonies. Tannins in arecanut are being used for dyeing clothes, ropes and for tanning leather. Plastic, hard boards and craft paper of satisfactory strength can be made from its husk. The leaf sheath is a good material for making throw-away cups and plates, plyboards, decorative veneer panels and picture mounds. Its stem forms a useful building material in the villages.

More than 10 million people depend on arecanut for their livelihood in India. World production of arecanut is about 0.610 million tonnes from an area of 0.476 million ha. The major areca nut growing countries in the world are India, Sri Lanka, Bangladesh, Malaysia, Indonesia and The Philippines. The current world productivity is 1.287 tonnes/ha. China has the highest arecanut productivity with 3.752 tonnes/ha. India ranks first in both area and production In the last four decades, productivity has increased from 845 kg/ha to 1243 kg/ha. Karnataka, Kerala, West Bengal, Assam and Tamil Nadu are major states producing arecanut. Andhra Pradesh, Goa, Maharashtra, Meghalaya, Mizoram, Nagaland, Tripura, Andaman and Nicobar Islands and Pondicherry are other states and union territories growing arecanut. Consumption of arecanut in the country increased to 0.336 from 0.114 million tonnes in a 50 year period from 1950. The projected demand of arecanut by 2020 is about 0.617 million tonnes. A small quantity, about 1502 tonnes, of processed arecanut in the form of panmasala, scented supari and gutka are exported.

The genus *Areca* has 76 species. Among these, *A. catechu* L. is the only cultivated species, though nuts of few other species like *A. triandra* Roxb. are also used for masticatory purposes. Five botanical varieties of *A. catechu*, namely, *A. communis, A. silvatica, A. batanensis, A. deliciosa* and *A. longicarpa* have been reported based on size and shape of fruits and kernels. Mostly bivalents with a rare quadrivalent in *A. catechu* and only bivalents in *A. triandra* and a secondary allotetraploid origin for *A. catechu* was reported. There

ARECANUT

Variety	Year of release	Breeding method	Pedigree/ parentage	Important traits	Institutes/ universities
Mangala	1972	Introduction, evaluation and selection	VTL 3 (China)	Semi Tall palm with partially drooping crown, earliness in bearing, more number of female flowers/inflorescence, higher nutset, quicker stabilization, round and medium-sized yellow coloured nuts. Average chali/dry kernel yield is 3.00 kg/palm/year	CPCRI, Kasaragod
Sumangala	1985	Introduction, evaluation and selection	VTL 11 (Indonesia)	Tall palm with partially drooping crown, oval to round-shaped deep yellow coloured nuts. High recovery of chali (26.50%) from fresh fruits. Average chali/dry kernel yield is 3.28 kg/palm/year	CPCRI, Kasaragod
Sreeman-gala	1985	Introduction, evaluation and selection	VTL 17 (Singapore)	Tall palm with sturdy stem, partially drooping crown. round and bold with deep yellow coloured nuts. Average chali/dry kernel yield is 3.18 kg/palm/ year	CPCRI, Kasaragod
Mohitnagar	1991	Introduction, evaluation and selection	VTL 60 (West Bengal, India)	Tall palm with medium thick stem, partially drooping crown, orange yellow coloured oval to round shaped nuts. The bunches are well placed and nuts loosely arranged on spikes which help	CPCRI, Kasaragod

ARECANUT : *Contd.*

Variety	Year of release	Breeding method	Pedigree/ parentage	Important traits	Institutes/ universities
				in uniform development. Average chali/ dry kernel yield is 3.67 kg/palm/year	
Cal-17/ Samrudhi	1995	Introduction, evaluation and selection	VTL 37 (Andamans and Nicobar Islands, India)	Tall palm with longer internodes, partially drooping crown. Elongated bold nuts with orange yellow in colour. Average chali/dry kernel yield is 4.34 kg/palm/year and recommended for Andaman and Nicobar group of Islands	CPCRI, Kasargod and CARI, Port Blair
Swarna-mangala	2006	Introduction, evaluation and selection	VTL 12 (Saigon)	Tall palm with medium thick stem and comparatively shorter internodes, partially drooping crown. Nuts are bold and heavier with high recovery of chali (26.40%). Average chali/dry kernel yield is 3.88 kg/palm/year	CPCRI, Kasaragod
VTLAH I	2006	Hybridization and evaluation	Hirehalli dwarf (VTL 56) and Sumangala (VTL 11)	Dwarf in nature. Sturdy stem with super imposed nodes, reduced canopy size, well spread leaves, medium sized oval to round shaped nuts and early stabilization and medium yielder. Average chali/dry kernel yield is 2.54 kg/palm/year. Advantages-reduced cost of cultivation in terms of harvesting and spraying	CPCRI, Kasaragod

ARECANUT : *Contd.*

Variety	Year of release	Breeding method	Pedigree/ parentage	Important traits	Institutes/ universities
SAS I	1995	Introduction, evaluation and selection	VTL 52 (Sirsi Local, India)	Tall palm with compact canopy, deep orange colour with round and even sized nuts. It is suitable for both tendernut and ripe nut processing. Average chali/dry kernel yield is 4.60 kg/palm/year and recommended for Sirsi hill zone of Karnataka	UAS, Dharwad

are many reports speculating the origin of arecanut palm. The suggested origins are in India, China, Malayan peninsula, East Indies and Indonesia (Sunda Islands). Prevalence of germplasm diversity, supports East Indies group of islands as the centre of origin.

COCOA

Botanical name: *Theobroma cacao* L.
Family: Sterculiaceae

Cocoa is a much recent introduction to India. It was only in the early years of 20th century that the cocoa cultivation was started in India. The *Theobroma cacao* is a native of Amazon region in South America. It was known and used by the natives in this region and was considered as the "food of the gods". The cocoa press was developed during early 17th century for the extraction of cocoa butter. In the later half of 17th century, the Swiss developed both milk chocolate and solid chocolate. The genus *Theobroma* has more than 20 species, but only *Theobroma cacao* is cultivated widely. There are three major varietal groups, namely, Criollo, Forestero and Trinitario. Forestero is the one that is commercially grown all over the world. It is high-yielding more resistant to pest and diseases and more tolerant to drought. Criollo produces fine and flavour beans. Trinitario variety is a cross from Criollo and Forastero. Tropical countries in the African continent are the major producers of cocoa. World production is about 30 lakh MT. Côte d'Ivoire, Ghana, Indonesia, Nigeria, Cameroon, Brazil, Ecuador and Malaysia are the major producers. These countries represent 90% of world production. India contributes only 10 thousand MT from an area of 27 thousand hectors.

In India, cocoa is cultivated as a mixed crop in the coconut and arecanut gardens. Though it comes under plantation crop, pure plantations of cocoa are rarely found in India. Commercial cultivation of cocoa was started in 1960s. Cocoa beans are the raw material for confectioneries, beverages, chocolates and other edible products. Kerala and Karnataka are the leading states in cocoa cultivation. Cocoa adapted well to the coconut and arecanut gardens of these states. In recent years, cocoa cultivation was started in states like Tamil Nadu and Andhra Pradesh. Cocoa has been known as the beverage crop even before tea or coffee. Today, cocoa bean is the major raw material for confectionery industries.

Based on pod and bean characters, three varieties are recognized, namely, Criollo, Forestero and Trinitario. Criollo cocoa pod is big with prominent beak, constriction and furrows. The finished product of whitish or light brown beans has a mild chocolate flavour. The

COCOA

Variety	Year of release	Breeding method	Pedigree/ parentage	Important traits	Institutes/ universities
VTLCC 1 (Vittal Cocoa Clone)	2006	Selection	NC-45/53 (Nigerian clone)	Early, heavy bearer, both self and cross compatible; color-green to yellow; pods – 75 tree/year; beans/pod – 37; single pod weight – 321 g; single dry bean weight – 1.05 g; dry bean yield/tree/year – 1.33 kg; yield/ha – 911 kg; fat content – 52.5%; shelling – 12%; suitable for areca and coconut gardens of Karnataka, Kerala and Tamil Nadu	CPCRI, Kasargod
VTLCH 1 (Vittal Cocoa Hybrid)	2006	Hybridization	II-67 × NC-42/94 (Malaysian and Nigerian)	Early, heavy bearer, tolerant to water stress; color-red to pink; pods – 40 tree/ year; beans/pod – 43; single pod weight – 430 g; single dry bean weight – 1.01 g; dry bean yield/tree/year – 1.25 kg; yield/ ha – 856 kg; shelling – 12%; stomatal resistance (s/cm) – 2.41; suitable for rainfed and irrigated areca, coconut gardens of Karnataka, Kerala, TN, AP, Maharashtra, Goa	CPCRI, Kasargod
CCRP 1	1998	Single tree selection from local population	M-16.9	Heavy-bearer, self-incompatible; Vascular Streak Dieback (VSD) tolerant; color-green to yellow, pods – 56.2/tree/year; beans/pod – 46.2; single pod weight – 384.7 g; single dry bean weight – 0.8 g; dry bean yield/tree/year – 2.5 kg	KAU, Vellanikara

COCOA : *Contd.*

Variety	Year of release	Breeding method	Pedigree/ parentage	Important traits	Institutes/ universities
CCRP 2	2001	Selection	M-13.12	Heavy-bearer, self-incompatible; Vascular Streak Dieback (VSD) tolerant; color-green to yellow, pods – 53.9/tree/year; beans/pod – 45.5; single pod weight – 311.3 g; single dry bean weight – 1.0 g; dry bean yield/tree/year – 2.4 kg	KAU, Vellanikara
CCRP 3	2001	Selection	GI-5.9	Heavy-bearer, self-incompatible; Vascular Streak Dieback (VSD) tolerant; color-green to yellow, pods – 68.5/tree/year; beans/pod – 42.3; single pod weight – 240.6 g; single dry bean weight – 1.0 g; dry bean yield/tree/year – 2.9 kg	KAU, Vellanikara
CCRP 4	1998	Selection	GII-19.5	Heavy-bearer, self-incompatible; Vascular Streak Dieback (VSD) tolerant; color-purple to yellow, pods – 66.2/tree/year; beans/pod – 45.4; single pod weight – 402.1 g; single dry bean weight – 1.1 g; dry bean yield/tree/year – 3.9 kg	KAU, Vellanikara
CCRP 5	1998	Selection	GIV-18.5	Heavy-bearer, self-incompatible; Vascular Streak Dieback (VSD) tolerant; color-green to yellow, pods – 37.9/tree/year; beans/pod – 45.25; single pod weight – 425 g; single dry bean weight – 0.8 g; dry bean yield/tree/year – 1.7 kg	KAU, Vellanikara

COCOA : *Contd.*

Variety	Year of release	Breeding method	Pedigree/ parentage	Important traits	Institutes/ universities
CCRP 6	1998	Selection	GVI-55	Heavy-bearer, self-incompatible; Vascular Streak Dieback (VSD) tolerant; color-green to yellow, pods – 50.1/tree/year; beans/pod – 48; single pod weight – 895 g; single dry bean weight – 1.9 g; dry bean yield/tree/year – 3.1 kg	KAU, Vellanikara
CCRP 7	1998	Selection	GVI-56	Heavy-bearer, self-incompatible; Vascular Streak Dieback (VSD) tolerant; color-green to yellow, pods – 78.1/tree/year; beans/pod – 46.9; single pod weight- 526.7 g; single dry bean weight – 0.9 g; dry bean yield/ tree/year – 4 kg	KAU, Vellanikara
CCRP 8	—	Hybridization	CCRP-1 × CCRP-7	Heavy bearer; Vascular Streak Dieback (VSD) tolerant; pods – 53.9/tree/year; beans/pod – 90.4; single pod weight – 389.3 g; single dry bean weight – 0.9 g; dry bean yield/tree/year – 3.5 kg	KAU, Vellanikara
CCRP 9	—	Hybridization	CCRP-1 × CCRP-4	Heavy bearer; Vascular Streak Dieback (VSD) tolerant; pods – 106.7/tree/year; beans/pod – 36.7; single pod weight – 370.7 g; single dry bean weight – 0.8 g; dry bean yield/tree/year – 3.1 kg	KAU, Vellanikara

COCOA : *Contd.*

Variety	Year of release	Breeding method	Pedigree/ parentage	Important traits	Institutes/ universities
CCRP 10	—	Hybridization	CCRP-3 × GVI-68	Heavy bearer; Vascular Streak Dieback (VSD) tolerant; pods – 79.6/tree/year; beans/pod – 41.6; single pod weight – 332.5 g; single dry bean weight – 1.1 g; dry bean yield/tree/year – 3.5 kg	KAU, Vellanikara

tree is very sensitive to stresses. Though, some plants of Criollo are found, the bulk of India cocoa belongs to the Forestero. The Forestero produces pod which is small with smoother surface and less prominent beak and Trinitario has evolved through genetic mixing of the other two types, thus, it has acquired characters of both. Within the varieties, a lot of variation exists.

CASHEW NUT

Botanical name: *Anacardium occidentale* L.
Family: Anacardiaceae

Cashew nut, a native of Eastern Brazil, was introduced to India just as other commercial crops like rubber, coffee, tea etc. by the Portuguese nearly five centuries back. The first introduction of cashew nut in India was made in Goa from where it spread to other parts of the country. In the beginning it was mainly considered as a crop for afforestation and soil binding to check erosions. Its nuts, apple and other byproducts are of commercial importance. Though its commercial exploitation began during early-60s, marginal lands and denuded forests were set apart for plantation development. Due to non-availability of high yielding-varieties and multiplication techniques, indiscript seeds and seedlings were used for planting purposes. Because of its adaptive ability in a wide range of agroclimatic conditions, it has become a crop of high economy, becoming an export-oriented commodity. India has a creditable record of attaining a good amount of foreign exchange by exporting its kernels. Cashew ranks second position in export. India is the largest area holder of cashew. In India, its cultivation confines mainly to the peninsular areas. It is grown in Kerala, Karnataka, Goa, Maharashtra, and Tamil Nadu, Andhra Pradesh, Orissa and West Bengal. It is also cultivated in Chattisgarh, north-eastern states (Assam, Manipur, Tripura, Meghalaya and Nagaland) and Andaman and Nicobar Islands.

A medium-sized tree, beautiful, and not unlike in appearance the walnut tree, with oval blunt alternate leaves and scented rose-coloured panicles of bloom, its tree produces a fleshy receptacle, commonly called an apple, at the end of which the kidney-shaped nut is borne; the end of it which is attached to the apple, is much bigger than the other. The outer shell is ashy colour, very smooth, the kernel is covered with an inner shell, and between the two shells is found a thick inflammable caustic oil, which raises blisters on the skin and be dangerously painful if the nuts are cracked with the teeth.

CASHEW

Variety	Year of release	Breeding method	Pedigree/ parentage	Important traits	Institutes/ universities
NRCC Sel-1 (VTH 107/3)	1989	Selection	3/8 Simhachalam	Yield 10.0 kg/tree, nut weight 7.6 g, 5 fruits/bunch, shelling 28.8%, W 210 grade	NRC Cashew, Puttur
NRCC Sel- (VTH 40/1)	1989	Selection	2/9 Dicherla	Yield 9.0 kg/tree, nut weight 9.2 g, shelling 28.6%, W 210 grade	NRC Cashew, Puttur
Bhaskara (Goa 11/6)	2006	Selection	Tree of seedling origin spotted from cashew plantation of Goa Forest Department	Yield 10.7 kg/tree, nut weight 7.4 g, shelling percentage 30.6% with 240 W grade, mid-season flowering with 8.6 fruits/ panicle	NRC Cashew, Puttur
Goa 1	1999	Selection	Selection from accession Balli 2 of Balli village	Yield 7.0 kg/tree, nut weight 7.6 g, shelling 30%, kernel weight 2.2 g, W 210 grade.	ICAR Research Complex for Goa, Panaji
BPP 4	1980	Selection	9/8 Epurupalam	Yield 10.50 kg/tree, nut weight 6.0 g, poor shelling 23%, W 400 grade	AICRP-Cashew, Bapatla
BPP 6	1980	Selection	T.No.56	Yield 10.5 kg/tree, nut weight 5.2 g, poor shelling 24%, W 400 grade	AICRP-Cashew, Bapatla

CASHEW : *Contd.*

Variety	Year of release	Breeding method	Pedigree/ parentage	Important traits	Institutes/ universities
BPP 8 (H 2/16)	1993	Hybrid	T.No.1 × T.No.39	Yield 14.5 kg/tree, nut weight 8.2 g, shelling 29%, W 210 grade.	AICRP Cashew, Bapatla
Bhubanes-war 1	1989	Selection	WBDC-V (Vengurla 36/3)	Yield 10.5 kg/tree, nut weight 4.6 g, cluster bearing, high shelling 32%, W 320 grade.	AICRP-Cashew, Bhubaneswar
Chinta-mani 1	1993	Selection	8/46 Taliparamba	Yield 7.2 kg/tree, nut weight 6.9 g, shelling 31%, W 210 grade.	AICRP-Cashew, Chintamani
Jhargram 1	1989	Selection	T.No. 16 of Bapatla	Yield 8.5 kg/tree, nut weight 5.0 g, shelling 30%, W 320 grade	AICRP-Cashew, Jhargram
Madakka-thara 1 (BLA-39-4)	1987	Selection	T.No. 39 of Bapatla	Yield 13.8 kg/tree, nut weight 6.2 g, shelling 26.8%, kernel weight 1.6 g, W 280 grade	AICRP-Cashew, Madakkat-hara
Madakka-thara 2 (NDR-2-1)	1987	Selection	Neduvellur material	Yield 17.0 kg/tree, nut weight 7.3 g, shelling 26.2%, kernel weight 2 g, W 240 grade	AICRP-Cashew, Madakkat-hara

CASHEW : *Contd.*

Variety	Year of release	Breeding method	Pedigree/ parentage	Important traits	Institutes/ universities
K-22-1	1987	Selection	Kottarakkara 22 (Layer 23)	Yield 13.2 kg/tree, nut weight 6.2 g, shelling 26.5%, kernel weight 1.6 g, W 280 grade	AICRP-Cashew, Madakkat-hara
Dhana (H-1608)	1993	Hybrid	ALGD-1 × K-30-1	Yield 17.5 kg/tree, nut weight. 9.5 g, shelling 28.0%, cluster bearing, W 210 grade	AICRP-Cashew, Madakkat-hara
Kanaka (H-1598)	1993	Hybrid	BLA-139-1 × H-3-13	Yield 19.0 kg/tree, nut weight 6.8 g, early variety, shelling 31.0%, W 210 grade	AICRP-Cashew, Madakkat-hara
Priyanka (H-1591)	1995	Hybrid	BLA-139-1 × K-30-1	Yield 16.9 kg/tree, nut weight 10.8 g, shelling 26.5%, kernel weight 2.2 g, W 180 grade	AICRP-Cashew, Madakkat-hara
Amrutha (H-1597)	1999	Hybrid	BLA-139-1 × H-3-13	Yield 18.4 kg/tree, nut weight 7.2 g, shelling 31.6%, kernel weight 2.2 g, W 210 grade	AICRP-Cashew, Madakkat-hara

CASHEW : *Contd.*

Variety	Year of release	Breeding method	Pedigree/ parentage	Important traits	Institutes/ universities
Vengurla 1	1974	Selection	Ansur 1	Yield 19.0 kg/tree, nut weight 6.2 g, shelling 31%, early flowering type, W 240 grade	AICRP-Cashew, Vengurla
Vengurla 4	1981	Hybrid	Midnapore Red × Vetore 56	Yield 17.2 kg/tree, nut weight 7.7 g, shelling 31%, cluster bearing, high sex ratio, W 210 grade	AICRP-Cashew, Vengurla
Vengurla 6	1991	Hybrid	Vetore-56 × Ansur 1	Yield 13.8 kg/tree, nut weight 8.0 g, shelling 28%, W 210 grade	AICRP-Cashew, Vengurla
Vengurla 7	1997	Hybrid	Vengurla 3 × M 10/4 (VRI-1)	Yield 18.5 kg/tree, nut weight 10.0 g, shelling 30.5%, kernel weight 2.9 g, W 180 grade	AICRP-Cashew, Vengurla
VRI-3 (M 26/2)	1991	Selection	M 26/2 – Edayanchavadi material	Yield 10.0 kg/tree, nut weight 7.2 g, shelling 29.1%, kernel grade W 210 grade	AICRP-Cashew, Vridha-chalam
Ullal 1	1984	Selection	8/46 Taliparamba	Yield 16.0 kg/tree, nut weight 6.7 g, shelling 30.7%, long duration of flowering and harvesting, W 210 grade	ARS, Ullal, Karnataka

CASHEW : *Contd.*

Variety	Year of release	Breeding method	Pedigree/ parentage	Important traits	Institutes/ universities
Ullal 3	1993	Selection	5/37 Manjeri	Yield 14.7 kg/tree, nut weight 7.0 g, shelling 30%, very short duration of flowering and fruiting, W 210 grade	ARS, Ullal, Karnataka
Ullal 4	1994	Selection	2/77 Tuni, Andhra	Yield 9.5 kg/tree, nut weight 7.2 g, shelling 31%, W 210 grade	ARS, Ullal, Karnataka
UN 50	1995	Selection	2/27 Nileshwar (T.No.25)	Yield 10.5 kg/tree, nut weight 9.0 g, shelling 32.8%, W 180 grade	ARS, Ullal, Karnataka

SPICES

Major Spices

BLACK PEPPER

Botanical name: *Piper nigrum* Linn.
Family: Piperaceae

Black pepper, the king of spices, is one of the oldest and most popular spices in the world. It is a perennial, climbing vine indigenous to the Malabar Coast of India. The hotly pungent spice made from its berries is one of the earliest spices known and is probably the most widely used spice in the world. It was mentioned as far back as 1000 BC in ancient Sanskrit literature. In early historic times, black pepper was widely cultivated in tropical areas of Southeast Asia, where it became an important article of overland trade between India and Europe. It became a medium of exchange, and tributes were levied in black pepper in ancient Greece and Rome. In the Middle Ages the Venetian and the Genoese became the main distributors, their virtual monopoly of the trade helping to instigate the search for an eastern sea route. The name pepper comes from the Sanskrit word pippali meaning berry.

Apart from India, black pepper is widely cultivated throughout Indonesia, Malaysia, Thailand, tropical Africa, Brazil, Sri Lanka, Vietnam and China also. It is a branching vine with a smooth, woody, articulate stem swollen at the joints. A woody climber, it may grow up to 10 m in height by means of its aerial roots. Its broad, shiny green, pointed, petiolate leaves are alternately arranged. The sessile, white, small flowers are borne in pendulous, dense, slender spikes of about 50 blossoms each. The berry-like fruits, or peppercorns, are round, about 0.5–1.0 cm in diameter and contain a single seed. They become yellowish-red at maturity and bear a single seed. The odour is penetrating and aromatic; the taste is hot, biting and very pungent.

Black pepper is used in almost all applications where spice is used, with exception of baked goods. It is used universally in sauces, gravies, processed meats, poultry, snack foods etc. Both black and white pepper are used in cuisine worldwide, at all stages of the cooking process and as a table condiment. White pepper has a distinguishably different flavour but is utilized to a lesser extent.

BLACK PEPPER

Variety	Year of release	Breeding method	Pedigree/ parentage	Important traits	Institutes/ universities
Panniyur 1	1967	Hybridization	Inter-cultivar hybrid of Uthirankotta × Cheriyakaniya-kadan	Vigorous growing vine. Medium maturity group. Long spikes, close setting of berries, bold berries, oleoresin 11.8%, piperine 5.3%, essential oil 3.5%, dry recovery 35.3% with a yield of 1242 kg/ha dry pepper	Pepper Research Station (KAU), Panniyur
Panniyur 2	1991	Selection	Clonal selection from open pollinated progeny of Balankotta	Shade tolerant, suitable for inter-cropping, medium maturity group, medium quality oleoresin 10.9%, high piperine (6.6%), essential oil 3.4%, dry recovery 35.7% with a yield of 2570 kg/ha dry pepper	Pepper Research Station (KAU), Panniyur
Panniyur 3	1991	Hybridization	Inter-cultivar hybrid of Uthirankotta × Cheriyankaniya-kadan	Suitable for all pepper growing regions, performs well under open situation. Late maturity group. Long spikes and bold berries, piperine 5.2%, oleoresin 12.7%, essential oil 3.1%, dry recovery 27.8% with a yield of 1953 kg/ha dry pepper	Pepper Research Station (KAU), Panniyur
Panniyur 4	1991	Selection	Clonal selection from Kuthiravally type II	Stable yield performs well under adverse condition also, tolerant to shade, late maturity, 4.4% piperine, 9.2% oleoresin, 2.1% essential oil and 34.7% dry recovery with a yield of 1277 kg/ha dry pepper	Pepper Research Station (KAU), Panniyur

BLACK PEPPER : *Contd.*

Variety	Year of release	Breeding method	Pedigree/ parentage	Important traits	Institutes/ universities
Panniyur 5	1996	Selection	Clonal selection from open-pollinated progeny of Perumkodi	Suitable for both mono and mixed cropping in coconut/arecanut gardens, shade tolerant, medium maturity, tolerant to nursery disease. Long spikes, piperine 5.3%, oleoresin 12.33%, essential oil 3.8% and dry recovery 35.7% with a yield of 1110 kg/ha dry pepper	Pepper Research Station (KAU), Panniyur
Panniyur 6	2000	Selection	Clonal selection from Karimunda type III	A vigorous vine. Tolerant to drought and adverse climatic conditions, stable and regular bearer, medium maturity group. Suitable for open condition as well as partial shade, spike 6–8 cm, more number of spikes/unit area, close setting and attractive bold berries, piperine 4.9%, oleoresin 8.27% and essential oil 1.33% and 33.0% dry recovery with a yield of 2117 kg/ha dry pepper	Pepper Research Station (KAU), Panniyur
Panniyur 7	2000	Selection	Open pollinated progeny of Kalluvally	Vigorous vine and a regular bearer, long spike, a hardy type vine, tolerates adverse climatic condition, suitable for open and shaded conditions, very long spike (16–24 cm) high piperine content (5.6%), oleoresin 10.6%, essential oil	Pepper Research Station (KAU), Panniyur

BLACK PEPPER : ***Contd.***

Variety	Year of release	Breeding method	Pedigree/ parentage	Important traits	Institutes/ universities
				1.5%, and 34.0% dry recovery with a yield of 1410 kg/ha dry pepper	
PLD–2	1996	Selection	Clonal selection from Kottanadan	Late maturity high quality cultivar, contains piperine 3.0%, oleoresin 15.45%, essential oil 4.8%. Suitable for plains and higher elevations with a yield of 2475 kg/ha dry pepper	NRC for Oil Palm, Regional Station, Palode
Sreekara	1990	Selection	Clonal selection from Karimunda (KS 14)	High yield and quality, Adaptable to various climatic conditions in all the pepper growing tracts, yield 2677 kg/ha dry pepper	IISR, Calicut
Subhakara	1990	Selection	Clonal selection from Karimunda	High yield and quality, wider adaptability, a yield 2352 kg/ha dry pepper	IISR, Calicut
Panchami	1991	Selection	Clonal selection from Aimpiriyan	A high yielding variety with excellent fruit set. Spike twisted in appearance due to high fruit set. Oleoresin content is high	IISR, Calicut
Pournami	1991	Selection	Clonal selection from germplasm (ottaplackal 1)	Tolerant to root knot nematode. A moderately high yielding vine with high oleoresin content, yield 2333 kg/ha dry pepper	IISR, Calicut

BLACK PEPPER : *Contd.*

Variety	Year of release	Breeding method	Pedigree/ parentage	Important traits	Institutes/ universities
IISR Thevam	2006	Selection	Clonal selection from Thevanmundi	Tolerant to foot rot diseases. Suited to high altitude areas, mean yield 5.17 kg per line (fresh)	IISR, Calicut
IISR Malabar Excel	2006	Hybridization	Cholamundi × Panniyur 1	Suitable for higher elevation and plains, mean yield 2.78 kg/vine (fresh)	IISR, Calicut
IISR Girimunda	2006	Hybridization	Narayakodi × Neelamundi	Medium maturing type, mean yield 6.14 kg per line (fresh)	IISR, Calicut
IISR Shakthi	2006	Selection	Open pollinated progeny of Perambramundi	Moderately resistant to *Phytophthora*	IISR, Calicut

CARDAMOM

Botanical name: *Elettaria cardamomum* Maton.
Family: Zingiberaceae

Cardamom, the 'queen of spices', is a rich spice culled from the seeds of a perennial plants, *Elettaria cardamomum*. It is highly prized spice in the world. Its original home is the mountains of South-Western parts of the Indian Peninsula. As early as the 4th century BC cardamom was used in India as a medicinal herb and was an article of Greek and Roman trade. It is a native of Western Ghats in India. India had a virtual monopoly of cardamom till recently. But now it is being cultivated in Guatemala, Sri Lanka, Thailand, Laos, Vietnam, Costa Rica, El Salvador and Tanzania. Its cultivation in India is confined to Kerala, Karnataka and Tamil Nadu.

Its seeds have a warm, slightly pungent and highly aromatic flavour. They are popular seasoning in Oriental dishes, particularly curries and in Scandinavian pasteries. In Middle East countries, it is used mostly in preparation of 'Gahwa', a strong cardamom-coffee concoction.

Cardamom is a perennial bushy herb. The tuberous underground rhizome is its real stem, while aerial shoot is a pseudostem and formed by the encircling leaf sheaths. The leaves are distichous, long, alternate and lanceolate acuminate in shape. Flowers are borne on panicles and they emerge directly from the underground stem on long floral stalks. They are hermaphrodite and zygomorphic. The corolla is tubular, 3-lobed, pale-green, androecium with petalloid labellum white in colour with pink or purplish veins, composed of three modified stamens with an undulated edge. There are two further rudimentary staminodes and one functional stamen. The fruits are tri-locular, ovoid or oblong, greenish-brown capsules containing about 15–20 seeds attached to axile placenta. Light reddish or dark reddish brown seeds are irregularily 3-sided, transversely wrinkled or furrowed and are covered by a membraneous aril.

GINGER

Botanical name: *Zingiber officinale* Rosc
Family: Zingiberaceae

Ginger is one of the earliest known oriental spices and is being cultivated in India both as a fresh vegetable and as a dried spice since time immemorial. Ginger is obtained from the rhizomes of *Zingiber officinale*. The ginger family is a tropical group especially abundant in Indo-Malaysia, consisting of more 1,200 species in 53 genera. The genus *Zingiber* includes about 85 species of aromatic

CARDAMOM

Variety	Year of release	Breeding method	Pedigree/ parentage	Important traits	Institutes/ universities
Mudigere 1	1984	Selection	Clonal selection from Malabar type	Erect and compact plant, short panicle, pale green, oval bold capsules, suitable for high-density planting, moderately tolerant to thrips, hairy caterpillar and white grubs, pubescent leaves. Contains 8.0% oil, 36.0% 1, 8 cineol, 42.0% α-terpenyl acetate, dry recovery 20.0% with 275 kg dry/ha	Regional Research Station, UAS, Mudigere
Mudigere 2	1996	Selection	Clonal selection from open pollination of Malabar type	Early-maturing, suitable for high density planting, round/oval and bold capsules, oil 8.0%, 1, 8 cineol 45.0%, α-terpenyl acetate 38.0%. with 475 kg dry/ha	Regional Research Station, UAS, Mudigere
PV 1	1991	Selection	A selection from Walayar collection, a Malabar type	An early-maturing type, short panicle, elongated slightly ribbed, light green capsules, essential oil 6.8%, 1, 8 cineol 33.0%, α-terpenyl acetate 46.0%, dry recovery 19.9% with 260 kg dry/ha	Cardamom Research Station, KAU, Pampadumpara
PV 2	2001	Selection	A selection from OP seedlings of PV-1, a Malabar type	Early-maturing, unbranched lengthy panicle, long bold capsules, high dry recovery percentage (23.8%), essential oil 10.45% field tolerant to stem borer and thrips. Suitable for elevation range of 1000–1200 m above MSL with 982 kg dry/ha	Cardamom Research Station, KAU, Pampadumpara

CARDAMOM : *Contd.*

Variety	Year of release	Breeding method	Pedigree/ parentage	Important traits	Institutes/ universities
ICRI 1	1992	Selection	Selection from Chakkupallam collection, a Malabar type	An early-maturing variety, medium-sized panicle with globose, round and extra bold dark green capsules contains oil 8.7%, 1,8 cineol 29.0%, α-terpenyl acetate 38.0% and dry recovery 22.9% with 325 kg dry/ha	ICRI (Spices Board), Myladum-para
ICRI 2	1992	Selection	Clonal selection from germplasm collection, Mysore type	Performs well under high altitude and irrigated condition, medium-long panicles, oblong bold and parrot green capsules, tolerant to azukkal disease. Dry recovery 22.5% with 375 kg dry/ha	ICRI (Spices Board), Myladum-para
ICRI 3	1993	Selection	Selection from Malabar type	Early-maturing, non-pubescent leaves, tolerant to rhizome rot disease, oblong, bold parrot green capsules, suitable for hill zone of Karnataka, oil 6.6%, 1,8 cineol 54.0%, a terpenyl acetate 24.0%, dry recovery 22.0% with 440 kg dry/ha	ICRI (Spices Board) Sakleshpur
ICRI 4	1997	Selection	Clonal selection from Vadagaraparai area of lower pullenys, a Malabar type	Early maturity, medium-sized panicle, globose bold capsules, oil 6.4%. Suitable for low rainfall areas, relatively tolerant to rhizome rot and capsule borer with 455 kg dry/ha	ICRI (Spices Board), Myladum-para

CARDAMOM : *Contd.*

Variety	Year of release	Breeding method	Pedigree/ parentage	Important traits	Institutes/ universities
ICRI 5	2006	Hybridization	MCC260 × MCC 49	First hybrid variety; early maturity, moderately tolerant to drought, high yield under intensive management (responds well to intensive management) capsule size – 68%; More than 7 mm volatile oil – 7.13%; dry recovery – 23.15%	ICRI (Spices Board), Myladum-para
ICRI 6	2006	Selection	Clonal selection	High yield; medium maturity, relatively tolerant to drought high percentage of bold capsules and volatile oil content capsule size – 71%; More than 7 mm volatile oil – 7.33%; dry recovery-19.0%	ICRI (Spices Board), Myladum-para
IISR Suvashini	1997	Selection	Clonal selection from OP progenies of Cl.37	Malabar type early maturing	IISR, CRC, Appangala
IISR Avinash	1999	Selection	Clonal selection from OP progenies of CCS-1	Malabar type tolerant to rhizome rot, shoot and capsule borer	IISR, CRC, Appangala
IISR Vijetha	2001	Selection	Clonal selection from Natural katte escapes (NKE 12)	Malabar type resistant to katte (Cardamom mosaic virus)	IISR, CRC, Appangala

GINGER

Variety	Year of release	Breeding method	Pedigree/ parentage	Important traits	Institutes/ universities
Suprabha	1988	Selection	Clonal selection from Kunduli local	Plumpy rhizome, less fibre, wide adaptability, suitable for both early and late sowing, duration 229 days. 8.9% oleoresin, 4.4% crude fibre, 1.9% essential oil and 20.5% dry recovery	High Altitude Research Station, OUAT, Pottangi
Suruchi	1990	Selection	Clonal selection from Kunduli local	Profuse tillering, bold rhizome, suitable for rainfed/irrigated conditions, duration 218 days, 23.5% dry recovery	High Altitude Research Station, OUAT, Pottangi
Surabhi	1991	Mutation	Induced mutant of Rudrapur local	Plumpy rhizome, dark skinned yellow fleshed, suitable for both irrigated/ rainfed, duration 225 days. 10.2% oleoresin, 2.1% essential oil, 4.0% crude fibre and 23.6% dry recovery	High Altitude Research Station, OUAT, Pottangi
Himgiri	1996	Selection	Clonal selection from Himachal collection	Best for green ginger, less susceptible to rhizome rot disease, suitable for rainfed condition, 4.29% oleoresin, 1.6% essential oil, 6.05% crude fibre, 20.2% dry recovery, 230 days duration	Department of Vegetable Crops, YSPUH&F, Solan

GINGER : ***Contd.***

Variety	Year of release	Breeding method	Pedigree/ parentage	Important traits	Institutes/ universities
IISR Varada	1996	Selection	Clonal selection from Sarjiguda collection	High yield and good quality, 20.7% dry recovery	IISR, Calicut
IISR Rejatha	2002	Selection	Clonal selection from Pampadumpara	High yield and plumy rhizomes, 19% dry recovery, 2.36% essential oil, 6.34% oleoresin and 4.0% fibre content	IISR, Calicut
IISR Mahima	2002	Selection	Clonal selection from Pottangi collection	High yield and plumy rhizomes, 23% dry recovery, 1.72% essential oil, 4.48% oleoresin and 3.26% fibre content	IISR, Calicut

herbs from East Asia and tropical Australia. The word ginger is derived from a Sanskrit word *singabera* meaning 'shaped like a deer's antlers (horn)'. Ginger is not known in a wild state and has been cultivated for so long in both China and India that its exact origin is unclear. It is believed to be a native of Southern Asia.

It is a slender herbaceous perennial herb. It forms a spreading, tuberous, underground stem or rhizome. The plant produces erect, tall and dark green leafy shoots (pseudstems) 30–100 cm of high. The aerial pseudostems ususally bear 8–12 distichous leaves. Leaves are alternate, lanceolate or linear-lanceolate, acute, smooth and with subsessile sheathing. They are 5–25 cm long and 1–3 cm wide. There is a broad, thin, glabrous ligule (about 5 mm long) and slightly bilobed. Flowers are borne in a bracteal, imbricated spike, terminating in a leafless stem reaching to about 15–25 cm in height. They are situated in the axils of large, greenish — yellow obtuse bracts. They are small, fraile, short lived, very a few, and usually arising one or two at a time. The evanoscent flowers are yellowish and speckled with a purplish lip. Calyx is thin, tubular, spathaceous, 1–1.2 cm long and is three-toothed. Corolla tube is 2–2.5 cm long with three lobes. The dorsal lobe is 1.5–2.5 cm long, 8 mm wide and is curved over the anther and narrowed to the tip. The labellum or lip corresponds to three stamens, is nearly circular, dull purple with cream blotches at base. The perfect stamen has a short filament; the anther is cream coloured and is prolonged into a beak-like appendage. The rhizome, protruding just below the apex of the appendage, has a circular apical aperture surrounded by stiff hairs. There are two slender free styloids. The inferior ovary is trilocular with several ovules per locule on axile placentation. Fruit is an oblong, thin walled, three — valved capsule but is rarely produced. The seeds are small, black and arillate. Rhizomes are fleshy sympodial, hard and thick, laterally compressed, often palmately branched with about 1.5–2.5 cm in diameter. The inner core is usually pale yellow while the outer is light yellow. They are covered with small distichous scales with an encircling insertion and fine fibrous roots in the top layer.

Ginger contains 1.5–3.0% volatile oil. The volatile oil is composed mainly of sesquiterpene hydrocarbons. Oxygenated sesquiterpenes are present up to 17% and the remainder is composed of monoterpene hydrocarbons and oxygenated monoterpenes. Of the sesquiterpene hydrocarbons, about 20-30% is (-)-α-zingeberene, up to 12% (-)-β-bisabolene, up to 19% (+)-ar-curcumene and up to 10% farnasene. The volatile oil possesses the aromatic odour and flavour but not the pungent flavour of the spice. Studies have shown that β-sesquiphellendrene and ar-curcumine were the major contributors to the 'ginger' flavour whereas α-terpenol and citral contributed a lemony flavour. The pungent components of ginger are three compounds: gingerol, shogoal and zingerone. Gingerols are a series of compounds with the general structure, 1-(4'-hydroxy-3'methoxyphenyl)-5-hydroxyalkan-3 ones. They are mainly condensation products of zingerone. The presence of zingerone and shogoals in fresh extract is non existant. They are formed during the drying process.

TURMERIC

Botanical name: *Curcuma longa* L.
Family: Zingiberaceae

Turmeric is a native to Asia and India. Its tuberous rhizomes or underground stems are used from antiquity as condiments, a dye and as an aromatic stimulant in several medicines. Turmeric is a very important spice in India, which produces nearly the whole world's crop and uses 80% of it. Presently, it is cultivated in China, Taiwan, Indonesia, Sri Lanka, Australia, Africa, Peru and the West Indies. Turmeric usage dates back nearly 4,000 years, to the Vedic culture in India, when turmeric was the principal spice and also of religious significance. It has been used in Indian systems of medicine for a long time.

It is a herbaceous perennial with a rhizome from which arises tufts of large, broad, lanceolate, bright green leaves acute at both ends. The plants grow up to 60–90 cm high. Leafy shoots are erect, bearing 6–10 leaves with the leaf sheath forming a pseudostem. The ligule is a small lobe (1 mm long). The sheath near the ligule has ciliate margins. The inflorescence is a cylindrical spike, 10–55 cm long, 5–7 cm wide and terminal on the leafy shoot. The flowers are yellow or pale yellow, borne in a spike. They arise from two buds situated in axils of bracts and mature successively. Bracts are greenish-white; the uppermost tinged with pink. The bracteoles are thin and elliptic. The calyx is short, unequally toothed and split nearly half way down on one side. The corolla is tubular at the base and the upper halfs cup-shaped. There are two lateral staminodes. The lip or labellum is obovate. The ovary is inferior and trilocular with a slender style held by anther lobes and passing between them. Fruits are seldom. The primary tuber at the base of the aerial stem is ellipsoidal bearing many rhizomes; straight or little curved, with secondary branches in two rows and further tertiary branches, the whole forming a dense clump. Rhizomes have a distinctive taste and smell, brownish and scaly outside and the inside is bright orange in colour. The roots are fleshy, often ending in a swollen starchy tuber.

TURMERIC

Variety	Year of release	Breeding method	Pedigree/ parentage	Important traits	Institutes/ universities
CO 1	1983	Mutation	Vegetative mutant by X-ray irradiation of Erode local	Bold and bright orange yellow rhizomes, curcumin 3.2%, oleoresin 6.7%, essential oil 3.7%, dry recovery 19.5%, suitable for drought prone, hilly areas saline and alkaline areas. Crop duration 270 days. Plants are robust, vigorous and taller	HC&RI, TNAU, Coimbatore
BSR 1	1986	Selection	Clonal selection from Erode local irradiated with X rays	Bright yellow rhizome, curcumin 4.2%, oleoresin 4.0%, essential oil 3.7%, dry recovery 20.5%, crop duration 285 days, suitable for drought-prone areas of Tamil Nadu	ARS, TNAU, Bhavani-sagar
BSR 2	1994	Mutation	Induced mutant from Erode local	A high-yielding short duration variety (245 days) with bigger rhizomes, resistant to scale insects	ARS, TNAU, Bhavani-sagar
Krishna	1983	Selection	Clonal selection from Tekurpeta collection	Plumpy rhizomes, curcumin 2.8%, oleoresin 3.8%, essential oil 2.0%, dry recovery 16.4%, duration 240 days. Moderately resistant to pests and diseases	Maharashtra Agrl. University, Kasba, Digraj
Sugandham	1984	Selection	Clonal selection from germplasm	Thick, round rhizomes with short internodes. Curcumin 3.1%, oleoresin 11.0%, essential oil 2.7%, dry recovery	Spices Research Station,

TURMERIC : *Contd.*

Variety	Year of release	Breeding method	Pedigree/ parentage	Important traits	Institutes/ universities
				23.3%, duration 210 days. Moderately tolerant to pest and diseases	GAU, Jagudan
Roma	1988	Selection	Clonal selection from T. Sunder	Suitable for both rainfed and irrigated conditions. suitable for hilly areas and late season planting. Curcumin 6.1%, oleoresin 13.2%, essential oil 4.2% and dry recovery 31.0%, duration 250 days	High Altitude Research Station, OUAT, Pottangi
Suroma	1989	Mutation	Clonal selection from T. Sunder by X-ray irradiation	Round and plumpy rhizome, field tolerance to leaf blotch, leaf spot and rhizome scale, curcumin 6.1%, oleoresin 13.1%, essential oil 4.4% and dry recovery 26.0%, duration 253 days	High Altitude Research Station, OUAT, Pottangi
Ranga	1992	Selection	Clonal selection from Rajpuri local	Bold and spindle-shaped mother rhizome, suitable for late planting and low lying areas. Moderately resistant to leaf blotch and scales, curcumin 6.3%, oleoresin 13.5%, essential oil 4.4% and dry recovery 24.8%, duration 250 days	High Altitude Research Station, OUAT, Pottangi
Rasmi	1992	Selection	Clonal selection from Rajpuri local	Bold rhizomes, suitable for both rainfed and irrigated conditions, early and late sown season, curcumin 6.4%, oleoresin 13.4%, essential oil 4.4% and dry recovery 23.0%, duration 240 days	High Altitude Research Station, OUAT, Pottangi

TURMERIC : *Contd.*

Variety	Year of release	Breeding method	Pedigree/ parentage	Important traits	Institutes/ universities
Rajendra Sonia	1939	Selection	Selection from local germplasm	Bold and plumpy rhizome, grows widely under all north Indian conditions. Curcumin 8.4%, essential oil 5.0% and dry recovery 18.0%, duration 225 days	Tirhut College of Agriculture, RAU, Dholi
Megha turmeric-1	1996	Selection	Selection form Lakadong type	Suitable for North-East hill and North West Bengal. Bold rhizomes, high curcumin content 6.8% and dry recovery 16.37%, duration 300–315 days	ICAR R.C. NEH Region, Shillong, Meghalaya
Suranjana	2000	Selection	Clonal selection from local types of West Bengal (TCP-2)	Suitable for open and shaded conditions, sole or intercrop, suitable for rainfed as well as high rain fall areas. Curcumin 5.7%, oleoresin 10.9%, essential oil 4.1%, dry recovery 21.2%, duration 235 days, tolerant to leaf blotch and rhizome rot. Resistant to rhizome scales and moderately resistant to shoot borer	UBKVV, Pundibari
Suguna	1991	Selection	Selection from germplasm collected from Assam	Short duration type (190 days), curcumin 4.9%, oleoresin 13.5%, essential oil 6.0% and dry recovery 20.4%, field tolerance to rhizome rot	IISR, Calicut

TURMERIC : *Contd.*

Variety	Year of release	Breeding method	Pedigree/ parentage	Important traits	Institutes/ universities
Suvarna	1991	Selection	Selection from germplasm collected from Assam	Bright orange coloured rhizome with slender fingers. Maturity 200 days, field tolerant to pest and diseases Curcumin 4.3%, oleoresin 13.5%, essential oil 7.0% and dry recovery 20.0%	IISR, Calicut
Sudhar-sana	1991	Selection	Selection from germplasm collected from Singhat, Manipur	High yielding variety, short duration type (190 days). Field tolerant to rhizome rot. Curcumin 5.3%, oleoresin 15.0%, essential oil 7.0% and dry recovery 20.6%	IISR, Calicut
IISR Prabha	1996	Selection	Open pollinated progeny selection	High yielding variety, curcumin content 6.5%, oleoresin 15.0%, essential oil 6.5% and dry recovery 19.5%, crop duration 205 days	IISR, Calicut
IISR Prathiba	1996	Selection	Open pollinated progeny selection	High quality line, 6.2% curcumin content with high yield, 16.2% oleoresin, 6.2% essential oil, 18.5% dry recovery, crop duration 225 days	IISR, Calicut
IISR Alleppy Supreme	2004	Selection	A clonal selection from Alleppy turmeric	Shows tolerance to leaf blotch disease. Rhizomes contain 5.55% curcumin, 16.0% oleoresin, 19.0% dry recovery, crop duration 210 days	IISR, Calicut

TURMERIC : *Contd.*

Variety	Year of release	Breeding method	Pedigree/ parentage	Important traits	Institutes/ universities
IISR Kedaram	2004	Selection	Clonal selection from Thodupuzha collection	Tolerant to leaf blotch disease, rhizomes contain 5.5% curcumin, 13.6% oleoresin, maturity 210 days and 18.9% driage	IISR, Calicut
Kanthi	1996	Selection	Clonal selection from Mydukur variety of Andhra Pradesh	Erect leaf with broad lamina, big mother rhizomes with medium bold fingers and closer internodes. Medium duration. Curcumin content (7.18%), oleoresin 8.25%, essential oil 5.15%, dry recovery 20.15%, duration 240–270 days	College of Hort., KAU, Vellanikkara, Trichur
Sobha	1996	Selection	Clonal selection from local type germplasm	Mother rhizome big with medium bold and closer internodes. Inner core of rhizomes is dark orange like Alleppey. More territory rhizomes. Dryage 19.38%, curcumin content (7.39%), oleoresin (9.65%), essential oil (4.24%), medium duration 240–270 days	College of Hort., KAU, Vellanikkara, Trichur
Sona	2002	Selection	Clonal selection from local germplasm	Orange-yellow rhizome, medium bold with no territory fingers. Best suited for central zone of Kerala. Rhizome medium bold, field tolerant to leaf blotch. Curcumin 7.12%, essential oil 4.4%, oleoresin 10.25%. 18.88% dry recovery, medium duration. 240–270 days	College of Hort., KAU, Vellanikkara, Trichur

TURMERIC : *Contd.*

Variety	Year of release	Breeding method	Pedigree/ parentage	Important traits	Institutes/ universities
Varna	2002	Selection	Clonal selection from local germplasm	Bright orange yellow rhizome, medium bold with closer internodes, tertiary fingers present. Suited to central zone of Kerala. Field tolerant to leaf blotch, curcumin 7.87%, essential oil 4.56%, oleoresin 10.8%. 19.05% dry recovery, medium duration 240–270 days	College of Hort., KAU, Vellanikkara, Trichur

Raw turmeric rhizomes have to be cured for both colour and aroma. For this, the fingers and bulbs are boiled separately in water for 30–45 min until the rhizomes are soft. This procedure gets rid of the 'raw' colour, reduces drying time, gelatinises the starch and gives the turmeric a more uniform colour. Water is then drained and the turmeric sun-dried for 10–15 days until it becomes dry and hard. For imparting orange-yellow colour, rhizomes are boiled in lime water or sodium bicarbonate solution. The dried produce is cleaned and polished mechanically in a drum rotated by hand or by power. Turmeric oleoresin is obtained by solvent extraction of the ground spice. It is orange-red in colour and consists of colouring matter, volatile oil, fatty oils and bitter principles. The volatile oil gives the turmeric its characteristic flavour. The important quality attributes of turmeric are size, physical form, colour, curcumin content, maturity, weight or bulk density, length and thickness, intensity of colour of the core and aroma.

Turmeric contains two primary constituents colouring matter and volatile oil. The volatile oil of turmeric is about 1.5–6.0% and is composed of a variety of sesquiterpenes, many of which are specific for the species. Several sesquiterpenes, germacrone, turmerone, ar-(+)-α-,β-turmerones; β-bisabolene; α-curcumene; zingiberene; β-sesquiphellandene, bisacurone; curcumenone; dehydrocurdione; procurcumadiol; *bis*-acumol; curcumenol; isoprocurcumenol epiprocurcumenol; procurcumenol; zedoaronediol; curlone; and turmeronol A and turmeronol B, are found in rhizomes. Most important for the aroma are turmerone (max. 30%), ar-turmerone (25%) and zingiberene (25%). A ketone, and an alcohol identified as p-tolylmethyl carbinol, are obtained. Conjugated di-arylheptanoids (1,7-diaryl-hepta-1,6-diene-3,5-diones, *e.g.* curcumin) are responsible for orange colour and probably also for pungent taste (3–4%). The crystalline coloring matter, curcumin, is a diferuloyl methane. It dissolves in concentrated sulphuric acid giving a yellow-red coloration. The rhizomes contain curcuminoids, curcumin, demethoxy curcumin, *bis*-demethoxycurcumin, 5'-methoxycurcumin and dihydrocurcumin which are found to be natural anti-oxidants. A new curcuminoid, cyclocurcumin, was isolated from nematocidally active fraction of turmeric. The fresh rhizomes also contain two new natural phenolics which possess antioxidant and anti-inflammatory activities and also two new pigments. The rhizomes also contain four new polysaccharides-ukonans having activity on reticuloendothelial system, along with stigmasterol, β-sitosterol, cholesterol and 2-hydroxymethyl anthraquinone.

TREE SPICES

CINNAMON

Botanical name: *Cinnamomum verum* Presl.
syn. *C. zeylanicum* Blume.
Family: Lauraceae

Cinnamon, a bushy evergreen tree, is native to Sri Lanka (Ceylon), the neighbouring Malabar Coast of India, and Myanmar (Burma). It is also cultivated in South America and the West Indies for the spice consisting of its dried inner bark. The spice is light brown in colour and has a delicately fragrant aroma and warm, sweet flavour. It is lighter in colour and milder in flavour than other related species.It was once more valuable than gold and has been associated with ancient rituals of sacrifice or pleasure. In Egypt, it was sought for embalming and witchcraft; in medieval Europe for religious rites and as flavouring. References to cinnamon are plenty throughout the Old Testament in the Bible. Later, it was the most profitable spice in the Dutch East India Company trade.

Cinnamon is cultivated as low bushes to ease the harvesting process. The leaves are long (10–18 cm), leathery and shining green on upper surface when mature. The flowers have a fetid, disagreeable smell. The fruit is a dark purple, one-seeded berry. It prefers shelter and moderate rainfall without extremes in temperature. About 8–10 lateral branches grow on each bush. After three years they are harvested. The Sri Lankan farmers harvest their main crop in the wet season, cutting the shoots close to the ground. In processing, the shoots are first scraped with a semicircular blade, then rubbed with a brass rid to loosen the bark, which is split with a knife and peeled. The peels are telescoped one into another forming a quill about 107 cm (42 inches) long and filled with trimming of the same quality bark to maintain the cylindrical shape. After 4 or 5 days of drying, the quills are rolled on a board to tighten the filling and then placed in subdued sunlight for further drying. Finally, they are bleached with sulphur dioxide and sorted into grades.

Cinnamon contains 0.5–1% essential oil, the principal component of which is cinnamic aldehyde (about 60%). Other components are eugenol, eugenol acetate, and small amounts of aldehydes, ketones, alcohols, esters and terpenes. The oil is distilled from fragments for use in food, liquor, perfume and drugs. The aldehyde can also be synthesized. Cinnamon leaf oil is unique in that it contains eugenol as its major constituent (70–90%).

CINNAMON

Variety	Year of release	Breeding method	Pedigree/ parentage	Important traits	Institutes/ universities
YCD1	1996	Selection	Clonal selection from open-pollinated seedlings progenies of Sri Lankan type	Good bark recovery, adapted to wide range of soil and rainfed conditions. Bark oil 2.8%, leaf oil 3.0%, bark recovery 35.3%	HRS, TNAU, Yercaud
PPI (C) 1	2003	Selection	Selection from open-pollinated seedlings progeny introduced from Sri Lanka seed supplied from IISR	Bark, higher oil recovery from bark (2.9%) and leaf oil recovery of 3.3%, bark oil 2.9%, leaf oil 3.3%, bark recovery 34.22%. Suitable for cultivation in high rainfall zones and hill regions of Tamil Nadu at an altitude range of 100–500 m MSL	HRS, TNAU, Pechiparai
Konkan Tej	1993	Selection	Seedling selection from progenies of Sri Lankan accessions	Superior qualities with 3.2% bark oil with bark recovery 29.16%, cuminal-dehyde in bark oil 70.23, eugenol in bark oil 6.93%, leaf oil 2.28%, eugenol in leaf oil 75.5%, yields 4.10 kg fresh bark. Bark recovery 51.78%	Regional Coconut Research Station, Dr BSKKV, Vengurle
Sugandhini	2000	Selection	Single tree selection from Wynad local collection. A Sri Lankan type	Recommended for cultivation for leaf oil production, bark oil 0.94%, leaf oil 1.6%, bark recovery 51.0%, cinnamal-dehyde in bark oil 45.0%, eugenol in leaf oil 93.7%; released mainly for leaf oil purpose. Densely foliage	Aromatic and Medicinal Plants Research Station, KAU, Odakkali

CINNAMON : *Contd.*

Variety	Year of release	Breeding method	Pedigree/ parentage	Important traits	Institutes/ universities
Navashree	1996	Selection	Seedling selection from Sri Lankan collection	Higher shoot regeneration	IISR, Calicut
Nithyashree	1996	Selection	Seedling selection from collection (In 189)	Higher cinnamaldehyde and oleoresin	IISR, Calicut

NUTMEG

Botanical name: *Myristica fragrans* Houtt.
Family: Myristicaceae

Nutmeg is the seed of an apricot-like fruit of nutmeg tree, while mace is its arillus, a thin leathery tissue between stone and pulp. Both spices are strongly aromatic, resinous and warm in taste. Mace is generally said to have a finer aroma than nutmeg, but difference is small. Nutmeg loses its fragrance quickly when ground. Naturally, nutmeg is limited to the Banda Islands, a tiny archipelago in Eastern Indonesia (Moluccas). Indonesia (East Indian Nutmeg) and Grenada (West Indian Nutmeg) are main nutmeg-producing countries. In many European countries, the name of nutmeg derives from Latin nux muscatus "musky nut; moschate nut"; the Middle English form is notemugge. Mace goes back to Greek makir, which was used to denote an oriental spice, though it is not clear whether this was identical to mace.

These spices have been appreciated since Roman times. Because of its very limited geographical distribution, nutmeg and mace became known in Europe comparatively late (11th century). Although nutmeg was available in Europe since the 13th century, significant trade started not before the 16th century, when Portuguese ships sailed to India and further, to the famed spice Islands (Moluccas). During 17th century, the Dutch succeeded in monopolizing the nutmeg trade, as they did with cloves. This situation changed only in the 18th century, when Frenchman Pierre Poivre succeeded in smuggling nutmeg trees from the Bandas to Mauritius and thereby broke the Dutch monopoly. The British East India Company introduced this tree to Penang, Singapore, India, Sri Lanka and the West Indies.

The nutmeg tree is a large, evergreen native to the Banda Islands in the Moluccas and grows to a height of about 18 m. It produces fruits 15–20 years after planting. The fruit of nutmeg tree, which is similar in colour and size to apricot, splits when ripe revealing the brilliant red arils encasing the brown nut. The red arils on drying become orange in colour and are the mace of commerce. The nut is also dried until the kernel inside rattles.

Nutmeg contains about 10% essential oil, which is mostly composed of terpene hydrocarbons (pinenes, camphene, p-cymene, sabinene, phellandrene, terpinene, limonene, myrcene, together 60–90%), terpene derivatives (linalool, geraniol, terpineol, together 5 to 15%) and phenylpropanes (myristicine, elemicine, safrol, together 2–20%). Of the latter group, myristicine (methoxy-safrol) is responsible for the hallucinogenic effect of large nutmeg dosages (typically, one or

NUTMEG

Variety	Year of release	Breeding method	Pedigree/ parentage	Important traits	Institutes/ universities
Konkan Sugandha	1998	Selection	Single plant selection from local seedling population	Adaptable in Konkan region. Tree canopy is conical and compact	Regional Fruit Research Station, Vengurla
Konkan Swad	2003	Selection	Selection from nutmeg seedlings from Ratnagiri dist	Adapted to Konkan region with warm, humid conditions as well as shade provision. Canopy erect, conical shape. Contain 39.8% essential oil in seed and 10.9% in mace. No incidence of pest and disease are noticed	Regional Coconut Research Station, Dr BVSKU, Bhatye
Vishwa-shree	2002	Selection	Clonal selection from elite trees (Mannoor, Calicut)	Bushy, compact canopy with high quality and low incidence of fruit rot	IISR, Calicut

more nuts). Oil of mace (up to 12% in the spice) contains the same aroma components in slightly different amounts. Mace is used to flavour milk-based sauces and is widely used in processed meats. It is also added sparingly to delicate soups and sauces with fish or seafood. Pickles or chutneys may be seasoned with mace. Nutmeg is a traditional flavouring for cakes, gingerbreads, biscuits and fruit or milk puddings. Today, nutmeg's popularity has shrunken and the spice is less used, still most in Arab countries, Iran and northern India, where both nutmeg and mace appear in delicately-flavoured meat dishes.

SEED SPICES

CORIANDER

Botanical name: *Coriandrum sativum* L.
Family: Umbelliferae

Coriander is an umbelliferous annual plant. Native to the eastern Mediterranean region and southern Europe, it is found in many other parts of the world. It is valued for dry ripe fruits, called coriander seeds and also fresh green leaves called cilantro. The herb is produced in Morocco, Romania, Mexico, Argentina, the People's Republic of China, Bangladesh, Bulgaria, Canada, Egypt, India, Indonesia, Nigeria, Poland, Syria, the United States, the USSR, and Yugoslavia. It is one of the oldest recorded spices, mentioned in the ancient Sanskrit texts and in Exodus. Seeds have been found in the tombs of the Pharaohs. The name originated from koris, the Greek word for a bed bug, so given because of the similarity between smell of coriander leaves and offending bug.

It is a rigid, strong-smelling annual with pronounced tap root, and slender branching stems up to 60 cm. Reaching to a height of 1 m, andromonoecious plant flowers in July and August. The plant has ferny, pinnately or ternately decompound leaves and produces compound umbels with small white or pinkish flowers that are attractive to bees. The seed capsules are round red-brown which are aromatic when ripe.

Since seeds shatter soon after maturity (about 90 days from planting), timeliness of harvesting and weather conditions greatly influence yield. The plants should be cut for seed when the fruits have turned brown. Young, immature fruits have a characteristic disagreeable odor and lack the desirable spicy aroma associated with mature fruits. Harvesting in early morning is ideal, while the dew is on its plants.The pleasing flavour in coriander fruits is not

CORIANDER

Variety	Year of release	Breeding method	Pedigree/ parentage	Important traits	Institutes/ universities
Guj. Cor 1	1974	Selection	Selection from germplasm	Suitable for early sowing, erect plant, moderately tolerant to wilt and powdery mildew, round bold grains, essential oil 0.35%, duration 112 days	Main Spices Research Station, GAU, Jagudan, Gujarat
Guj. Cor 2	1985	Selection	Reselection from CO. 2	Semi spreading type, suitable for early sowing, moderately tolerant to powdery mildew, oblong, essential oil 0.40%, lodging and shattering resistant	Main Spices Research Station, GAU, Jagudan, Gujarat
CO 1	1977	Selection	Selection from germplasm Koilpatti local	A variety with small statured plant, duration 110 days, suitable for rainfed areas and for greens and grains, small grain, essential oil 0.27%	HC and RI, TNAU, Coimbatore, Tamil Nadu
CO 2	1982	Selection	Reselection from culture P2 of Gujarat	A dual purpose variety, suitable for saline, alkaline and drought prone areas, maturity 90–100 days, seeds oblong, medium 0.40% essential oil	HC and RI, TNAU, Coimbatore, Tamil Nadu
CO 3	1991	Selection	Reselection from Acc.695 of IARI,	A dual purpose variety, good yielder, medium sized grains, suitable for	HC and RI, TNAU,

CORIANDER : *Contd.*

Variety	Year of release	Breeding method	Pedigree/ parentage	Important traits	Institutes/ universities
			New Delhi	rainfed and irrigated conditions, *rabi* as well as *kharif* season. Field tolerant to powdery mildew, wilt and grain mould. Seed oil content ranges from 0.38 to 0.41%, 85-95 days duration	Coimbatore, Tamil Nadu
CO(CR) 4	2002	Selection	Reselection from germplasm ATP 77 Guntur collection	Early maturing variety (65-70 days), suitable for both rainfed and irrigated conditions; grains oblong and medium; essential oil content is 0.4%; field tolerant to wilt & grain mould	HC and RI, TNAU, Coimbatore, Tamil Nadu
Rajendra Swati	1988	Pure line selection	Pure line Selection from Muzaffarpur collection	Medium sized plant with fine, aromatic round grains, essential oil 0.65%. Suitable for intercropping, field tolerance to aphids	Dept. of Hort., Tirhut College of Agriculture, RAU, Dholi-843 121, Muzaffarpur, Bihar
RCr 41	1988	Recurrent selection	Recurrent half sib selection from local type from 'Kota'	A tall erect plant type with thick stem. Grows well under irrigated condition, resistant to stem gall, wilt and moderately resistant to powdery mildew; small seeds (9.3 g/1000 seed),	SKN College of Agri-culture, RAU, Jobner, Rajasthan

CORIANDER : *Contd.*

Variety	Year of release	Breeding method	Pedigree/ parentage	Important traits	Institutes/ universities
				essential oil 0.25%. Long duration variety (130–140 days)	
RCr 20	1997	Recurrent selection	Recurrent half sib selection from Jaipur local	Medium sized bush plant, suitable for rainfed crop or limited moisture conditions and heavy soils of south Rajasthan. Moderately resistant to stem gall, bold grains, essential oil 0.25%. Early maturity, duration 100–110 days	SKN College of Agri-culture, RAU, Jobner, Rajasthan
RCr 435	2003	Selection	Half sib selection from local germ-plasm from Jalore.	Plants are bushy, erect, bold seeds, medium size, contain essential oil 0.33%; medium maturity; adapted for irrigated condition, moderately resistant to root knot and powdery mildew. Medium maturity, duration 110–130 days	SKN College of Agri-culture, RAU, Jobner, Rajasthan
RCr 436	2001	Selection	Half sib selection from local germ-plasm from 'Kota'	Plants semi dwarf, bushy type; with quick early growth and bold big seeds, resistant to root rot and root knot nematodes; most suitable for limited moisture condition and heavy soils of south Rajasthan, essential oil 0.33, early maturing (90–100 days)	SKN College of Agri-culture, RAU, Jobner, Rajasthan

CORIANDER : *Contd.*

Variety	Year of release	Breeding method	Pedigree/ parentage	Important traits	Institutes/ universities
RCr 684	1999	Mutation	Mutation breeding by gamma rays. Induced mutant of Rcr-20	A variety, resistant to stem gall and less susceptible to powdery mildew. Adapted to medium heavy textured soil, essential oil 0.32% and sandy loam soil under irrigations. Seeds of the variety are bold. Plants are tall and erect with higher number of seeds/umbel. Medium maturity (110–120 days)	SKN College of Agri-culture, RAU, Jobner, Rajasthan
RCr 446	2001	Selection	Half sib selection from local type from Jaipur local	Plants tall, are leafy erect with higher number of seeds per umbel. Seeds are medium in size and 0.33% volatile oil content. Moderately resistant to stem gall and wilt under rainfed, limited moisture or irrigated conditions, medium maturity (110–130 days)	SKN College of Agri-culture, RAU, Jobner, Rajasthan
Sadhana	1989	Selection	Mass selection from local Alur collection	Dual purpose, semi-erect variety; suitable for rainfed condition, field tolerance to diseases and white fly, mites and aphids, essential oil 0.2%. A mid late variety 95–110 days duration withstands moisture stress, responded well to input management under optimum moisture level	RARS, ANGRAU, Lam, Guntur, Andhra Pradesh

CORIANDER : *Contd.*

Variety	Year of release	Breeding method	Pedigree/ parentage	Important traits	Institutes/ universities
Swathi	1989	Selection	Mass selection from Nandyal germplasm	Plants medium-sized, semi erect type. Early-maturing variety (80–85 days), suitable for rainfed condition and late sown season, essential oil 0.30%. Medium size oval grain. Field tolerant to white fly, moderately tolerant to disease. It suits well to the areas where the soil moisture retentiveness is comparably less. Being early maturity, it escapes powdery mildew disease	RARS, ANGRAU, Lam, Guntur, Andhra Pradesh
Sindhu	1991	Selection	Mass selection from germplasm Warangal local	Oval medium breakable grains, essential oil (0.4%); suitable for rainfed areas, tolerant to wilt, powdery mildew as well as drought condition, medium duration (100–110 days)	RARS, ANGRAU, Lam, Guntur, Andhra Pradesh
Hisar Anand	1994	Selection	Mass selection from Haryana collection	A medium tall, dual purpose variety, oval medium size seeds; essential oil 0.35%, wider adaptability to different soil conditions. Resistant to lodging due to spreading habit	CCSHAU, Hisar, Haryana
Hisar Sugandh	2001	Selection	Mass selection from indigenous germplasm	Suitable for irrigated conditions. Resistant to stem gall disease	CCSHAU, Hisar, Haryana

CORIANDER : *Contd.*

Variety	Year of release	Breeding method	Pedigree/ parentage	Important traits	Institutes/ universities
Hisar Surabhi	2004	Selection	Mass selection from local germplasm	Bushy erect plant type, seed medium, oblong; oil content 0.425%, tolerant to frost, less susceptible to aphids, medium duration (130–140 days)	CCSHAU, Hisar, Haryana
Azad Dhania-1	1996	Selection	Mass selection from Kalyanpur germplasm collection	Erect, early branching, number of umbellates per umbel 5, essential oil 0.29%, tolerant to moisture stress, tolerant to powdery mildew and aphids, maturity 120–125 days	CSAUAT, Kanpur, Uttar Pradesh
Pant Haritima	1993	Selection	Selection from local type Pant Dhania	A tall erect plant, a dual purpose type, good yielder of leaf, smaller seeds with high oil (0.4%), resistant to stem gall, maturity 155–160 days	GBPUAT, Udham Singh Nagar, Uttarakhand
DWA 3	1999	Pure line selection	Pure line selection from Karnataka collection	A dual purpose variety and for seed production in *rabi* crop, essential oil 0.27%. Moderately tolerant to powdery mildew, black clay soils are best suited	UAS, Dharwad, Karnataka
NRCSS ACr 1	2006	Selection	Selection from exotic collection	It is a variety from late maturity group developed at NRCSS, Ajmer is suitable for leaves and seeds production, and grows well under irrigated conditions. The grains are medium to small, having	NRC Seed Spices, Ajmer

CORIANDER : ***Contd.***

Variety	Year of release	Breeding method	Pedigree/ parentage	Important traits	Institutes/ universities
				essential oil content up to 0.6% and average yield 12.5 q seeds/hectare. The plants are tall erect, seeds are medium in size, round in shape and suitable for export purpose. The plants are resistant to stem gall and have tolerance to powdery mildew. This variety is suitable for growing as greens during off-season also	

thoroughly developed until it is completely dry. The whole plant may be tied in bundles or spread on screens to dry. As soon as dry, the fruits should be separated by threshing and winnowing. The clean seeds should be stored in bags or closed containers.

For essential oil extraction, the seeds are ground immediately before distillation to increase oil yield and minimize distillation time. The essential oil content of dried fruit ranges from 0.1 to 1.5% and the oil contains d-linalool (also known as coriandrol), camphor, pinenes, camphene, sabinene, myrcene, terpinenes, limonene, and other constituents. Coriander fruits also contain a fixed or fatty oil. Coriander leaf or cilantro contains about 4% volatiles, on a wet leaf basis, primarily 2-decenal and 2-dodecenal.

CUMIN

Botanical name: *Cuminum cyminum* L.
syn. *Cuminum odorum* Salisb.
Family: Umbelliferae

Cumin is a small annual plant of the parsley family, widely cultivated in the Mediterranean region of Europe and in India. Primary cultivation of cumin is in Europe, Asia, the Middle East, and North Africa with India and Iran as the largest cumin exporters. The valued portion of the plant is the dried fruit called cumin seed, which is esteemed as a condiment. Cumin was known to the Egyptians 5,000 years ago and it was found in the pyramids. In ancient times cumin was a symbol of greed and meanness. Curiously, by the Middle Ages it was regarded as a symbol of faithfulness.

A small, annual herbaceous plant, it grows to a height of about 25 cm. It flourishes best in sunny places with some rainfall. The small white or pink flowers grow on small compound umbels. The small, boat-shaped seed has nine ridges and it is brown-yellow in colour. The reported life zone of cumin is 9–26°C with an annual precipitation of 0.3–2.7 m and a soil pH of 4.5–8.3. Cumin thrives on rich, well-drained sandy loam soil. The plant, which needs mild temperatures during 3–4 months growing season, is intolerant of long periods of dry heat. Plant the seeds in a sunny location after the soil has become warm in spring. If the rows are 2 feet apart, with 16–20 seeds to the foot, no thinning is necessary.

The odour and flavour of cumin is derived largely from the essential oil, which contains cumaldehyde or cuminic aldehyde as the main constituent. Other ingredients of oil are dihydrocuminaldehyde, dl-pinene, d-pinene, para-cymene, dipentene, and cuminyl alcohol.

CUMIN

Variety	Year of release	Breeding method	Pedigree/ parentage	Important traits	Institutes/ universities
5-404	1952	Selection	Selection from local germplasm	An erect plant, medium size fruit, contain 2.2% essential oil, 7.7% crude fibre. Moderately tolerant to powdery mildew	Spices Research Station, GAU, Jagudan, Gujarat
Mc 43	1970	Selection	Selection from germplasm	Plants semi spreading, grains bold lustering withstand lodging and shattering moderately tolerant to *Fusarium* wilt, Alternaria blight and powdery mildew. Essential oil content 2.7%, crude fibre 15.5%	Spices Research Station, GAU, Jagudan, Gujarat
Guj. Cumin 1	1983	Selection	Selection from local germplasm (Vijaypur-5)	Plants bushy and spreading, grains bold, linear oblong; withstand shattering and lodging, moderately tolerant to wilt, powdery mildew and blight, 3.6% essential oil, 14.25% crude fibre	Spices Research Station, GAU, Jagudan, Gujarat
Guj Cumin 2	1992	Pure line selection	Pure line selection from M2 irradiated seeds of MC 43	Bushy plant, good branching habit, grains bold, lustrous medium sized, tolerant to wilt and blight and powdery mildew, 4.0% essential oil, 22.1% crude fibre	Spices Research Station, GAU, Jagudan, Gujarat

CUMIN : *Contd.*

Variety	Year of release	Breeding method	Pedigree/ parentage	Important traits	Institutes/ universities
Guj. Cumin 3	2000	Recurrent selection	Recurrent selection derived from W. German entry EC-232689	Bushy dwarf plant, fruit medium sized, frost wilt resistant variety suitable for winter season limited irrigation. Higher essential oil (4.4%) content	Spices Research Station, GAU, Jagudan, Gujarat
Guj. Cumin-4	2003	Selection	Natural crossing followed by selection	The first wilt resistant variety, normal, seed appearance and no splitting seed habbit	Spices Research Station, GAU, Jagudan, Gujarat
RZ 19	1988	Recurrent selection	Recurrent single plant progeny selection from collection of Ajmer	Erect plant, pink flowers, bold, lustrous grain, grey pubescent, tolerant to wilt and blight suitable for late sowing season. Duration 140–150 days	SKN College of Agriculture, RAJAU, Jobner, Rajasthan
RZ 209	1995	Recurrent selection	Recurrent single plant progeny selection from Jalore	A variety shown some resistance with blight disease. Maturity 120–130 days	SKN College of Agriculture, RAJAU, Jobner, Rajasthan

CUMIN : *Contd.*

Variety	Year of release	Breeding method	Pedigree/ parentage	Important traits	Institutes/ universities
RZ 223	1999	Mutation	Mutation breeding in UC-216	Wider adaptability, resistant to wilt, superior in yield and seed quality over RZ-19. Plants bushy, semi erect, long bold attractive seeds, with 3.0 to 3.5% volatile oil content, medium duration (120–130 days)	SKN College of Agriculture, RAJAU, Jobner, Rajasthan

Synthetic cuminaldehyde is an adulterant to cumin oil and is very difficult to detect chemically. The dried seed of cumin has 2.5–5% essential oil on a dry-weight basis and is obtained by steam distillation. Cumin is used as a flavoring agent in many ethnic products such as cheeses, pickles, sausages, soups, stews, stuffings, rice and bean dishes, and liqueurs. It is an essential component of Mexican foods, along with chilli pepper and oregano. Its use is prevalent in many Latin American cuisines. Cumin is the key ingredient of Indian cooking like all types of curries and chilli powders. Oil of cumin is used in fragrances. As a medicinal plant, cumin has been utilized as an antispasmodic, carminative, sedative, and stimulant. Cumin oil has been reported to have antibacterial activity. Distinct phototoxic effects have been reported from undiluted cumin oil. It is also used in veterinary medicines and perfumes.

FENNEL

Botanical name: *Foeniculum vulgare* Mill.
Family. Umbelliferae

Fennel is an erect growing perennial herb native to southern Europe and the Mediterranean area, commonly grown for flavoring purposes. The mature seeds are the dried fruits of this herb which are used commercially. The young tender shoots and leaves are used in foods in European countries. This plant has a thickened base of leaves, which can be blanched or boiled and eaten as a vegetable. Principal fennel production areas are located in India, the People's Republic of China, Egypt, Argentina, Indonesia, and Pakistan. Fennel has been known to herbalists and doctors since time immemorial. It was believed to be the total cure and have the power to make people young, strong and healthy. It is one of nine Anglo-Saxon sacred herbs. It was often hung over doorways to ward off evil spirits. Fennel was used in ancient Greece and Rome. The name comes from the Latin *foenum*, a variety of fragrant hay, reflecting the plant's odour.

It is a greyish-green, strong-smelling herbaceous perennial, with slim stems, bearing soft lacy, dark green leaves with thread-like lobes and swollen bases. Creeping rootstock gradually extends plant into a sparse clump. Reaching a height of 1.5 m, its plants have small mustard-yellow flowers on a compound umbel. The fruits split into two seeds, which are oval shaped with five ridges. The seed is light green to gray, about 0.75 cm long, and curved.

Fennel contains 1–3% of a volatile oil which is composed of about 50–60% anethole and about 20% d-fenchone. Other compunds present

FENNEL

Variety	Year of release	Breeding method	Pedigree/ parentage	Important traits	Institutes/ universities
S-7-9	1956	Selection	—	A bushy plant with big umbel, moderately tolerant to blight, contain 1.2% essential oil. Duration 210 days, 24.0% crude fibre	Spices Research Station, GAU, Jagudan, Gujarat
PF 35	1973	Selection	Selection from local germplasm	Plant tall and spreading, moderately tolerant to leaf spot, leaf blight and sugary diseases	Spices Research Station, GAU, Jagudan, Gujarat
Guj. Fennel 1	1984	Pure line selection	Pure line selection from Vijaypur local	Plant tall and bushy, shattering and lodging, suitable for early sowing and as rabi crop, reasonably tolerant to drought, moderately tolerant to sugary disease, oblong, medium bold and dark green seeds	Spices Research Station, GAU, Jagudan, Gujarat
Guj. Fennel 2	1997	Selection	Pedigree selection from local germ-plasm	Plants bushy, bold grains, rich in volatile oil (2.4%) suitable for both rainfed and irrigated conditions	Spices Research Station, GAU, Jagudan, Gujarat

FENNEL : *Contd.*

Variety	Year of release	Breeding method	Pedigree/ parentage	Important traits	Institutes/ universities
Guj. Fennel 3	2003	Selection	Selection based on individual plant progeny performance from local germplasm	A medium maturity type (148 days) adopted to rabi season under irrigation; seeds medium bold, contains 1.8% volatile oil	Spices Research Station, GAU, Jagudan, Gujarat
CO 1	1985	Selection	Reselection from PF 35	Medium statured, diffuse branching, suitable for intercropping and border cropping with chilli and turmeric. Suitable for drought, prone, water logged saline and alkaline conditions. Duration 220 days	HC and RI, TNAU Coimbatore, Tamil Nadu
RF 101	2001	Selection	Recurrent half sib selection from local germplasm collection from Jobner	Erect, medium tall nature, medium maturity type (150–160 days) with long bold grains, most suitable for loamy and black cotton soil	SKN College of Agriculture, RAJAU, Jobner, Rajasthan
RF 125	2003	Selection	Recurrent half sib selection is an exotic collection EC 243380 from Italy	Plants are short statured with compact umbels and long bold seeds. Duration 110–130 days, tolerant to sugary disease	SKN College of Agriculture, RAJAU, Jobner, Rajasthan

FENNEL : *Contd.*

Variety	Year of release	Breeding method	Pedigree/ parentage	Important traits	Institutes/ universities
Hisar Sawrup	2004	Selection	Mass selection from indigenous germ-plasm of Haryana	Plants grow upright, spreading, gives a bushy appearance. A late maturity type (175–185 days) grain long and bold with 1.6% volatile oil content, resistant to lodging, no shattering of grains	Department of Vegetable Crops, CCS HAU, Hisar, Haryana
NRCSS-AF 1	2006	Selection	Selection from local collection	The variety is suitable for growing both for early-sowing and as *rabi* crop. The plants are erect and tall, bearing large size umbels. The seeds are attractive bold, medium sized, fragrant with volatile oil content upto 1.6%. The variety is from late maturity group and produces average yield of 19.5 q/ha during *rabi* season and 25.1 q/ha when grown as an early crop. This variety has tolerance to Ramularia and Alternaria blight	National Research Centre on Seed Spices, Ajmer

in fennel are d-α-pinene, d-α-phellandrene, dipentene, methyl chavicol, feniculun, anisaldehyde, and anisic acid. Oil of fennel is commonly available. An oleoresin is also available with a volatile oil content of only about 3–6%. Bitter fennel oil is thought to contain more fenchone (a bitter mixture with a camphor-like odor and flavor) and less anethole than sweet fennel oil. Sweet fennel oil is of a superior quality with a more pleasing aroma and flavour. Some analyses have indicated a lack of fenchone in sweet fennel and high concentrations of limonene in bitter fennel.

Both the seeds and the oil distilled from them are used for flavoring. Fennel seed is used in the food and flavour industry for addition to meats, vegetable products, fish sauces, soups, salad dressings, stews, breads, pastries, teas, and alcoholic beverages. Crushed seed are used in salad dressings, in mayonnaise, in savoury and sweet baking and as a substitute for juniper in flavoring gin. Ground fennel is used in many curry powders. The essential oil and the oleoresin of fennel are used in condiments, soaps, creams, perfumes, and liqueurs. Several types of fennel differing in morphology and leaf color are available for ornamental use and as a fresh vegetable. Soft growing tips are widely used to flavour and garnish fish dishes, soups and baked foods.

As a medicinal plant, fennel seed has been used as an antispasmodic, carminative, diuretic, expectorant, laxative, stimulant, and stomachic. Roots once boiled as a vegetable, and used as an expectorant in cough mixtures. Fennel has also been used to stimulate lactation, as a remedy against colic, and to improve the taste of other medicines. Chinese herbal medicine includes the use of fennel for gastroenteritis, hernia, indigestion, abdominal pain, and to resolve phlegm and stimulate milk production. Fennel is known to provoke both photodermatitis and contact dermatitis in humans. The volatile oil may cause nausea, vomiting, seizures, and pulmonary edema. The essential oil has been reported to stimulate liver regeneration in rats. It has antibacterial properties.

FENUGREEK

Botanical name: *Trigonella foenum-graecum*
Family: Leguminosae

Fenugreek is a slender annual herb of the pea family (Fabaceae). Its dried seeds are used as a food, a flavouring, and a medicine. The seeds' aroma and taste are strong, sweetish, and somewhat bitter, reminiscent of burnt sugar. They are farinaceous in texture and may be mixed with flour for bread or eaten raw or cooked. It was used by the ancient

FENUGREEK

Variety	Year of release	Breeding method	Pedigree/ parentage	Important traits	Institutes/ universities
CO 1	1982	Selection	Reselection from TG-2336 introduced from north India	A quick growing short, dual purpose, early maturing variety (80–95 days) tolerant to root rot disease. Seed contain 21.7% protein	HC&RI, TNAU, Coimbatore, Tamil Nadu
CO 2	1999	Selection	Selection from CF 390	Short duration dual purpose variety, field tolerant to *Rhizoctonia* root rot disease, suitable for both *kharif* and *rabi* season. Early maturity, short duration (85–90 days)	HC&RI, TNAU, Coimbatore, Tamil Nadu
Rajendra Kanti	1988	Pure line selection	Pure line selection from Reghunath-pur collection	Medium sized bushy plant; early maturity (65–90 days), suitable for intercropping in both *kharif* and *rabi* season; field tolerant to *Cercospora* leaf spot, powdery mildew and aphids. Seed contains 9.5% protein	Tirhut College of Agriculture, RAU, Dholi, Bihar
Rajendra Abha (Kasuri methi)	—	—	—	—	Tirhut College of Agriculture, RAU, Dholi, Bihar
Hisar Sonali	1994	Pure line selection	Pure line selection from local germplasm	Tall and bushy vigorous growing variety, dual purpose variety, late maturity (140–145 days), suitable for cultivation under	Dept. of Vegetable

FENUGREEK : *Contd.*

Variety	Year of release	Breeding method	Pedigree/ parentage	Important traits	Institutes/ universities
				irrigated condition. Moderately resistant to root rot and aphids	Crops (CCS HAU), Hisar, Haryana
Hisar Suvarna	2001	Pure line selection	Pure line selection from local germplasm	A quick growing, erect and tall, dual purpose, medium maturity (130–140 days), moderately resistant to cercospora and powdery mildew. Wider adaptability suitable for cultivation through out the country	Dept. of Vegetable Crops (CCS HAU), Hisar, Haryana
Hisar Madhavi	2001	Pure line selection	Pure line selection from local germplasm of Uttar Pradesh	A quick growing, erect and tall dual purpose, medium maturity (130–140 days) variety, moderately resistant to powdery mildew and to downy mildew. A variety with wider adaptability suitable for both irrigated and rainfed conditions	Department of Vegetable Crops (CCS HAU), Hisar, Haryana
Hisar Mukta	2001	Pure line selection	Pure line selection natural green seed coated mutant line from Uttar Pradesh	A quick growing seed type variety medium maturity (135–140 days), moderately resistant to downy mildew and powdery mildew. Erect and tall plants. Wide adaptability, suitable for both irrigated and rainfed conditions	Department of Vegetable Crops (CCS HAU), Hisar, Haryana

FENUGREEK : *Contd.*

Variety	Year of release	Breeding method	Pedigree/ parentage	Important traits	Institutes/ universities
Guj Methi 1	1999	Recurrent selection	Recurrent selection based on pure line selection from J. Fenu 102	The first variety from Gujarat released for the state. Plant dwarf	Spices Research Station, GAU, Jagudan-382 710, Mehsana Dist, Gujarat
RMt 1	1990	Pure line selection	Pure line selection from Nagaur local	Vigorous semi erect medium size moderately branched growth habit, maturity 140–150 days, medium sized bold and attractive typically yellow coloured grains, moderately resistant to root-knot nematode and powdery mildew and aphids. Contain 0.2% diogenin, 21.0% seed protein	SKN College of Agriculture, Jobner, Rajasthan
RMt 143	1997	Pure line selection	Pure line selection of local collection of Jodhpur	Moderately resistant to powdery mildew, seeds bold yellow colour 140–150 days maturity, suitable for heavier soils	SKN College of Agriculture, Jobner, Rajasthan

FENUGREEK : *Contd.*

Variety	Year of release	Breeding method	Pedigree/ parentage	Important traits	Institutes/ universities
RMt 303	1999	Mutation breeding	Mutation breeding from variety RMt 1	Medium maturity (145–150 days) variety, seeds bold, less susceptible to powdery mildew	SKN College of Agriculture, Jobner, Rajasthan
RMt 305	—	Mutation breeding	Mutation breeding from variety RMt 1	First determinant type, multipoded, early maturing wider adaptability, resistant to powdery mildew and root knot nematodes, seeds bold, attractive and yellow, contain 21.7% protein, duration 120–125 days	SKN College of Agriculture, Jobner, Rajasthan
Lam sel 1	1992	Selection	Selection from germplasm collection of Uttar Pradesh	Dual purpose early maturing variety (85–90 days), bushy type with medium height (40 cm), more number of branches and green matter. When cultivated for green leaves purpose it gives an average green yield of 12 tonnes/ha, field tolerant to major pests and diseases, seeds contain 53% protein	RARS, ANGRAU, Guntur, Andhra Pradesh
Pant Ragini	2001	Selection	Selection from local germplasm	A dual purpose tall bushy type resistant to downy mildew and root rots, medium maturity (170–175 days), seeds contain 2.0 to 2.5% essential oil	GBPUAT, Pantnagar, Uttaranchal

FENUGREEK : *Contd.*

Variety	Year of release	Breeding method	Pedigree/ parentage	Important traits	Institutes/ universities
NRCSS AM 1	2006	Selection	Selection from local collection	The variety has been developed through pure line selection from local germ-plasm at NRCSS, Ajmer. The variety is medium in height bears broad leaves with less bitterness. The seeds are bold and large. The number of seeds per plant range from 17–20 with 17–20 g test weight. The crop takes 137 days to mature and gives seed yield of 27.2 q/ha. The crop grown exclusive for leaf purpose yields 76 q/ha green leaves from 3 cuttings	NRC Seed Spices, Ajmer
NRCSS AM 2	2006	Selection	Selection from local collection	The variety has been developed through pure line seletion from local germplasm at NRCSS, Ajmer. The variety is medium in height bear broad leaves with more bitterness. The seeds are small in size. The number of seeds per pod range from 16–18. The crop takes 138 days to mature and gives average seed yield of 18.1 q/ha. The crop grown exclusively for leaf purpose yields 72 q/ha green leaves from 3 cuttings	NRC Seed Spices, Ajmer

Egyptians and is mentioned in medical writings in their tombs. The Romans grew it as fodder for their cattles. Historically, the main usage of fenugreek was medicinal rather than as a labour. The botanical name *Trigonella* refers to the angular seeds, while *foenum-graecum* translates as 'Greek hay', which explains its use as cattle feeds.

Native to India and Southern Europe and the Mediterranean region, its plants are cultivated in central and southeastern Europe, western Asia, India, and Northern Africa. The plants are erect, loosely branched, less than 3 feet (1 m) tall with trifoliate, light green leaves and small white flowers. The slender pods are up to 6 inches (15 cm) long, curved and beaked, and contain 10–20 yellow-brown seeds—flat rhomboids characterized by a deep furrow, less than 0.2 inch (1 cm) long. The seeds are about 0.5 cm in diameter and are irregularly shaped, very hard, and tan or mustard coloured.

The volatile oil content of fenugreek is very small (less than 0.02%). It also contains fixed oils at about 5–7%. They contain the alkaloids trigonelline and choline and a yellow colouring matter. It is rich in proteins, minerals and vitamins.

AJOWAN

Botanical name: *Carum copticum* (L.) Benth. et Hook.
(syn. *Carum ajowan, Trachyspermum ammi*)
Family: Umbelliferae; Apiaceae

Ajowan looks like wild parsley (similar to caraway, celery and cumin seeds). It is a native to India. It is grown throughout the country in Madhya Pradesh, Andhra Pradesh, Gujarat, Maharashtra, Uttar Pradesh, Rajasthan, Bihar and West Bengal, and also grown in Pakistan, Afghanistan, Iran and Egypt. The striped seeds are used as the spice. It is an erect, glabrous or minutely pubescent, branched annual that grows up to 90 cm. Stems are striate and leaves are distant and pinnately divided. Small white flowers are on terminal or seemingly lateral pedunculate, compound umbels. The fruits are ovoid, greyish brown, aromatic cremocarps with single seed.

It grows on all kinds of soil but does well on loams or clayey loams, both as a dry crop and under irrigation. Seeds are sown from September to November. The plants flower in about two months and the fruits become ready for harvesting when their flower heads turn brown. They are pulled out, dried on mats and the fruits are separated by rubbing by hands or feet. The sensoric quality of ajowan is similar to thyme, but stronger and less subtle. The essential oil (2.5–5% in the dried fruits) is dominated by thymol (35–60%); furthermore, α-pinene, β-cymene, limonene and ε-terpinene have been found.

AJOWAN

Variety	Year of release	Breeding method	Pedigree/ parentage	Important traits	Institutes/ universities
NRCSS-AA 1	2006	Selection	Selection from local collection	It is suitable for cultivation both under irrigated and rainfed situations. The average height of plant is 112 cm which bears average number of 219 umbels/ plant. It is a late variety and takes about 165 days to maturity. The variety possesses high yield potential and gives an average yield of about 14.26 q/ha under irrigated condition and 5.8 q/ha under rainfed condition. The seeds yield essential oil content of 3.4%	NRC Seed Spices, Ajmer
NRCSS-AA 2	2006	Selection	Selection from local collection	This variety matures in 147 days. It is suitable for cultivation under rainfed condition. The average height of plant is 80 cm and bears average number of 185 umbels/plant. It gives an average yield of about 12.83 q/ha under irrigated condition and 5.2 q/ha under rainfed condition. It is resistant to powdery mildew. The seeds yield essential oil content of about 3%	NRC Seed Spices, Ajmer

Usage of ajowan is almost confined to Central Asia and northern India. Ajowan is particularly popular in savoury Indian recipes like savoury pastries, snacks and breads. For example, the Bengali spice mixture panch phoron is sometimes enhanced with ajowan. Ajowan enjoys, however, some popularity in the Arabic world and is found in berebere, a spice mixture of Ethiopia which shows both Indian and Arabic heritage. In Southern Indian cuisine (which is predominantly vegetarian), tadka-like preparations are not only applied to dried lentils and beans, but also to green vegetables. Ajowan is much used as a medicinal plant in ayurvedic medicine for its antispasmodic, stimulant, tonic and carminative properties. The seeds are used to ease asthma and indigestion. It is also widely used to treat diarrhoea and flatulence. In the West, thymol is used in medicines against cough and throat irritation. The thymol content makes ajowan a potent fungicide.

DILL

Botanical name: *Anethum graveolens* L.
Family: Umbelliferae

Dill is a fennel-like annual or biennial herb of parsley family. Native to Mediterranean countries and South-Eastern Europe, dill is now widely cultivated in Europe, India, and North America. The name dill comes from an Old Norse word, dilla, meaning "lull" since they used it to quiet crying babies. Dill was widely used in Greek and Roman times. In the Middle Ages, it was thought to have magical properties and was used in witchcraft, love potions and as an aphrodisiac. The whole plant is aromatic. The young leaves and fully developed green fruits are used for flavoring purposes. Native to Southern Europe and Western Asia, dill grows wild in Spain, Portugal and Italy. It is now cultivated in India, Germany, Rumania, and England and to some extent in North and South America as a commercial crop.

Dill is an annual herb, growing 45–75 cm in height, with finely feathered blue-green fern-like leaves and hollow stems. It produces small open umbels of creamy-yellow flowers in summer followed by dark brown seeds. The fruit, or seed, is broadly oval in shape, about 0.14 inch (3.5 mm) long, with three longitudinal dorsal ridges and two wing-like lateral ridges.Its seeds contains 2–5% volatile oil. Its main constituent is carvone and other components are d-limonene and phellandrene. A recent study also found eugenol and vanillin present in the seed. Dill weed or leaves contains 0.3–1.5% volatile oil, the chief constituent also being carvone.

DILL

Variety	Year of release	Breeding method	Pedigree/ parentage	Important traits	Institutes/ universities
NRCSS-AD 1	2006	Selection	Selection from local collection	It is a variety of European type of dill, grows to a height of 134 cm and takes about 142 days to reach seed maturity. The leaves are dark green in colour with good herbal quality. It is suitable for cultivation under irrigated conditions and seeds are dillapiole less, a toxic substance as compared to Indian dill varieties. It is susceptible to powdery mildew but possesses field resistance to blight. The variety is suitable for export purpose and seeds contain about 3.5% essential oil. The average seed yield of 14.7 q/ha can be obtained under irrigated conditions	NRC Seed Spices, Ajmer
NRCSS-AD 2	2006	Selection	Selection from local collection	It is an Indian dill type, grows to a height of 90 cm and is suitable for cultivation under both irrigated and rainfed conditions. Suitable for cultivation under both early and *rabi* seasons. 135 days to reach maturity. Seeds are bold and dark brown in colour. The seeds contain about 3.2% essential oil content. An average yield of about 14.6 q under irrigated conditions and 5.8 q/ha under rainfed conditions can be obtained	NRC Seed Spices, Ajmer

NIGELLA

Botanical name: *Nigella sativa*
Family: Ranunculaceae

The seeds of *N. sativa*, known as kalonji, black cumin (though this can also refer to *Bunium persicum*) or just nigella, are used as a spice in Indian and Middle Eastern cuisine. The dry roasted nigella seeds flavor curries, vegetables and pulses. The black seeds taste like oregano and have a bitterness to them like mustard-seeds. It can be used as a "pepper" in recipes with pod fruit, vegetables, salads and poultry. The *Nigella sativa* is an annual flowering plant, native to southwest Asia. It grows to 20–30 cm tall, with finely divided, linear (but not thread-like) leaves. The flowers are delicate, and usually coloured pale-blue and white, with 5–10 petals. The fruit is a large and inflated capsule composed of 3–7 united follicles, each containing numerous seeds. The seed is used as a spice.

ANISE

Botanical name: *Pimpinella anisum* Linn.
Family: Umbelliferae

Anise, a herbaceous annual native to the Mediterranean region and Egypt, is cultivated in Europe, the Middle East, Mexico, North Africa, India and Russia chiefly for its fruits, called aniseed, the flavour of which resembles that of licorice. Anise was well-known to the ancient Egyptians and Romans. Its plants reach a height of about 0.75 m and require a warm and long frost-free growing season of 120 days. It has long-stalked basal leaves and shorter, stalked stem leaves. Its small and yellowish-white flowers form loose umbels. The fruit is nearly ovoid in shape, about 3.5 mm long, and has five longitudinal dorsal ridges. The fruit consists of two united carpels each containing an anise seed. The seed is small and curved, about 0.5 cm long and grayish brown. Its usually contains hair-like protrusions from each end.

Fresh leaves may be used in salads, especially apple; seeds in cookies and candies. While the entire plant is fragrant, it is the fruit of anise, commercially called anise seed, that has been highly valued since antiquity. The delicate fragrance is widely used for flavouring curries, breads, soups, cakes, candies, desserts, non-alcoholic beverages, and such liqueurs as anisette and arak. Aniseed is widely used to flavour pasteries; it is the characteristic ingredient of a German bread called Anisbrod. In the Mediterranean region and in Asia, aniseed is commonly used in meat and vegetable dishes.

NIGELLA

Variety	Year of release	Breeding method	Pedigree/ parentage	Important traits	Institutes/ universities
NRCSS-AN 1	2006	Selection	Selection from local collection	This variety has been developed through selection from a local collection. It is suitable for cultivation under semi-arid region under irrigated conditions. The plants are 32 cm in height. This variety takes 135 days to reach seed maturity and has resistance to root rot. The ovary is pentamerous and each capsule contains 65 seeds. Its seeds contain about 0.7% of essential oil content. The average seed yield is 8 q/ha with potential yield of 12 q/ha under semi-arid conditions	NRC Seed Spices, Ajmer

ANISE

Variety	Year of release	Breeding method	Pedigree/ parentage	Important traits	Institutes/ universities
NRCSS-ANi 1	2006	Selection	Selection from local collection	It is the first variety of anise in India and identified for release very recently under AICRP on Spices. It is a high-yielding variety, bears attractive seeds and high volatile oil content of 3.2%. This variety is suitable for cultivation in semi-arid region under irrigated conditions. It gives an average yield of 7.3 q/ha with potential yield of 11.5 q/ha under semi-arid conditions	NRC Seed Spices, Ajmer

It is used in Italian sausage, pepperoni, pizza topping and other processes meat items.The volatile or essential oil, obtained by steam distillation of crushed anise seed, is valuable in perfumery and soaps and has been used in toothpastes, mouthwashes and skin creams. The essential oil is used to flavour absinthe, and Penod liqueurs. Anise oil is sometime used as an adulterant in the essential oil of licorice. The oil is also sometime used as sensitizer for bleaching colors in photography. The seeds are chewed after a meal in India to sweeten the breath.

It makes a soothing herbal tea and has been used medicinally from prehistoric items. As a medicinal plant, anise has been used as a carminative, antiseptic, antispasmodic, expectorant, stimulant, and stomachic. In addition, it has been used to promote lactation in nursing mothers and as a medicine against bronchitis and indigestion. Oil of anise is used today as an ingredient in cough medicine and lozenges and is reported to have diuretic and diaphoretic properties. If ingested in sufficient quantities, anise oil may induce nausea, vomiting, seizures and pulmonary edema. Contact of the concentrated oil with the skin can cause irritation.

CELERY

Botanical name: *Apium graveolens* Linn.
Family: Umbelliferae; Apiaceae

Celery is a hardy biennial, occasionally annual, widely cultivated for its fleshy leafstalk, which is used as a vegetable. Its seeds produced in the second season, are commonly used as a flavouring agent. Separate varieties have been developed for production of seeds for condiment use and for vegetable celery. There are also root-producing varieties that are used for flavouring soups and stews. It is a herbaceous annual or biennial and is native to southern Europe. In India, it grows wild and also is cultivated in foothills of the northwestern Himalayas and outlying hills of Punjab, Himachal Pradesh and Uttar Pradesh. Roots are succulent, well-developed and numerous. Stems are branching, angular, jointed and reaching a height of 2–4 m. Leaves are oblong, 7–18 cm long, pinnate or trifoliate. It throws up a flower head in the second year producing masses of fruits. The flowers are white or greenish-white, very small on sessile compound umbels. The fruit itself is two united carpels which each produce a single seed. The seed is small, about 1–2 mm in length, oval and greenish brown.

The seeds can be used in pickling fish and in salads, salad dressings, and other dishes where celery flavour is desired. Ground celery is

CELERY

Variety	Year of release	Breeding method	Pedigree/ parentage	Important traits	Institutes/ universities
NRCSS-ACel 1	2006	Selection	Selection from local collection	The variety has been developed through selection from Punjab Local at NRCSS, Ajmer. This is suitable for cultivation in semi-arid regions under irrigated conditions. It produces an average yield of 8 q/ha, the potential being 12 q/ha. The variety yields essential oil content of 2.4% from seeds	NRC Seed Spices, Ajmer

used in a large variety of products like meat dishes, snack foods, gravies and sauces to provide a flavour enhancing effect. The leafstalks and roots give flavour as well as food value to soups and salads.

Blanched celery leaves are eaten raw or cooked as a vegetable. Whole seeds can be added to bread dough or when making cheese biscuits. Celery salt and celery pepper are both made by grinding the seeds with either salt or peppercorns in the required proportions.

Celery seeds have stimulant and carminative properties. Since they have tranquilizing effect it is prescribed as a decoction in psychiatric, epilepsy like diseases. The fatty oil from seeds is antispasmodic and nerve stimulant. The roots possess diuretic property. The oleoresin is also used in a large variety of food items. The oil from the seeds is used medically to treat asthma, flatulence and bronchitis.

MEDICINAL PLANTS

OPIUM POPPY

Botanical name: *Papaver somniferum*
Family: Papaveraceae

Poppy is an ornamental flowering plant of the poppy family. Red-flowered and double and semi-double strains are garden ornamentals. About 50 other species of *Papaver* are grown for their attractive papery flowers or interestingly cut foliage. Opium, from which morphine, heroin, codeine, and papaverine are derived, comes from the milky fluid in the unripe seed capsule of opium poppy. Ancient Greeks have long been aware of medicinal and narcotic properties of poppy. Poppy seed capsules have been found in Switzerland in the remains of prehistoric lake dwellings. Through Arab traders and the spread of Islam to the East, the opium poppy was introduced to Persia, South East Asia and India. *Papaver somniferum* means sleep inducing poppy.

It is an herbaceous annual native to Greece and the Orient. It plants have blue-green stems, lobed or dissected leaves, milky sap and nodding buds on solitary stalks. An annual plant, it bears 12.7 cm wide blue-purple or white flowers on plants 1–5 m tall. They have bisexual, regular, cup-shaped, four- to six-petaled flowers with one superior pistil (female structure) and numerous stamens surrounding the ovary. The two sepals drop off as the petals unfold. The fruit is a capsule, the leaves are usually deeply cut or divided into leaflets,

OPIUM POPPY

Variety	Year of release	Breeding method	Pedigree/ parentage	Important traits	Institutes/ universities
Jawahar Aphim 16	1984	Selection	Local race	White flowered with deeply serrated petals; 2–3 capsules/plant; 60–65 kg/ ha latex yield; 12.5% morphine	AINRP on M&AP, Jawaharlal Nehru Krishi Vishwa Vidyalaya, Mandsaur
Kirtiman	1990	Selection	Local race	One m tall, white-flowered, serrated broad leaf, flat capsule 45–50 kg/ha latex yield	AINRP on M&AP,ND University of Agriculture and Technology, Faizabad
Jawahar Opium 539	1997	Selection	Local race	Deeply serrated leaf with 2 capsules/ plant; 65 kg/ha latex yield; 14.85% morphine	AINRP on M&AP, Jawaharlal Nehru Krishi Vishwa Vidyalaya, Mandsaur
Jawahar Opium 540	1998	Selection	Local race	75–80 cm tall; serrated leaf; 75 kg/ha latex yield; 13% morphine	AINRP on M&AP, Jawaharlal Nehru Krishi Vishwa Vidyalaya, Mandsaur
Chetak Aphim	1994	Selection	Local race	75–82 cm tall, waxy capsule and leaves, large sized round and smooth capsule, 12.6% morphine	AINRP on M&AP, Rajasthan Agricultural University, Udaipur
Trishna	1989	Selection	Local colletion (IC-42)	Flower colour pink; 5–7 capsules; morphine 14.78; latex yield 49–53 kg/ha	NBPGR, New Delhi

and the sap is coloured. The ovary develops into a short, multi-seeded capsule that opens in dry weather, permitting small seed to escape when it is shaken by the wind. The seeds are small (about 1 mm in length), kidney-shaped, and grayish-blue to dark blue in colour.

The opium poppy is also grown for its non-narcotic ripe seeds, which are used for seasoning, oil, and birdseed. Tiny dried seeds of opium poppy are used as food, food flavouring, and the source of poppy-seed oil. Poppy seeds have no narcotic properties, because the fluid contained in buds, that becomes opium, is present only before the seeds are fully formed. They have a faint nut like aroma and a mild, nutty taste, especially popular in breads and other baked goods. Poppy seeds are more common in India, where they are ground and used as a thickening agent in curries and sauces. They are also sprinkled over cooked noodles, or sweetened with honey and made into a dessert dip or sauce. Dry-fried seeds are added to salads and salad dressings.

Poppy seed contains from 44–50% fixed oil, the principal components of which are linoleic and oleic acids. The poppy seed oil is used by artists as a drying oil. The seeds are used in painkillers, cough mixtures, syrups and as an expectorant. An infusion of seeds can provide relief to toothache and earache.

ASALIO

Botanical name: *Lepidium sativum*
Family: Brassicaceae

Garden cress or Asalio, grown in India, Europe and US is an underutilized crop. The edible whole seed is known to have health promoting properties. Hence, it was assumed that these seeds can be a functional food. It is a fast-growing, edible plant botanically related to watercress and mustard and sharing their peppery, tangy flavor and aroma. In some regions garden cress is known as garden pepper cress, pepper grass or pepperwort. It is a perennial plant, and an important green vegetable consumed by human beings, most typically as a garnish or as a leaf vegetable. Garden cress is found to contain significant amounts of iron, calcium and folic acid, in addition to vitamins A and C. The garden cress produces an orange flower suitable for decorative use and also produces fruits which, when immature, are very much like caper berries. Garden cress is also used as a medicine in India in the system of ayurveda. It is used to prevent postnatal complications; the seeds of this plant perform as an aperient when boiled with milk. In Ethiopia, seeds are offered for sale on almost every market and the plants are grown in gardens and fields in most provinces.

ASALIO

Variety	Year of release	Breeding method	Pedigree/ parentage	Important traits	Institutes/universities
GA-1	1998	Selection	Local collection	High yielder	Gujarat Agricultural University, Anand

SAFED MUSLI

Variety	Year of release	Breeding method	Pedigree/ parentage	Important traits	Institutes/ universities
Jawahar Safed musli 405 (JSM-405)	2004	Single plant selection and multi-plication	Selection from local genetic stock	Narrow dense green leaves, anther yellow in colour, fasiculated root and in shape, blunt tips, purple cylindrical colour and seed dark black in colour	AINRP on M&AP, Jawaharlal Nehru Krishi Vishwa Vidyalaya, Mandsaur

PERIWINKLE

Variety	Year of release	Breeding method	Pedigree/ parentage	Important traits	Institutes/universities
Prabhat	2002	Selection	Local race	High-yielding	AINRP on M&AP,CCS Haryana Agricultural University, Hisar

An erect glabrous annual grows 15–45 cm in height. The leaves are entire, the upper ones are sessile whereas lower petiolate. The flowers are white in clour, small and are in long racemes. The fruits are small pods, obovate, with two seeds per pod. The seeds are brownish red in colour and become slimy when soaked in water. The seeds contain a volatile essential aromatic oil, the active principle and a fatty oil. A glycoside is present in seeds, but absent in seed-oil. A new alkaloid — lepidine isolated from seeds along with sinapic acid ethyl ester.

SAFED MUSLI

Botanical name: *Chlorophytum borivilianum*
Family: Liliaceae

It is a herb with linear leaves appearing over ground with the advent of summer rains. Flowers white. It penetrates by fleshy roots/root-tubers.

Located in foot-hills of Uttaranchal, Himachal Pradesh and Uttar Pradesh, Madhya Pradesh, Tamil Nadu, Kerala, Karnataka, Rajasthan, Gujarat and Maharashtra. Safed Musli requires well drained loamy to sandy loam soils rich in organic matter. Warm and humid climatic condition with good amount of soil moisture during the growing season favour luxuriant vegetative growth and facilitate fleshy root development.

The crop matures in about 90 days under cultivation. At maturity the leaves start yellowing and ultimately dry up from the collar part and fall down. The crop could thus be harvested when leaves have dried which occurs in the months of September/October. During digging of plants, fleshy root bunches should be lifted form the soil. The harvested fleshy roots are cleaned and is removed and white musali tubers are dried spread in the shade for about 4–7 days to dry-out its moisture. About one tonne of fleshy roots can be collected per ha. The yield after processing and drying is about 200 kg/ha.

PERIWINKLE

Botanical name: *Catharanthus roseus*
Family: Apocyanaceae

Periwinkle or sadabahar is a tropical, perennial herb. It has long been cultivated as a pot herb in India because of its pink or white flowers, produced almost throughout the year. Being pantropic in distribution, it is found growing all over the country but its commercial cultivation is done in 3,000 ha area in Tamil Nadu and Karnataka and to a small extent in Gujarat and Madhya Pradesh. It is a unique medicinal plant. Its leaves contain VLB alkaloids with anti-neoplastic properties used in curing blood cancer, whereas roots are rich in ajmalicine, raubacine and reserpine used in controlling high blood pressure. There is a good demand for both these group of alkaloids in world trade.

It is an erect, branched, herb growing around 1m tall. Its plants bear oblong-obovate, shortly stalked, opposite leaves. Flowers are borne in axil in pairs with long cylindrical corolla tube. The flowers are pink or white. Fruit is a follicle, 2.5 cm long and 2.5–3 mm broad containing up to 30 black seeds. The fruits split along the length at maturity. All parts of the plant contain alkaloids.

The foliage are picked up (twice) at 120 and 150 days after planting. The third picking is obtained at uprooting of plants for root crop 200 days after planting. The root growth is high between 150 and 200 days when root alkaloids are maximum. At picking, a few lower leaves are left over the plants to allow continued growth, whereas root is uprooted at maturity after giving a light irrigation.

The foliage contain 0.003–0.004% VLB alkaloid. The crop is dried in shade for 7–10 days, frequently turning it all over to avoid mould formation. The foliage, stem and roots lose 80, 70 and 60% of moisture respectively during drying. Root is cut into 15–20 cm pieces before drying. The dry roots are wrapped in gunny bags and stored in a cool, dry, ventilated godown up to 1–2 years. On an average, a leaf yield of 14 q/ha together with 4q/ha of roots and basal stems are obtained from irrigated crop. Rainfed crop produces 1 tonne/ha of foliage and 2–2.5 q/ha of root and basal stems.

SENNA

Botanical name: *Cassia angustifolia*
Family: Leguminaceae

Senna is a small, perennial, branched under-shrub.It is cultivated traditionally over 10,000 ha in semi-arid lands in coastal districts of Tirunelvelli, Ramnathapuram and Madurai in Tamil Nadu. Although, successful cultivation has been demonstrated in many parts in western India, its commercial cultivation has recently come up in Kutch (Gujarat) and Jodhpur division of Rajasthan. It can grow over sand-dunes after rainy season and can be maintained as a perennial crop for 2–3 years. Since its leaves and pods are common laxatives,

SENNA

Variety	Year of release	Breeding method	Pedigree/ parentage	Important traits	Institutes/ universities
Anand Late Selection	1989	Selection	Local race	75% higher leaf dry matter than the local check	AINRP on M&AP, Anand Agricultural University, Anand
KKM. Se. 1	2001	Selection	Thenkalam local	The yield of dried leaves is 712 kg/ha with a pod yield of 266 kg/ha, which is 47.7 and 53.8% respectively higher than local variety. This variety possesses good rejuvenation after stripping once in 30 days. Pods and leaves are good laxatives with medicinal property. It possesses good export potential. Duration is 135–140 days with a sennoside content of 2.54%	AC&RI, TNAU, Killikulam

they are widely used in medicine and as a household remedy for constipation all over the world. India is the main producer and exporter of senna leaves, pods and sennosides concentrate to world market.

Its plants grow up to 1m tall over marginal lands in subtropical climate in peninsular India. It sheds leaves with the onset of cold weather in north-western India. The crop thrives over well-drained, sandy loam lateritic soils of 7–8.5 pH, though fertile fields and irrigated crop support better growth and produce higher yields. It needs an alround warm and dry weather. Even temporary stagnation of water in fields can cause loss to the crop. As an annual crop, it remains in field for 110–130 days. The plants bears compound leaves, made up of 5–8 pairs of shortly stalked oval-lanceolate leaflets (2.5 cm × 1.5 cm) and produce successive flush of flowering shoots both in axillary and subterminal position 60–70 days after sowing. The flowers are large and brilliant yellow in colour, producing medium-sized pods (3.5–6.5 cm × 1.5 cm) after 90 days. They contain 5–8 yellowish, flat seeds. It is predominantly self-pollinated crop but outcrossing could be high (20%) through beetles.

Three pickings of leaves and pods are usually taken 50, 90 and 130 days after sowing. Usually, flowering shoots are chopped off initially to increase branching and to allow its plants to put up more vegetative growth. Mature leaves and 15–25 days old pods are harvested. On an average, 1.2–1.5 tonnes/ha of dry leaves and 3.5–4 q pods/ha are obtained.

The dry foliage and pods should possess a minimum of 2.5 and 3.0% total sennosides respectively. The harvested produce should be spread in thin layer in open sun for 6–10 h. It is further dried in well-ventilated drying sheds for 3–5 days. It should have not more than 8.0% moisture at storage. The colour of dry leaves and pods should be ensured to remain light green to yellow. The sennosides are soluble in water and exposure of the produce to rain water during drying can reduce these contents. The produce is liable to storage loss up to 30% in content. Therefore, it is recommended to store in cool, dry place after reducing the bulk under hydraulic press and wrapped in gunny bags lined with polythene, particularly for distant transportation. Grading of the produce is common for marketing. The extra large, bold, yellowing-green leaves and pods are placed in first grade and sell at a premium.

MEDICINAL YAM

Botanical name: *Dioscorea floribunda* Mart. & Gal.
Family: Dioscoreaceae

Diosgenin, obtained from dioscorea tubers, is the major base chemical for several steroid hormones including sex hormones, cortisone, other corticosteriods and is the active ingredient in the oral contraceptive pill. The growing need for steroid drugs and the high cost of obtaining them from animal sources led to a widespread search for plant sources of sterodal sapogenins, which ultimately led to the identification of the genus *Dioscorea* as the most promising one.

This genus *Dioscorea* belonging to family Dioscoreacea with over 600 species is widely distributed in tropical world. Some of the species like *D. alata* and *D. esculenta* are under cultivation for long time for their edible tubers. There are about 15 species of this genus containing diosgenin.Among this *D. floribunda* and *D. compostia* are widely grown for diosgenin production.

The major Dioscorea producing countries are Mexico, Guatamala, Costa Rica, India and China. The estimated world consumption of diosgenin is somewhere between 800 to 1000 tonnes per year. India produces only 25 tonnes of diosgenin annually mainly from the natural sources and to be self sufficient, India has to step up its production to the tune of sbout 200 tonnes of diosgenin per year.

The best adapted species in Karnataka, Assam, Meghalaya, Tamil Nadu, Goa and Andaman is *D. floribunda*, an introduction from Mexico. It produces compact tubers at a shallow depth. The diosgenin content varies from 2–7% depending on the age of the tubers.

FOXGLOVE

Botanical name: *Digitalis lanata* Ehrh.
Family: Scrophularaceae

The genus *Digitalis* includes several species which are either medicinally or ornamentally important. *D. purpurea* and *D. lanata* are the two species recognized medicinally important as they contain three lanatoside A, B, C, which on hydrolysis yield digitoxin, gitoxin and digoxin, of which digoxin is the most important. The main use of these is in heart diseases and these glycosides have so far not been synthesized and hence cultivation of this crop is the only source of production of these glycosides. India requires about 3000 kg of this drug annually. The hill stations of India offers excellent scope for the cultivation of this crop.

YAM

Variety	Year of release	Breeding method	Pedigree/ parentage	Important traits	Institutes/ universities
FB(C)-1	1974	Selection	Local collections	High steroid content, 2.5% diosgenin content	IIHR, Bangalore
Arka Upakar	1980	Selection	Local collections	High steroid content, 3.5% diosgenin content	IIHR, Bangalore

FOXGLOVE

Variety	Year of release	Breeding method	Pedigree/ parentage	Important traits	Institutes/ universities
D.76	1991	Selection	Local race	Petal dull white with purple to yellow veins, 0.9–0.97% glycoside content	AINRP on M&AP, Y.S. Parmar University of Horticulture and Forestry, Nauni Solan

Digitalis is native of Europe and is a tall herbaceous biennial or perennial plant. The leaves of *D. purpurea* is broadly lanceolate than that of *D. lanata* while the glycoside content is 2–3 times more in *D. lanata* (1–1.4%), *D. pupurea* has 0.2 and 0.4% glycoside content. *D. purpurea* produces white or purple coloured much elongated raceme while *D. lanata* produces relatively shorter inflorescence with smaller flowers having cream or yellow colours.

LIQUORICE

Botanical name: *Glycyrrhiza glabra*
Family: Fabaceae

Liquorice or *mulhati* is a perennial, branched, bushy under-shrub. It has recently been introduced into cultivation in India. Its root is a commercial part traded both in un-peeled and peeled condition besides its solid aquous extract is sold in the form of stick, block and spray dried powder. The roots contain a sweet substance glycyrrhizin, which is 50 times sweeter than sugar besides liquiritoside and isoliquril. The isoliquril imparts yellow colour to its roots. It has emollient and demulsive properties, useful in soar throat, cough and catarrhal affections. It is also used as taste moderator in pharmacy. Itsanti-inflammatory and spasmolytic activity are useful in treating peptic ulcers. By far, its proportionately larger demand is for flavouring and blending of tobacco products besides its common use in confectionery. After extraction, the spent root is used in production of mushroom compost, insulating mill boards and fire-extinguishing boards. At present, India imports a large quantity of these roots annually.

Its roots are dug out in dry autumn season. These are cleaned, cut into 15–20 cm long pieces and split if more than 2 cm thick. These are dried in the open sun for 7–10 days when 85% of their moisture is lost. The root yield is 2.5–4.0 tonnes from 20 month-old crop and 7 tonnes/ha for 30–36 month-old crop. Older roots possess high (6–8%) glycyrrhizin content. Fresh roots are sometimes decorticated. The peeled roots are smooth, yellow in colour with faint pleasant odour and sweet in taste. Artificial drying through hot blast of air (40°C) is ideal. The produce is packed in gunny bags for storage in dry cool place. It can be stored for a long period without loss in quality.

LIQUORICE

Variety	Year of release	Breeding method	Pedigree/ parentage	Important traits	Institutes/ universities
Haryana Mulhatti 1	1989	Selection	Local race	Erect and tall growing with broad and dark green leaves, 6–7% glycyrrhizin content	AINRP on M&AP, CCS Haryana Agricultural University, Hisar

YELLOW HORNED POPPY (*Glaucium flavum*)

Variety	Year of release	Breeding method	Pedigree/ parentage	Important traits	Institutes/universities
H 47-3	1991	Selection	Local race	Leaves semi-spreading, pods are longer, 2.26–2.3% glaucine content	AINRP on M&AP, Dr YSPUHF, Nauni, Solan

SARPAGANDHA

Botanical name: *Rauvolfia serpentina*
Family: Apocynaceae

Sarpagandha is a perennial and native Indian herb. Its roots are used for controlling high blood pressure and certain forms of insanity in Ayurvedic system of medicine since ancient times. It has received worldwide recognition after isolation of its bioactive reserpine alkaloid in allopathic medicines. In addition, its sedative property is utilized by Ayurvedic physicians in treatment of insomnia, epilepsy and asthama. It is cultivated commercially in Madhya Pradesh, Uttar Pradesh, West Bengal, Assam and Orissa. The root is its economic part, containing 55 alkaloids, of which reserpine, deserpidine, ajmalicine, serpentine and yohimbine being pharmacologically active and important. The commercial root crop contains 1.4–3.0% of total alkaloids depending upon age, locality and drying of roots. Of these, ajmalicine, reserpine and serpentine content are highly potent in medicine.

Sarpagandha is an erect, sparsely branched, evergreen herb of humid tropics, attaining 0.75–1.0 m height under cultivation. It gives out 2–6 branches from lower part of the stem, bearing simple, shortly-stalked, leathery, elliptic-lanceolate leaves, arranged 3–5 in a whorl. Flowers are borne in both terminal and axillary compact cymes. It is 1.2–2.5 cm long with dark pink thalamus rendering it an attractive appearance. It has white, long tubular corolla. Fruit is a drupe, ovoid in shape, purplish black at maturity containing ovoid, hard seeds. The root is tuberous, 40–60 cm deep, branched, with corky bark, having irregular longitudinal fissures. The bark, very bitter in taste, has most of its alkaloids.

The roots are dug out in autumn (November–February) during second year and both tap root and its lateral branches besides attached fibrous roots are collected. These are washed and dried in the sun. Since its fibrous roots are rich in alkaloids, they should not be left out. It loses 60% of its weight on drying due to loss of moisture. An average yield of 1.6–2.0 tonnes/ha of dry root is obtained. The roots are cut into 12–15 cm pieces for convenience of drying and storage. It can be stored for 2–3 years in godown without loss in potency.

LONG PEPPER

Botanical name: *Piper longum*
Family: Piperaceae

Pipali or long pepper is a herbaceous, trailing dioecious plant of humid tropics. It is distributed along water courses and over shola

SARPAGANDHA

Variety	Year of release	Breeding method	Pedigree/ parentage	Important traits	Institutes/universities
RI 1	—	Selection	Local race	Medium plant height 60–80 cm, profuse branching, 1.4–1.7% alkaloid content	AINRP on M&AP, Jawaharlal Nehru Krishi Vishwa Vidyalaya, Indore

LONG PEPPER

Variety	Year of release	Breeding method	Pedigree/ parentage	Important traits	Institutes/universities
Viswam	1996	Clonal selection	Local collection	High-yielding, 2.8% alkaloid content, 25% dry matter content of dry fruits	AINRP on M&AP, College of Horticulture, Trichur

lands in Assam, Kerala and Karnataka. The unripe fruits (female spikes) and to a smaller extent, root and thick basal stems constitute commercial produce. These are traded all over South-East Asia for their use as a stimulant, appetizer and general tonic, also included as a food aromatic. In Ayurveda and Unani medicines, it is given in bronchital asthama, insomnia, jaundice and viral hepatitis. Eaten with ginger and honey, it is given in rheumatism and as a nervine tonic. When added to food, it improves bioavailability of nutrients and body resistance. It is cultivated commerically in Akola-Amravati tract of Maharashtra, whereas small and scattered cultivation is done in foothills of Assam, Meghalaya, West Bengal and Orissa besides humid tracts of south Gujarat, Kerala and Karnataka.

Pipali is an erect, twinning, perennial under-shrub, with slender branches trailing or twinning over support. It has thick, jointed and branched rootstock, bearing, numerous, simple, alternate, succulent broadly ovate dark green, shining leaves. The vines produce flowers almost throughout the year. The inflorescence is a spike of diocious nature. The female spikes have short and thick stalk 1.5–2.5 cm long and 0.5–0.7 cm thick, whereas male spikes are elongate and withers early. The unripe female spikes are collected, dried in sun and traded commercially.

An irrigated crop produces 400 kg of dry spikes during first year and up to 1–1.3 tonnes/ha third year, whereas rainfed crop has shorter flush of fruiting, producing 150 kg during first and 1 tonne/ha third year. The plants are dug out after third year. The roots and thick basal stems are separated, cut into 2–5 cm pieces and dried. These are called pipalamool in trade. The spikes contain volatile oil, resin, piperine and piperlonguminine alkaloids, whereas roots and basal stems contain piperine, piperlonguminine and dihydrostigmasterol in small quantity. The dry spikes can be stored in moisture-proof containers for a few years. India is currently importing a large quantity of pipali and pipalamool from Sri Lanka, Malaysia and Indonesia, where the produce is obtained from alien species and has low pungency.

ISABGOL

Botanical name: *Plantago ovata*
Family: Plantiginaceae

Isabgol or psyllium a small, stemless, annual herb, grows mainly in north Gujarat and south-western Rajasthan. A native of Persia, it is cultivated in 50,000 ha in Mehsana, Banaskantha and Sabarkantha districts in Gujarat, and Jalore, Pali, Jodhpur, Barmer, Nagour and

ISABGOL

Variety	Year of release	Breeding method	Pedigree/ parentage	Important traits	Institutes/ universities
Gujarat Isabgol 1	1976	Selection	Local race	About 11% higher seed yield than the local check	AINRP on M&AP, Anand Agricultural University, Anand
Gujarat Isabgol 2	1983	Selection	GI 1	20% higher seed yield than GI 1	AINRP on M&AP, Anand Agricultural University, Anand
Haryana Isabgol 5	1989	Mutation	Local race breeding	Profuse tillering, long and compact spikes and moderatly resistant to downy mildew	AINRP on M&AP, CCS Haryana Agricultural University, Hisar
Jawahar Isabgol 4	1996	Selection	Local race	High-yielding	AINRP on M&AP, Jawaharlal Nehru Krishi Vishwa Vidyalaya, Mandsaur

Sirohi districts in Rajasthan. Recently its cultivation has been extended to Mandsaur district of Madhya Pradesh. India is the largest producer of isabgol seed and seed-husk, 90% of the total being exported all over the world. The seed husk is rosy-white membranous covering which constitutes the drug in commerce. It is a safe laxative which is beneficial in habitual constipation, chronic diarrhoea and dysentery. It is also used in dyeing and calico printing, ice cream-making as a stabilizer, cosmetics and confectionery industries. The leftover seeds, after removal of husk, is a rich source of protein and is used as a poultry feed.

It grows 30–50 cm tall and remains in field for about 120 days. It gives out a large number of tillers from the base, 60 days after sowing, which in turn bear a rosette of long, narrow, 3-nerved, distantly toothed leaves covered with soft hair, 7.0–20 cm × 0.6 cm in dimension. It produces 10–15 ovoid or cylindrical, terminal 1.2–4 cm long spikes, containing 45–69 colourless bisexual tetramerous protogynous flowers, subtended by a bract. The mature capsules are formed in next 2 months. Brown-purplish in colour, these capsules are oval-shaped, each containing two seeds. The seed is long, elliptical-ovate, 2–3.5 mm × 1–1.5 mm in dimension and dull-white in colour.

KHASI KATERI

Botanical name: *Solanum viarum*
Family: Solanaceae

Khasi kateri is a steroid-bearing perennial, tall, bush distributed all over Assam (Khasi, Jayanti and Naga hills), Sikkim and Manipur up to 1,600–2,000 m elevation. It is a naturalized plant in several other parts of India occurring at lower elevations. Its berry pulp is rich in solasodine alkaloid, which is a starting chemical for production of steroids. It is used in production of contraceptive pills, corticosteroids and sex hormones. It is cultivated in 3,000–5,000 ha, mainly in Maharashtra and Andhra Pradesh, and West Bengal, Assam, Madhya Pradesh and Gujarat on a small scale.

It is a large, semi-woody and hardy plant, growing 75–150 cm tall. It is covered with straight and curved prickles over stem and foliage. The leaves are large, ovate and lobed, covered with hirsute and prickle spines on both surfaces. Flowers are white, 1–4 in group in axillary racemes, producing during most part of the year, except autumn. Fruit is a globose, green berry, 2.4 cm across containing numerous, smooth, brown compressed seeds. The berries possess higher solasodine content when its green colour changes to pale on gradual ripening over the plants. The berries of cultivated crop yield

KHASI KATERI

Variety	Year of release	Breeding method	Pedigree/ parentage	Important traits	Institutes/ universities
Arka Sanjeevani	1989	Selection	—	Spine free stem and petiole, high alkaloid content, sutable for high-density planting	IIHR, Bangalore
Arka Mahima	1992	Mutation breeding	Arka Sanjeevani	About 2% solasodine content, spineless variety	IIHR, Bangalore

SOLANUM LACINIATUM

Variety	Year of release	Breeding method	Pedigree/ parentage	Important traits	Institutes/ universities
NH 88-12	1991	Selection	—	Dry berry yield = 31.38 q/ha solasodine content = ~ 4.18 % (in berries) and 1.74% (in leaves). Suitable for sub-temperate conditions	AINRP on M&AP, Y.S. Parmar University of Horticulture and Forestry, Nauni Solan

2.5–3.0% total solasodine content which mainly resides in pulp — 60% around seeds and 40% below the pericarp.

The berries are harvested when they attain yellowish colour. About 2–4 pickings are done at weekly intervals. These fresh berries contain 70–75% water content and need to be dried in the sun very fast since they are prone to fungal infection. The drying of berries should be protected from rain water as solasodine is soluble in water and is lost. Drying through hot air blast at 50°C is preferred. The dry berry yield is 1.8–2 tonnes/ha and contains 2.5% solasodine. Usually khasi kateri is cultivated under contract buying with processing industries as there are a few buyers of berry crop and long storage of dry berry crop is not advisable. The solvent extracted and purified solasodine, its 16-DPA and its higher upstream formulations have market in export trade.

ASWAGANDHA

Botanical name: *Withania somnifera*
Family: Solanaceae

Asgand is a pantropic native medicinal plant growing all over north-western and central India. Its roots are employed as an ingredient in a large number of Ayurvedic medicines. It has adaptogenic, immuno-modulator, aphrodisiac, anti-stress and mildly sedative properties. It provides remedy for general debility, fatigue, stress-induced disorders, dropsy, dyspepsia, hiccup, joint pains, improves male potency, reduces neurosis and inflammation in the lungs. It contains a large number of alkaloids and withaniols (0.13–0.68%). These are found in the bark. Withaniol group of alkaloids (withanine and pseudo-withanine) are therapeutically important. The roots of cultivated crop contain up to 50% of these alkaloids, whereas roots of wild plants are rich in 3-tropy1 tigolate. Leaves contain anaferine but are not marketed commercially. At present, 4,000–5,000 ha area is under its cultivation mainly in Madhya Pradesh and neighbouring districts (Kota and Churu) of Rajasthan.

It is a perennial, branched, evergreen under-shrub, 60–120 cm tall plants. The plants are large,bear simple, opposite, ovate leaves and green to dull yellow flowers. The fruit is a small, globose berry, orange in colour but turning red on maturing. The seeds are small, light, flat and light yellow. The roots are dug out during early-February–mid-March when lower leaves begin to dry up. It gives 6–7 q of dry roots/ha in well-managed fields. The roots are dried in the sun (8% moisture). These roots are wrapped in gunny bags and stored in well-ventilated cool godowns for a long period without loss in quality. The roots are graded for marketing.

ASHWAGANDHA

Variety	Year of release	Breeding method	Pedigree/ parentage	Important traits	Institutes/ universities
Jawahar Asgand 20	1989	Pure line selection	Selection from local bulk	Spreading plant type, medium in height, ovate yellow-green leaf, surface non-hairy, Berries are yellow	AINRP on M&AP, Jawaharlal Nehru Krishi Vishwa Vidyalaya, Mandsaur
Jawahar Asgand 134	1998	Pedigree method	Selection from cross JA 20 × Wild type of Asgandha (1987)	Plant eract, tall, leaf cordate, dark green, surface hairy, berries yellow, yellow-brown	AINRP on M&AP, Jawaharlal Nehru Krishi Vishwa Vidyalaya, Mandsaur

BETELVINE

Botanical name: *Piper beetle*
Family: Piperaceae

Betelvine is a perennial, dioecious, evergreen creeper grown in India. It is a chewing stimulant. The crop is highly labour-intensive and particularly suited to small holdings. Once established, it becomes a perennial source of employment and cash flow for day-to-day income of farmers. It is an important cash crop in Andhra Pradesh, Assam, Bihar, Karnataka, Kerala, Madhya Pradesh, Maharashtra, Orissa, Tamil Nadu, Tripura, Uttar Pradesh and West Bengal, with an annual turnover of about Rs 700 crore. Its leaves are exported to Pakistan, Bangladesh, Indonesia, Malaysia, Burma and Thailand.

Mature leaves are plucked along with a portion of petiole. They are plucked by hand without any aid. However, in certain areas iron nail is used to facilitate plucking. In Karnataka and Tamil Nadu, leaves are plucked from side shoots. In south India, comparatively tender leaves are preferred in the market. After plucking, they are washed thoroughly and made into bundles according to the prevailing custom of the area. On an average, about 60–80 lakh leaves are harvested annually from one hectare garden.

EGYPTIAN HENBANE

Botanical name: *Hyoscyamus muticus*
Family: Solanaceae

The Egyptian Henbane is one of the most important medicinal herbs produced in Egypt and is a valuable source of the alkaloids hyoscyamine, and traces of hyoscine and atropine. These alkaloids can be extracted from the entire aboveground plant parts. The drug obtained from this plant provides relief from the painful spasmodic condition of the nonstriated muscles, irritation of hysteria and irritable cough. A cataplasm of fresh leaves is used as a pain killer and dried leaves are smoked as cigarettes against asthma.

There are two forms of the plant, an annual and a biennial. The annual grows during the summer to a height of 30 to 60 cm and then flowers and sets seed. The biennial produces during the first season only a tuft of basal leaves, which disappear in winter, leaving underground a thick fleshy root, from the crown of which arises in spring a branched flowering stem, usually much taller and more vigorous than the flowering stems of the annual plants. The whole henbane plant has a powerful, nauseous odour.

BETELVINE

Variety	Year of release	Breeding method	Pedigree/ parentage	Important traits	Institutes/ universities
SGM 1 (spb 10)	1994	Clonal selection	Clonal selection from Phalghat, Kerala	Erect bushy female vine short statured multilateraled vine (25–31 vine) with thick stem and attractive yellowish green (cucumber) leaves. The internodes are shortened with profuse leaf production, pungent leaves. Field tolerant to pest and diseases. Ideal homestead variety, leaf yield 110 lakh/ha in 2 years. Grown on live supports of *Sesbania grandiflora* in open condition with double row system having irrigation channels on both sides of row or raised bed	ARS (TNAU), Sirugamani
SGM 2 (spb 12) (Pancha mapatti)	2004	Clonal selection	Clonal selection from Panchampatti Dindugal type	Multilateral vine (17–20/vine) with thick stem and attractive dark green leaves, with long petiole (4–5 cm). The internodes are shortered with profuse leaf production. It can yield 100 lakh leaves per ha in 2 years. The leaves are slightly pungent. SGM 2 is tolerant to various pests and bacterial leaf spot disease. It is suited for open conditions with *Sesbania grandiflora* as live support	ARS (TNAU), Sirugamani
Utkal Sudam (Kpb 2)	1999	Clonal selection	West Bengal Ghanagate local cultivar	Short internode. Tolerant to major diseases. Suitable for Bareja type cultivation	OUAT, Bhubaneswar

BETELVINE : *Contd.*

Variety	Year of release	Breeding method	Pedigree/ parentage	Important traits	Institutes/ universities
Bidhan pan	2002	Clonal selection	A clonal selection of IC 360989	The variety is Bangla type having cordate leaves and very short internodes. This variety produces higher leaf yield with tolerance to pest and diseases	BCKV, Kalyani
Krishna Pan	2001	Clonal selection	Local clone Shirpurkanta (from) District Dhule	Having more number of *kali and phaphada* leaves. The leaves are oblong, thicker and light green in colour. Moderately resistant to foot rot, however, susceptible to betelvine bug and nematodes	MPKV, Sangli

EGYPTIAN HENBANE

Variety	Year of release	Breeding method	Pedigree/ parentage	Important traits	Institutes/ universities
HMI-80-1	—	Selection	Local race	Leaf leathery and succulant, stem smooth, flower violet; 120–125 days of maturity, height 40–50 cm; 0.4–0.7% alkaloid content	AINRP on M&AP, Jawaharlal Nehru Krishi Vishwa Vidyalaya, Indore

THYME

Botanical name: *Thymus vulgaris*
Family: Lamiaceae

Thyme is a native to Southern Europe from Spain to Italy. Apart from Europe it is grown in Australia, North Asia, North Africa, Canada and USA. In India, it is cultivated in the Western temperate Himalayas and Nilgiris. It prefers a mild climate, a mallow upland soil and grows best in the hills.

Thyme is used to season, tomato soups, fish and meat dishes, liver and pork sausages, headcheese, cottage and cream cheese. Thyme oil is used in treatment of bronchitis. It has anti-spasmodic and carminative properties. It possesses anti-oxidant and anti-microbial properties.

Thyme contains 0.8–2.0% volatile oil of which the main component is thymol. Other major components are p-cymene and d-linalool. Appreciable amounts of camphene, γ-terpinene and carvacrol are also reported.

Thyme is an aromatic, small shrub-like perennial growing 30 to 45 cm in height. It has gnarled thin, square stems, woody at the base and small, about 0.5 cm in length, grayish green, elliptical and very narrow leaves. The flowers are small, lilac or white and fragrant.

AROMATIC PLANTS

LEMON GRASS

Botanical name: *Cymbopogon flexuosus*
Family: Gramineae

Lemon grass is a stemless, perennial sedge, which grows wild in tropical southern states in India. It is now commercially cultivated in many parts of India. Its leaves contain an aromatic oil, with a characteristic lemon like odour, containing 75–80% citral. The oil is used in scenting of soaps, cosmetics and as a disinfectant besides its aroma compound; citral is a starting material for manufacturing of ionones and vitamin A. It is commercially grown in 30,000 ha in Kerala and Assam, mainly as a rainfed crop, whereas it is grown on a smaller scale in Orissa, Andhra Pradesh and Karnataka. India has been a large and traditional producer and exporter of its oil, citral and its upstream aroma compounds.

THYME

Variety	Year of release	Breeding method	Pedigree/ parentage	Important traits	Institutes/ universities
Ooty (TV) 1	—	Selection	Selection from germplasm	Average yield 10.7 tonnes green leaves/ha, high oil per cent (0.7), thymal 23.63%, tolerant to root rot, nematodes and white fly	HRS (TNAU), Ooty

LEMON GRASS

Variety	Year of release	Breeding method	Pedigree/ parentage	Important traits	Institutes/ universities
NLG-84	1994	Selection	Selection from germplasm	High herbage, oil content (0.4%) and citrol content (84%)	NDUAT, Faizabad

PALMAROSA

Variety	Year of release	Breeding method	Pedigree/ parentage	Important traits	Institutes/ universities
Rosha Grass 49	1989	Selection	—	About 200 cm tall, long and dense inflorescence, long and broad leaves with thick stem, 90% geraniol	AINRP on M&AP, CCS Haryana Agricultural University, Hisar
CI-80-68	—	Selection	—	About 175–200 cm tall, medium tillering capacity, 1–1.5% essential content	AINRP on M&AP, Jawaharlal Nehru Krishi Vishwa Vidyalaya, Indore

It grows up to 2 m tall under cultivation. It produces flowers and sets seeds luxuriantly during autumn under short day conditions only. The inflorescence is a long, terminal, branched panicle, bearing paired spikes, subtended by a leafy bract. A spike consists of 5–11 spikelets in pairs, of which one is sessile and the other is stalked. The sessile spikelet is awned, holds bisexual florets with 4 glumes while stalked ones are awnless with 3 glumes and a staminate floret.

It is harvested after 90 days during first year and after 60–65 days thereafter except in dry summer season. It is cut at 20 cm above the ground and left for withering for 4–6 h in field. It is cut into small pieces later and distilled in a steam distillation unit. It has 0.5–0.8% oil. It yields 18–20 tonnes/ha of herbage in rainfed condition and 25–30 tonnes/ha in irrigated one, producing 80–100 kg and 150–180 kg oil/kg respectively from well-managed fields. Average citral content varies from 80 to 86%.

PALMAROSA

Botanical name: *Cymbopogon martini* var. *motia*
Family: Gramineae

Palmarosa oil grass is a tall, perennial sedge which grows wild in dry, open, scrub forests of central India. The flowering tops and foliage contain a sweet oil emitting rose-like aroma, rich in geraniol. It is widely used in scenting of soaps, cosmetics and tobacco blending. Its geraniol content is extensively used in perfumery industry. India is the principal producer and exporter of its oil and geraniol. It is grown in 5,000–6,000 ha area in Maharashtra, Andhra Pradesh, Madhya Pradesh, Uttar Pradesh and Orissa.

It is a branched, profusely-tillering, drought hardy crop growing up to 1.75–2.0 m in height. It has cylindrical terete stem bearing long (35 cm × 1.8 cm), linear lanceolate leaves, sheathing the stem. The root system is fibrous, shallow and establishes easily in submarginal shallow soils and eroded river banks. Inflorescence is given out in autumn and is made up of branched terminal compound panicles of fawn colour, bearing white hairy star-like spiked flowers. The spikelets are heterogamous. Its lower pair is oblong-elliptical in shape containing bisexual florets. Seed is very light which remains attached to the glume for a long period. It can be removed through beating; seed loses much of its viability after one year of storage.

Its crop has maximum oil content at flower opening stage but the oil quality in terms of odour value is superior at early seeding stage. This coincides with high concentration of primary geraniol content in oil which gives highly sweet scented rosaceous green aroma. The first coppicing of herbage is obtained in October–November (early-autumn season), whereas second one is taken in May–June and third harvesting is done in next September–October. The plants are cut 20 cm above the ground. The herbage is wilted in sun for 6 h before chopping into small pieces for steam distillation. Distillation takes 3 h to exhaust the semi-dry herb. By and large, 15–20 tonnes/ha of herbage is obtained in 3 pickings, taken during late-November, June and October. The produce gives 150–200 kg of oil/ha, besides 5–8 q of pigeonpea as intercrop. The oil is light yellow, contains 90% total alcohol, calculated as geraniol.

SPEARMINT

Botanical name: *Mentha spicata*
Family: Labiatae

Spearmint is another important mint. Its oil is rich in carvone (65%) content and emits caraway like odour. The oil is useful in dentrifice, confectionery and pharmaceutical products. It bears lanceolate stalkless, light green leaves and narrow, long, terminal flowering spikes with lilac flowers, attaining a height of up to 60 cm. Punjab Spearmint is an erect-growing with quadrangular purple-green, hairy stem, producing 20 q/ha of fresh herb. It contains 0.57% oil, oil yield being 120 litres/ha containing 68% carvone.

MUSHAKBALA

Botanical name: *Valeriana jatamansi*
Family: Valerianaceae

Valeriana jatamansi (locally known as Mushakbala, Tagar) is a highly valuable medicinal and aromatic plant species used both in traditional and modern system of medicine. The plant starts sprouting just after snow melt in April. Aromatic rhizomes and roots yield a sweet smelling essential oil, containing valerenal, valerenic acid, eugenyl esters, actinide, valeriamine, valerine and chatinine. The plant is used in Ayurvedic, herbal and allopathic medicines, cosmetics and perfumery. The main uses are as a sedative, carminative and antispasmodic against hysteria, epilepsy, neurosis, and insomnia, but also against constipation and scorpion poisoning.

SPEARMINT

Variety	Year of release	Breeding method	Pedigree/ parentage	Important traits	Institutes/universities
Punjab Spear-mint-1	1991	Selection	—	Stem purple-green, semi-erect, 0.57% essential oil, 60–70% carvone	AINRP on M&AP, Dr Y.S. Parmar University of Horticulture and Forestry, Nauni, Solan

MUSHAKBALA

Variety	Year of release	Breeding method	Pedigree/ parentage	Important traits	Institutes/universities
Dalhousi Clone	1994	Selection	—	Dry rootstock yield is 8–9 q/ha, vale-potriates content is ~ 4.0 %, suitable for sub-temperate region	AINRP on M&AP, Dr Y.S. Parmar University of Horticulture and Forestry, Nauni, Solan

GERANIUM

Variety	Year of release	Breeding method	Pedigree/ parentage	Important traits	Institutes/ universities
KKL 1	1987	Selection	Clonal selection from an Algerian variety	It is vigorous in growth. It yields 45.2 tonnes/ha of green leaves from which 54.4 kg of essential oil can be extracted. The oil recovery is 0.16%. First harvesting can be made eight months after planting and continued thereafter once in three months. The leaf oil contains 60.56% geraniol against 54% in local type	TNAU, AC&RI, Killikulam

GERANIUM

Botanical name: *Pelargonium graveolens*
Family: Geraniaceae

It is an aromatic perennial herb. It is grown in cooler, subtropical climate of Mysore and Bangalore (Karnataka), and extended to Hyderabad (Andhra Pradesh). Its oil has a refreshingly delicate long-lasting rosy odour with fruity under-note, containing 66–78% primary alcohols (rhodinol). It is used in manufacturing of perfumes, creams, talcum powder and body lotion. It is stable in alkaline medium and is therefore used for scenting of soaps. China, Egypt, Re-Union Islands, and Malagasy Republic are major producers and exporters of its oil, whereas Algeria and Morocco also maintain sizeable plantations. We import part of our requirement annually.

It is a bushy plant with cylindrical stem and large cordate-ovate deeply lobed pubescent leaves. The plants grow up to 1m in height under cultivation. In hills, it flowers in February–March and September. These are bisexual, pentamerous flowers with pink corolla and pale-yellow shrivelled anthers, devoid of pollens. They drop out soon, producing no seed. There are 2 genotypes suitable for cultivation in India. Algerian type forms slender, erect plants with dark pink flowers. From this, PG 7 is selected for growing in hills. It contains 0.3% oil with 57% l-citronellol. The other culture, *Re-union type*, produces more bushy growth with light pink flowers. It grows well at lower elevation but it is more susceptible to wilt. It gives oil at par but the oil has marginally higher l-citronellol (59.4%) content. An Egyptian culture has recently been introduced in the country. It is drought hardy, tolerant to high summer temperature and relatively more tolerant to wilt disease. It grows well on lateritic soils around Hyderabad in Andhra Pradesh.

List of Selected Photographs

Garden Pea : Arka Ajit; Arka Karthik; Arka Sampoorna
Potato : Kufri Alankar; Kufri Anand; Kufri Arun; Kufri Ashoka; Kufri Badshah; Kufri Bahar; Kufri Chamatkar; Kufri Chandermukhi; Kufri Chipsona-1; Kufri Chipsona-2; Kufri Chipsona-3; Kufri Dewa; Kufri Girdhari; Kufri Giriraj; Kufri Himalini
Potato : Kufri Himsona; Kufri Jawahar; Kufri Jeevan; Kufri Jyoti; Kufri Kanchan; Kufri Khasigaro; Kufri Khyati; Kufri Kuber; Kufri Kumar; Kufri Kundan; Kufri Lalima; Kufri Lauvkar; Kufri Megha; Kufri Muthu; Kufri Naveen; Kufri Neela; Kufri Pukhraj; Kufri Pushkar; Kufri Red; Kufri Sadabahar; Kufri Safed; Kufri Shailja; Kufri Sheetman; Kufri Sherpa; Kufri Sindhuri; Kufri Surya; Kufri Sutlej; Kufri Swarna
Brinjal : Arka Anand; Arka Keshav; Arka Kusumakar; Arka Navneeth; Arka Neelkanth; Arka Nidhi; Arka Shirish
Chillies : Arka Harita; Arka Lohit; Arka Meghana; Arka Suphal; Arka Sweta
Tomato : Arka Abha; Arka Abhijit; Arka Alok; Arka Ananya; Arka Ashish; Arka Meghali; Arka Saurabh; Arka Shreshta; Arka Vikas; Arka Vishal
Capsicum : Arka Basanth; Arka Gaurav; Arka Mohini
Okra : Arka Anamika; Arka Abhay

MEDICINALAND AROMATIC CROPS **541-542**

Asalio : *Lepidium sativum*
Aswagandha : *Withania somnifera*
Betelvine : *Piper betle*
Glaucium : *Glaucium flavum; Glycyrrhiza glabra*
Isabgol : *Plantago ovata*
Lemongrass : *Cymbopogon flexuosus*
Long Pepper : *Piper longum*
Muskdana : *Abelmoschus moschatus*
OpiumPoppy : *Papaver somniferum*
Periwinkle : *Catharanthus roseus*
Palmarosa : *Cymbopogon martinii*
Safed musli : *Chlorophytumborovilianum*
Sarpagandha : *Rauvolfia serpentina*
Senna : *Cassia angustifolia; Solanumlaciniatum; Solanumviarum*
Vetiver : *Vetiver*

SPICES **543-556**

Black Pepper : IISR Girimunda; ISR Malabar Excell; IISR Panchami; IISR Pournami; IISR Shakthi; IISR Sreekara; IISR Subhakara; IISR Thevum; Panniyur-1; Panniyur-2; Panniyur-3; Panniyur-4; Panniyur-5; PLD-2
Ginger : IISR Mahima; IISR Rajatha; IISR Varada; Suprabha; Suruchi
Turmeric : IISR Alleppey Supreme; IISR Kedaram; ISR Prabha; IISR Prathibha; IISR Sudarsana; IISR Suguna; IISR Suvarna
Cardamom : IISR Avinash; IISR Suvasini; IISR Vijetha
Cinnamom : IISR Navashree; IISR Nithyashree
Nutmeg : IISR Viswashree
Ajowain : NRCSS AA-1; NRCSS AA-2
Coriander : NRCSS ACar-1; Rcr 435; Rcr 436
Cumin : RZ 19; RZ 209; RZ 223; RZ 341; UC 198
Dill : NRCSS AD-1; NRCSS AD-2
Fennel : NRCSS AF-1; RF-101; RF-125; RF-143; RF-178; RF-205
Fenugreek : NRCSS AM-1; NRCSS AM-2; Rmt 1; Rmt 143; Rmt 303; Rmt 305; Rmt 351
Anise : NRCSS AANI-1
Celery : NRCSS ACel-1
Nigella : NRCSS AN-1

FRUITS

Aonla — Goma Aishwarya

Ber — Goma Kirti

Ber — Thar Bhubhraj

Mango — Ambika

Mango — Amrapali

Mango — Mallika

Pomygranate — Bhagwa

Pomygranate — Ganesh

Banana — CO 1

Banana — Udayam

Guava — Lalit

Guava — Shweta

Papaya — CO 1

Papaya — Pusa Delicious

Papaya — CO 7

Plantation Crops

Arecanut — VTLAH-1
Arecanut — Mangala
Arecanut — Mohitnagar
Arecanut — Sreemangala
Arecanut — Sumangala
Arecanut — Swarnamangala
Cocoa — CCRP-1
Cocoa — CCRP-2
Cocoa — CCRP-3
Cocoa — CCRP-4
Cocoa — CCRP-5
Cocoa — CCRP-6
Cocoa — CCRP-7
Cocoa — CCRP-8
Cocoa — CCRP-9

Plantation Crops

Cocoa — CCRP-10

Cocoa — VTLCC-1

Cocoa — VTLCC-1

Cashew — Bhaskara

Cashew — BPP-8

Cashew — Dhana

Cashew — Ullal-3

Cashew — Vengurla-4

Cashew — Vengurla-7

Cashew — VRI-3

Coconut — Ananda Ganga

Coconut — Chandra Kalpa

Coconut — Chandra Laksha

Coconut — Chandra Sankara

Coconut — Chowghat Orange Dwarf

Plantation Crops

Coconut — Kalpa Dhenu | Coconut — Kalpa Mitra | Coconut — Kalpa Pratibha

Coconut — Kalpa Raksha | Coconut — Kera Chandra | Coconut — Kera Ganga

Coconut — Kera Sankara | Coconut — Kera Sowbhgya | Coconut — Kera Sree

Coconut — Laksha Ganga | Coconut — Pratap | Coconut — VHC-1

Coconut — VHC-2 | Coconut — VHC-3 | Coconut — VPM-3

Vegetables

Cauliflower

Arka Kanti

Amaranthus

Arka Arunima

Amaranthus

Arka Suguna

Spinach

Arka Anupama

Garlic

Agrifound Parvati (G-313)

Garlic

Agrifound White (G-41)

Garlic

Godavari

Garlic

Yamuna Safed-2 (G-50)

Garlic

Yamuna Safed-3 (G-282)

Garlic

Yamuna Safed-4 (G-323)

Garlic

Yamuna Safed (G-1)

Onion

Agrifound Dark Red

Onion

Agrifound Light Red

Onion

Agrifound White

Onion

Arka Bindu

Vegetables

Onion — Arka Kalyan
Onion — Arka Keerthiman
Onion — Arka Lalima
Onion — Arka Niketan
Onion — Arka Pitamber
Onion — Arka Pragathi
Onion — B-780
Onion — N-2-4-1
Onion — N-53
Onion — Pusa Madhvi
Onion — Pusa Red
Carrot — Arka Suraj
Radish — Arka Nishant
Tapioca — Sree Padmanabha
Greater Yam — Sree Shilpa

Vegetables

Sweet Potato

Sree Arun

Bittergourd

Arka Harit

Kakdi

AHC-2

Kakdi

AHC-13

Pumpkin

Arka Chandan

Pumpkin

Arka Suryamukhi

Snapmelon

AHS-10

Snapmelon

AHS-82

Kachari

AHK-119

Kachari

AHK-200

Muskmelon

Arka Jeet

Muskmelon

Arka Rajahans

Ridgegourd

Arka Sujat

Ridgegourd

Arka Sumeet

Watermelon

Arka Manik

Vegetables

Watermelon

Thar Manak

French Bean

Arka Anoop

French Bean

Arka Bold

French Bean

Arka Komal

French Bean

Arka Suvidha

Vegetable Cowpea

Arka Garima

Vegetable Cowpea

Arka Samurdhi

Vegetable Cowpea

Arka Suman

Vegetable Dolichos Bean

Arka Jay

Vegetable Dolichos Bean

Arka Vijay

Garden Pea

Arka Ajit

Garden Pea

Arka Karthik

Garden Pea

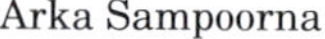

Arka Sampoorna

Potato

Kufri Alankar

Potato

Kufri Anand

Vegetables

Potato

Kufri Arun

Potato

Kufri Ashoka

Potato

Kufri Badshah

Potato

Kufri Bahar

Potato

Kufri Chamatkar

Potato

Kufri Chandermukhi

Potato

Kufri Chipsona-1

Potato

Kufri Chipsona-2

Potato

Kufri Chipsona-3

Potato

Kufri Dewa

Potato

Kufri Girdhari

Potato

Kufri Giriraj

Potato

Kufri Himalini

Potato

Kufri Himsona

Potato

Kufri Jawahar

Vegetables

Potato — Kufri Jeevan
Potato — Kufri Jyoti
Potato — Kufri Kanchan
Potato — Kufri Khasigaro
Potato — Kufri Khyati
Potato — Kufri Kuber
Potato — Kufri Kumar
Potato — Kufri Kundan
Potato — Kufri Lalima
Potato — Kufri Lauvkar
Potato — Kufri Megha
Potato — Kufri Muthu
Potato — Kufri Naveen
Potato — Kufri Neela
Potato — Kufri Pukhraj

Vegetables

Potato

Kufri Pushkar

Potato

Kufri Red

Potato

Kufri Sadabahar

Potato

Kufri Safed

Potato

Kufri Shailja

Potato

Kufri Sheetman

Potato

Kufri Sherpa

Potato

Kufri Sindhuri

Potato

Kufri Surya

Potato

Kufri Sutlej

Potato

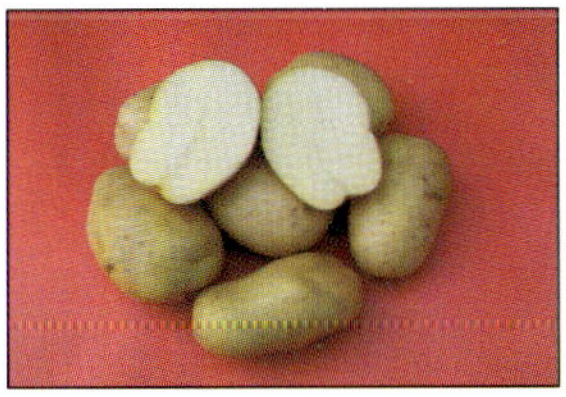

Kufri Swarna

Brinjal

Arka Anand

Brinjal

Arka Keshav

Brinjal

Arka Kusumakar

Brinjal

Arka Navneeth

Vegetables

Brinjal — Arka Neelkanth

Brinjal — Arka Nidhi

Brinjal — Arka Shirish

Chillies — Arka Harita

Chillies — Arka Lohit

Chillies — Arka Meghana

Chillies — Arka Suphal

Chillies — Arka Sweta

Tomato — Arka Abha

Tomato — Arka Abhijit

Tomato — Arka Alok

Tomato — Arka Ananya

Tomato — Arka Ashish

Tomato — Arka Meghali

Tomato — Arka Saurabh

Vegetables

Tomato — Arka Shreshta

Tomato — Arka Vikas

Tomato — Arka Vishal

Capsicum — Arka Basanth

Capsicum — Arka Gaurav

Capsicum — Arka Mohini

Okra — Arka Abhay

Okra — Arka Anamika

Medicinal and Aromatic Crops

Asalio — *Lepidium sativum*

Aswagandha — *Withania somnifera*

Betelvine — *Piper betle*

Medicinal and Aromatic Crops

Glaucium flavum

Glycyrrhiza glabra

Isabgol

Plantago ovata

Lemongrass

Cymbopogon flexuosus

Long Pepper

Piper longum

Muskdana

Abelmoschus moschatus

Opium Poppy

Papaver somniferum

Periwinkle

Catharanthus roseus

Palmarosa

Cymbopogon martinii

Safed musli

Chlorophytum borovilianum

Sarpagandha

Rauvolfia serpentina

Senna

Cassia angustifolia

Solanum laciniatum

Solanum viarum

Vetiver

Spices

Black Pepper

IISR Girimunda

Black Pepper

IISR Malabar Excell

Black Pepper

IISR Panchami

Black Pepper

IISR Pournami

Black Pepper

IISR Shakthi

Black Pepper

IISR Sreekara

Black Pepper

IISR Subhakara

Black Pepper

IISR Thevum

Black Pepper

Panniyur-1

Black Pepper

Panniyur-2

Black Pepper

Panniyur-3

Black Pepper

Panniyur-4

Black Pepper

Panniyur-5

Black Pepper

PLD-2

Ginger

IISR Mahima

Spices

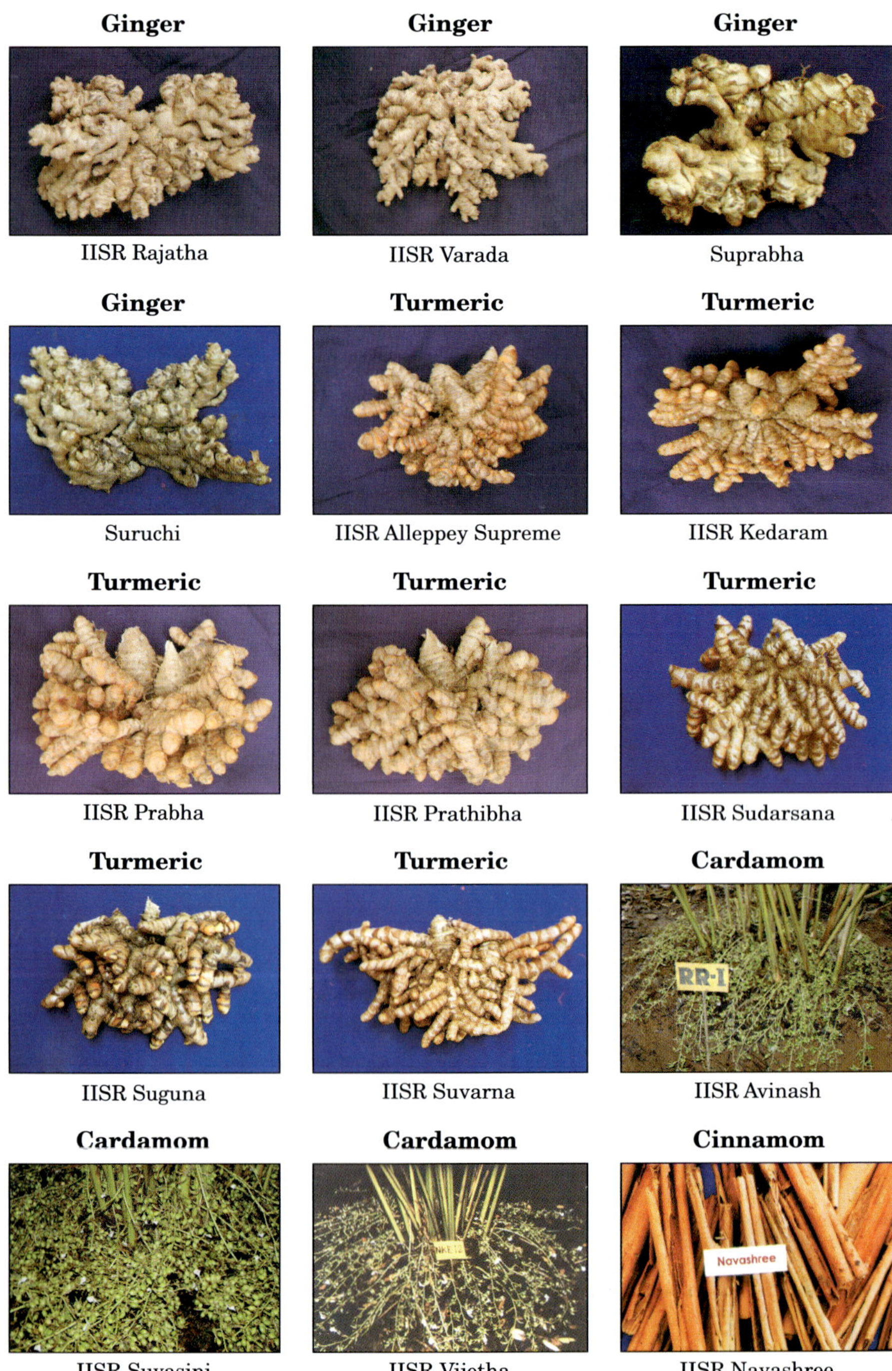

Ginger — IISR Rajatha

Ginger — IISR Varada

Ginger — Suprabha

Ginger — Suruchi

Turmeric — IISR Alleppey Supreme

Turmeric — IISR Kedaram

Turmeric — IISR Prabha

Turmeric — IISR Prathibha

Turmeric — IISR Sudarsana

Turmeric — IISR Suguna

Turmeric — IISR Suvarna

Cardamom — IISR Avinash

Cardamom — IISR Suvasini

Cardamom — IISR Vijetha

Cinnamom — IISR Navashree

Spices

Cinnamom — IISR Nithyashree

Nutmeg — IISR Viswashree

Ajowain — NRCSS AA-1

Ajowan — NRCSS AA-2

Coriander — NRCSS ACar-1

Coriander — Rcr 435

Coriander — Rcr 436

Cumin — RZ 19

Cumin — RZ 209

Cumin — RZ 223

Cumin — RZ 341

Cumin — UC 198

Dill — NRCSS AD-1

Dill — NRCSS AD-2

Fennel — NRCSS AF-1

Spices

Fennel — RF-101

Fennel — RF-125

Fennel — RF-143

Fennel — RF-178

Fennel — RF-205

Fenugreek — NRCSS AM-1

Fenugreek — NRCSS AM-2

Fenugreek — Rmt 1

Fenugreek — Rmt 143

Fenugreek — Rmt 303

Fenugreek — Rmt 305

Fenugreek — Rmt 351

Anise — NRCSS AANI-1

Celery — NRCSS ACel-1

Nigella — NRCSS AN-1

Subject Index

A

C

O

S

T